10th International Conference on Mechatronics and Manufacturing (ICMM 2019)

IOP Conference Series: Materials Science and Engineering
Volume 635

Bangkok, Thailand
21 – 23 January 2019

ISBN: 978-1-5108-9837-0
ISSN: 1757-8981

Printed from e-media with permission by:

Curran Associates, Inc.
57 Morehouse Lane
Red Hook, NY 12571

Some format issues inherent in the e-media version may also appear in this print version.

Institute of Physics
Dirac House, Temple Back
Bristol BS1 6BE UK

Phone: 44 1 17 929 7481
Fax: 44 1 17 920 0979

techtracking@iop.org

Additional copies of this publication are available from:

Curran Associates, Inc.
57 Morehouse Lane
Red Hook, NY 12571 USA
Phone: 845-758-0400
Fax: 845-758-2633
Email: curran@proceedings.com
Web: www.proceedings.com

Table of contents

Volume 635

10th International Conference on Mechatronics and Manufacturing (ICMM 2019)

21–23 January 2019, Bangkok, Thailand

Accepted papers received: 29 August 2019
Published online: 28 October 2019

Preface

Papers

Engineering Material Design and Performance Analysis

Mechatronics and Control Automation

Mechanical Engineering and Advanced Manufacturing Technology

IOP Conf. Series: Materials Science and Engineering **635** (2019) 011001 doi:10.1088/1757-899X/635/1/011001

Preface

2019 10th International Conference on Mechatronics and Manufacturing (ICMM 2019) was held at the Chulalongkorn University, Bangkok, Thailand during January 21-23, 2019. The conference is intended to offer a primary environment to encourage discussion and exchange of ideas leading to advanced methods for Mechatronics and Manufacturing.

The key focus of ICMM 2019 is to promote worldwide awareness of mechatronics and manufacturing among different industry and academic professionals. It provides a platform for the scientists, researchers, scholars, engineers, students as well as marketing professionals from all over the world to present their experiences, cases and ongoing research activities through presentation, discussions, and publications and exchanging the ideas and recent technical advances. The aim is to foster research relations between academia and the industry. It brings together academics and other professionals, for the presentation and exchange their thoughts and practices in mechatronics and manufacturing.

Another purpose of ICMM 2019 is to allow university instructors, scholars, researchers to present and publish their cutting edge research results. All papers in the conference proceedings have been undergone an intensive review process performed by the international technical committee, and only accepted papers are included. These papers are scholarly articles and representation of research from all over the world.

We would like to acknowledge and give special appreciation to our keynote speakers for their valuable contributions, our delegates for being with us and sharing their experiences, and our invitees for participating in ICMM 2019. We would also like to extend my appreciation to the Program Committee and Technical Committee for the devotion of their precious time, advice and hard work to prepare for this conference. The last but not least, I would like to thank those who have contributed to the success of ICMM 2019. We hope that all participants and other interested readers benefit scientifically from the proceedings.

Ratchatin Chancharoen
Chulalongkorn University, Thailand
Conference Chair

IOP Conf. Series: Materials Science and Engineering **635** (2019) 011001 doi:10.1088/1757-899X/635/1/011001

Statement of Peer Review

All papers published in this volume of *IOP Conference Series: Materials Science and Engineering* have been peer reviewed through processes administered by the proceedings Editors. Reviews were conducted by expert referees to the professional and scientific standards expected of a proceedings journal published by IOP Publishing.

IOP Conf. Series: Materials Science and Engineering **635** (2019) 011002 doi:10.1088/1757-899X/635/1/011002

Conference Committee

Conference Chairs

Prof. Yoshihiko Uematsu, Gifu University, Japan

Assoc. Prof. Ratchatin Chancharoen, Chulalongkorn University, Thailand

Program Chairs

Prof. Mahir Dursun, Gazi University, Turkey

Prof. Vadim R. Gasiyarov, South Ural State University, Russia

Technical Committee

Prof. Prakash.S, Sathyabama University, India

Prof. Osman Adiguzel, Firat University, Turkey

Prof. N.ETHIRAJ, Dr. M.G.R Educational and Research Institute, Chennai, India

Prof. Bandit Suksawat, King Mongkut's University of Technology North Bangkok, Thailand

Prof. Hamid Reza Karimi, Polytechnic University of Milan, Italy

Prof. Ranganath M. S, Delhi Technological University, India

Prof. Anna A. Sandulyak, Moscow technological university, Russia

Prof. Zhao-Jun YANG, Jilin University, China

Prof. Yusri Yusof, Universiti Tun Hussein Onn Malaysia (UTHM), Malaysia

Assoc. Prof. Yung-Chang Cheng, National Kaohsiung First University of Science and Technology, Taiwan

Assoc. Prof. Ghassan A. Al-Kindi, Sohar University - Sultanate of Oman

Assoc. Prof. Shih-Chuan Yeh, Hungkuo Delin University of Technology, Taiwan

Assoc. Prof.Shailendra Kumar, S.V.National Institute of Technology, India

IOP Conf. Series: Materials Science and Engineering **635** (2019) 011002 doi:10.1088/1757-899X/635/1/011002

IOP Conf. Series: Materials Science and Engineering **635** (2019) 011003 doi:10.1088/1757-899X/635/1/011003

Peer review statement

All papers published in this volume of *IOP Conference Series: Materials Science and Engineering* have been peer reviewed through processes administered by the proceedings Editors. Reviews were conducted by expert referees to the professional and scientific standards expected of a proceedings journal published by IOP Publishing.

IOP Conf. Series: Materials Science and Engineering **635** (2019) 012001 doi:10.1088/1757-899X/635/1/012001

Preparation and characterization of silk fibroin from four different species of Thai-local silk cocoon for Bone implanted applications

S Pitjamit[1, a], K Thunsiri[1,2, b], W Nakkiew[1*] and P Pothacharoen[3, c]

[1] Advanced Manufacturing Technology Research Center (AMTech), Department of Industrial Engineering, Faculty of Engineering, Chiang Mai University, Chiang Mai, 50200, Thailand

[2] Biomedical Engineering Institute, Chiang Mai University, Chiang Mai, 50200, Thailand

[3] Department of Biochemistry, Faculty of Medicine, Chiang Mai University Chiang Mai, 50200, Thailand

*Corresponding author e-mail: wasawat@eng.cmu.ac.th, [a]siwasit_p@cmu.ac.th, [b]kittiya.thunsiri@hotmail.com, [c]peraphan.pothacharoen@gmail.com

Abstract. The metal interlocking nail is normally used in diaphyseal fractures. The bio-composite interlocking nail will be fabricated in the near future by using local silk fibroin reinforce Polylactic acid filament for 3D printing. Four species of local Thai Bombyx mori silk cocoons were selected to extract silk fibroin. The silk cocoon species consist of 1) Nangnoi Srisaket-I (NN), 2) Nanglai (NL), 3) Luang Saraburi (LS), and 4) J108. After the extraction, fibroins of each silk cocoon species were characterized and compared the physical property by using Scanning Electron Microscopy (SEM), Energy Dispersive X-ray (EDS) and Fourier Transform Infrared Spectroscopy (FT-IR). Then, the biological test was performed on cell viability and cytotoxicity with human fetal osteoblast cell line. The result presents that all of local silk cocoons species presented non-cytotoxicity ability which can be used in human or animal body without endangerment.

1. Introduction

Diaphyseal fractures are common orthopedic injuries which can be treat by using interlocking nails. The developments of interlocking nails were performed to increase the treatment abilities especially in femur, tibia, and humerus fractures [1]. However, the removal of interlocking nails is still required due to the complications in long-term implantation [2]. According to the materials of interlocking nails, titanium and stainless steel did not provided biological environment to the living cells. This research looking for the biomaterials which can eliminated the limitations of metal interlocking nails. Currently, the use of three-dimensional printing (3DP) in biomedical technology is dramatically increased [3]. According to the rapidly process of 3DP, there were several studies of bone implant applications using 3DP for emergency cases [4, 5]. Fused filament fabrication (FFF) technique is the most satisfied 3DP technique. Thermoplastic filament will be feed through the heating nozzle. The melted filament will be printed layer by layer until the end. Material costs, production speed, and design flexibility are the advantage of FFF technique. There were several types of biomaterials which can be printed by using

FFF technique. Thus, biopolymer such as Polylactic acid (PLA) will be used as a main structure for 3DP. In bone tissue engineering (TE), the choice of appropriate biomaterials is one of particular challenge to mimic mechanical and biological characteristics of the native tissue [6]. Several types of biomaterials were tested and implanted to the animal models to confirm the materials abilities for bone TE [7]. Silk products have widely been used as an implantation biomaterial such as silk sutured, silk proteins for wound healing or TE application [8, 9]. Fibroin is a well-known biomaterial which could be extracted from silk cocoon, spider web and red ant nests [10]. It could be fabricated into a structure or combined with other biomaterials to increase the biological and mechanical properties [11]. Previous studies have proved that fibroin has good biological and mechanical properties such as biocompatibility, biodegradability, water permeability, non-cytotoxicity and the strength and resiliency of silk fibers [12-14]. Silk fibers has an ultimate tensile strength 740 MPa while collagen and polylactic acid has an ultimate tensile strength only 0.9 to 7.4 and 28 to 50 MPa, respectively [15]. Therefore, the silk fibroin will be used as a reinforcing material for PLA to create bio-composite filament for interlocking nail 3DP.

In this study, four species of local silk cocoon were provided by the Queen Sirikit Department of Sericulture (QSDS) in Chiang Mai. After silk fibroin was extracted from each species, the characterization will be performed and compared the biological ability by using cell viability test of osteoblast-like cells. The most effective silk cocoon specie will be used to reinforce PLA for 3DP in the future study.

2. Materials and methods

2.1. Silk Fibroin Preparation

Four species of local silk cocoons consist of Nangnoi Srisaket-I (NN), Nanglai (NL), Luang Saraburi (LS) and J108 (Figure 1) were selected for silk fibroin extraction. The silk cocoons were cut into small pieces. Then, the cut silk cocoons were degummed in 100 °C of Na_2CO_3 solution for 30 minutes and rinsed the degummed silks by 80-100 °C of DI water and repeated this process. The degummed silks were dried in 37 °C hot air oven for 24 hours. Dried degummed silk was dissolved in 70 °C of ternary solvent ($CaCl_2/CH_3CH_2OH/H_2O$, in mole ratio 1:2:8) for 6 hours. The silk solution was packed in dialysis tube and soaked in 4 °C of DI water for 3 days (change the DI water every day). The dialyzed silk solution was filtered and frozen at -80 °C freezer. Finally, the frozen silk solution was lyophilized to generated silk fibroin sponges as shown in Figure 2 [16-18].

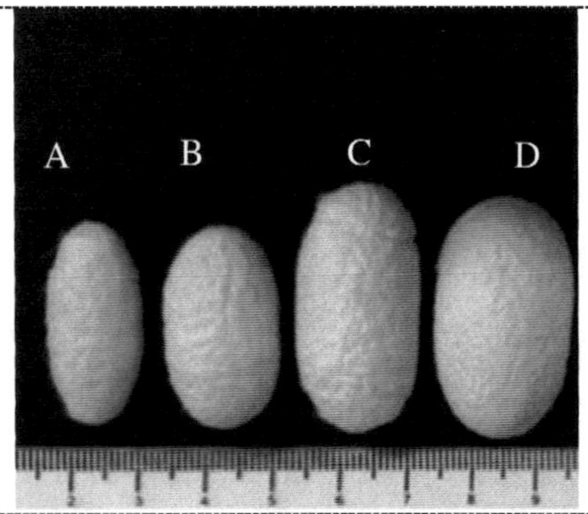

Figure 1. Different species of 4th silk cocoons in Chiang Mai, Thailand (A) NN (B) NL (C) LS and (D) J108.

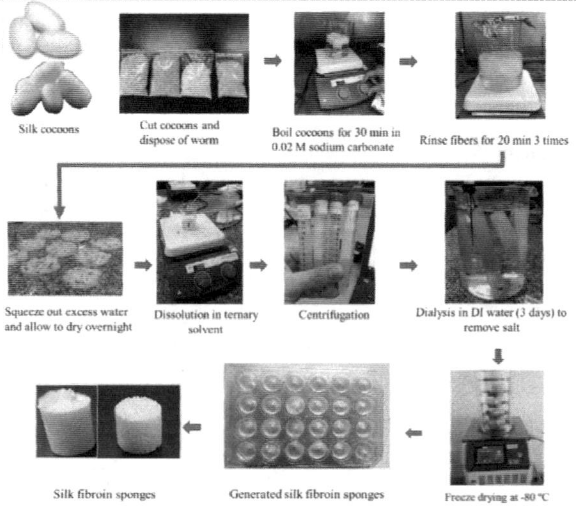

Figure 2. Schematic of the silk fibroin extraction procedure

2.2. Fourier-transform infrared spectroscopy (FTIR)

FTIR spectra of lyophilized fibroin sample from each silk species were obtained from a Thermo Fisher FTIR spectrometer equipped with an ATR (attenuated total reflectance) accessory. Sixty-four repetitive scans from 800-2500 centimetre^{-1} were averaged and presented.

2.3. Scanning Electron Microscopy (SEM) and Energy Dispersive X-ray Spectrometer (EDS)

The structure of lyophilized fibroin from each silk species was observed under a JSM -IT300 scanning electron microscope. The surface picture will be captured and Energy Dispersive X-ray Spectrometer (EDS) was performed.

2.4. Fibroin solution preparation

Prior cell viability and cytotoxicity testing, lyophilized fibroin from each silk species were dissolved in DMEM/Ham's F-12 medium (0.5 g/10 ml) and mixed well.

2.5. Cell viability and cytotoxicity

Human fetal osteoblast cell line (hFOB1.19, CRL NO.11372) was purchased from ATCC and expanded in a 1:1 mixture of phenol red-free DMEM/Ham's F-12 medium supplemented with 10% fetal bovine serum (FBS), 100 U/ml penicillin and 100 μg /ml streptomycin (basal media). The culture cells were incubated in 37 °C with 5% CO_2 until reach the requirement number. The hFOB1.19 cells were seeded in 96-well plates at a density of 5 x 10^3 cells per well. After 24 hours incubation, the cells were cultured with 100 μl of each silk species fibroin solution for another 24 hours. At indicated time of treatment [19], Alamar blue dye (10% v/v in PBS) was added to each well and incubated at 37 degree Celsius for 4 hours before the fluorescence measured at excitation 530 nanometres and emission 590 nanometres. The percentage of cell viability was calculated by the following equation 1.

$$\%\, Cell\ viability = \frac{OD\ of\ treated\ cell}{OD\ of\ control\ cell} \cdot 100 \quad (1)$$

2.6. Statistical Analysis

All data were showed mean ±SD. P-value < 0.05 was measured as statistically significant. A paired t-test with two-tailed distribution was used to determine the different between control group and each silk species group.

3. Results and Discussion

3.1. Silk fibroin characterization

FTIR spectroscopy was used to investigate the conformations of fibroin protein in regenerated each silk scaffolds. The FTIR spectra of the regenerated each silk fibroin species shown at Figure 3. For C=O stretching (amide I) at 1630 to 1650 centimetres^{-1}, For secondary NH bending (amide II) at 1540 to 1520 centimetres^{-1}, and for C-N stretching (amide III) at 1270 to 1230 centimetres^{-1} in their FTIR spectra show characteristic vibrational bands of protein materials [15].Furthermore, the conformations of the protein materials were indicated by the positions of these bands; Beta-sheet at 1630 centimetres^{-1} and random coil at 1650 centimetres^{-1} for amide I, Beta-sheet at 1520 centimetres-1 and random coil at 1540 centimetres^{-1} for amide II, Beta-sheet at 1270 centimetres^{-1} and random coil at 1230 centimetres^{-1} for amide III. The four species of silk scaffolds show typical Beta-sheet structures and random coil structures respectively. These band shifts occur because of the distinct hydrogen bonding states produced by different conformations adopted by the protein chains.

3.2. Scanning Electron Microscopy (SEM) and Energy Dispersive X-ray Spectrometer (EDS)

The silk fibroin structures were observed under SEM. Structure of all silk fibroin species provided interconnect pore. The pore size of each silk species was not significantly different as shown in Figure 2A-2D. The average of each silk scaffolds pore size was about 10 to 60 microns. Each silk species scaffolds were indicated by the area EDS taken at the centre, and its consistent EDS graph is presented in the figure. As shown in Figure 4, silk fibroin weight percentage consist of Carbon (C), Nitrogen (N) and Oxygen (O) element only which symbolize the proteinaceous compounds originating from Silk Bombyx mori.

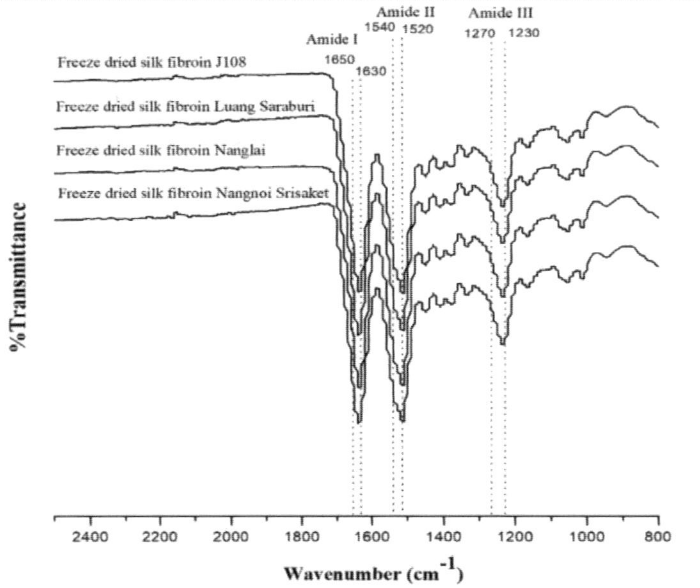

Figure 3. FTIR spectra of four types local Thai silk fibroin scaffolds.

3.3. Cell viability and cytotoxicity test using Alamar blue assay.

After 24 hours of cell culture in 100 µl of each silk species fibroin solution, the Alamar blue assay was performed on the set to observe percent cell viability and confirm non-toxicity ability of silk fibroin solution. The comparison of each silk species with control presented that cell viability percentage all scaffolds were not significantly the control (p-value> 0.05) as shown in Figure 5 and Table 1. Thus, all species of silk fibroin were non-cytotoxic materials.

Table 1: Cell viability percentage

	Mean (%)	SD	P-value (vs. control)
Control	100.00	1.11	-
NS	97.69	14.83	0.730
NL	106.73	8.26	0.168
LS	111.12	7.26	0.111
J108	110.41	3.62	0.051

IOP Conf. Series: Materials Science and Engineering **635** (2019) 012001 doi:10.1088/1757-899X/635/1/012001

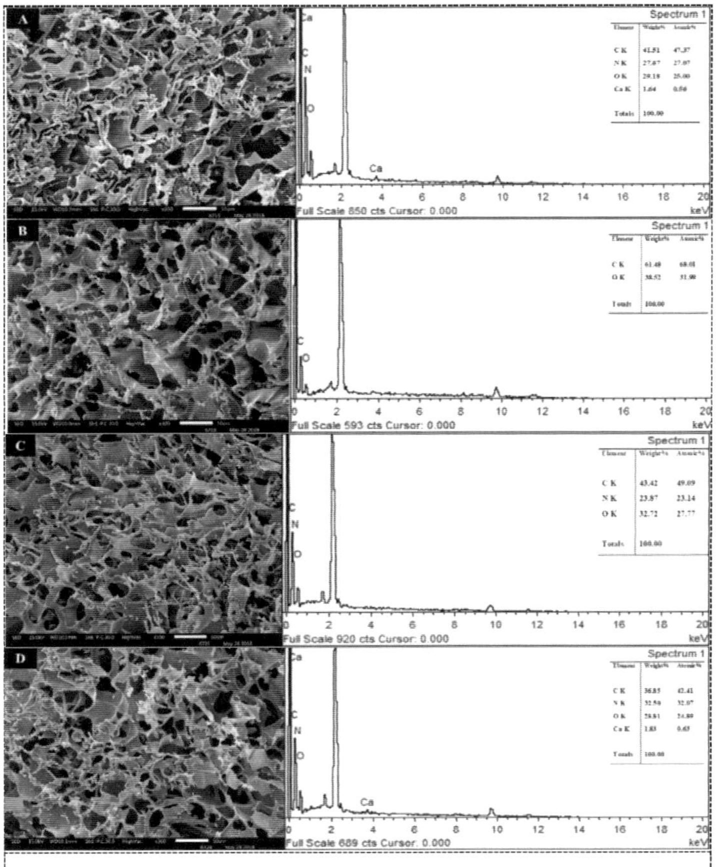

Figure 4. SEM images and Field emission scanning microscopy equipped with EDS results. (A) NN (B) NL (C) LS and (D) J108 silk fibroin.

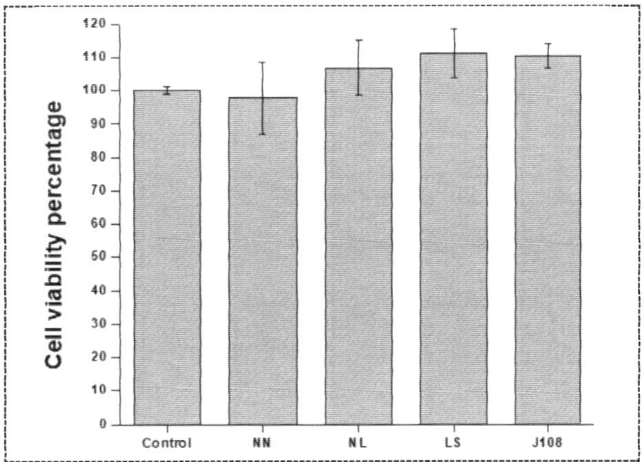

Figure 5. Cell viability percentage of each silk species in 24 hours.

4. Conclusion

According to the purpose of the end study, silk fibroin will be used to reinforce PLA interlocking nails which will be fabricated by 3DP process [15]. Silk fibroin should provide good biological environment to the fracture area to accelerate bone healing process by providing good surface for cell adhesion, migration, and proliferation [20]. Furthermore, silk fibroin and PLA can be degraded by biologically enzyme [18]. So, the 3DP silk fibroin reinforcing PLA interlocking nails will eliminate the removal process and provided good biological environment for bone regeneration. Therefore, four species of local silk cocoon which provided by the QSDS were used as raw materials to extract silk fibroin and performed biological test to compared and confirmed the biological property of each silk species. The best silk species from biological performance will be used to reinforce PLA interlocking nails using 3DP process in the future study. From the result, all of local silk cocoons species present non-cytotoxicity ability which can be used in human or animal body without endangerment. For future work, bio-composite filament for 3DP from silk fibroin reinforcing PLA will be tested and observed the other abilities such as cell proliferation ability, mechanical properties and printing morphology.

5. Acknowledgments

The authors of this study would like to acknowledge the Faculty of Engineering, Chiang Mai University under the research assistant scholarship ID RA/007/2559 for financial sponsorship of research. The authors are also grateful to Advanced Manufacturing Technology Research Centre (AMTech), Department of Industrial Engineering, Faculty of Engineering, Biomedical Engineering Institute, Department of Biochemistry, Faculty of Medicine, Graduate School, Chiang Mai University and Queen Sirikit Department of Sericulture for their facility support.

6. References

[1] Brumback, R. J. (1996). The rationales of interlocking nailing of the femur, tibia, and humerus: an overview. Clinical Orthopaedics and Related Research®, 324, 292-320.

[2] Seyhan, M., Guler, O., Mahirogullari, M., Donmez, F., Gereli, A., & Mutlu, S. (2018). Complications during removal of stainless steel versus titanium nails used for intramedullary nailing of diaphyseal fractures of the tibia. Annals of medicine and surgery, 26, 38-42.

[3] Ventola, C. L. (2014). Medical applications for 3D printing: current and projected uses. Pharmacy and Therapeutics, 39(10), 704.

[4] Tai, B. L., Kao, Y. T., Payne, N., Zheng, Y., Chen, L., & Shih, A. J. (2018). 3D Printed composite for simulating thermal and mechanical responses of the cortical bone in orthopaedic surgery. Medical engineering & physics.

[5] Barba, A., Maazouz, Y., Diez-Escudero, A., Rappe, K., Espanol, M., Montufar, E. B., ... & Franch, J. (2018). Osteogenesis by foamed and 3D-printed nanostructured calcium phosphate scaffolds: effect of pore architecture. Acta Biomaterialia.

[6] Hutmacher, D. W. (2006). Scaffolds in tissue engineering bone and cartilage. In the Biomaterials: Silver Jubilee Compendium (pp. 175-189).

[7] Pearce, A. I., Richards, R. G., Milz, S., Schneider, E., & Pearce, S. G. (2007). Animal models for implant biomaterial research in bone: a review. Eur Cell Mater, 13(1), 1-10.

[8] Nazarov, R., Jin, H. J., & Kaplan, D. L. (2004). Porous 3-D scaffolds from regenerated silk fibroin. Biomacromolecules, 5(3), 718-726.

[9] Kundu, B., Rajkhowa, R., Kundu, S. C., & Wang, X. (2013). Silk fibroin biomaterials for tissue regenerations. Advanced drug delivery reviews, 65(4), 457-470.

[10] Kamhaengpol, A., Maensiri, S., & Siri, S. (2010). Fibroin protein extract from red ant nests for a production of electrospun nanofibers. Asia-Pacific Journal of Science and Technology, 15(10), 919-929.

[11] Tamada, Y. (2005). New process to form a silk fibroin porous 3-D structure. Biomacromolecules, 6(6), 3100-3106.

[12] Xie, M. B., Li, Y., Li, J. S., Chen, A. Z., Zhao, Z., & Li, G. (2014). Biomedical applications of silk fibroin.
[13] Murphy, A. R., & Kaplan, D. L. (2009). Biomedical applications of chemically-modified silk fibroin. Journal of materials chemistry, 19(36), 6443-6450.
[14] Liu, T. L., Miao, J. C., Sheng, W. H., Xie, Y. F., Huang, Q., Shan, Y. B., & Yang, J. C. (2010). Cytocompatibility of regenerated silk fibroin film: a medical biomaterial applicable to wound healing. Journal of Zhejiang University SCIENCE B, 11(1), 10-16.
[15] Koh, L. D., Cheng, Y., Teng, C. P., Khin, Y. W., Loh, X. J., Tee, S. Y., ... & Han, M. Y. (2015). Structures, mechanical properties and applications of silk fibroin materials. Progress in Polymer Science, 46, 86-110.
[16] Rockwood, D. N., Preda, R. C., Yücel, T., Wang, X., Lovett, M. L., & Kaplan, D. L. (2011). Materials fabrication from Bombyx mori silk fibroin. Nature protocols, 6(10), 1612.
[17] Wattanutchariya, W., & Thunsiri, K. (2015). Effects of Fibroin Treatments on Physical and Biological Properties of Chitosan/Hydroxyapatite/Fibroin Bone's Scaffold. In Applied Mechanics and Materials (Vol. 799, pp. 488-492). Trans Tech Publications.
[18] Thunsiri, K., Oonjai, A., & Wattanutchariya, W. (2018). Characterization of Hydroxyapatite/Silk Fibroin/Chitosan Scaffold for Cartilage Tissue Engineering. In Key Engineering Materials (Vol. 775, pp. 120-126). Trans Tech Publications.
[19] Keogh, M. B., O'Brien, F. J., & Daly, J. S. (2010). A novel collagen scaffold supports human osteogenesis—applications for bone tissue engineering. Cell and tissue research, 340(1), 169-177.
[20] Qi, Y., Wang, H., Wei, K., Yang, Y., Zheng, R. Y., Kim, I. S., & Zhang, K. Q. (2017). A review of structure construction of silk fibroin biomaterials from single structures to multi-level structures. International journal of molecular sciences, 18(3), 237.

IOP Conf. Series: Materials Science and Engineering **635** (2019) 012002 doi:10.1088/1757-899X/635/1/012002

Optimization of electrospinning parameters and degradation properties for Gelatin/CMC nanofibers

K.Jongwuttanaruk[1], P.Surin[1] and F.Wiwatwongwana[1, 2]

[1]Department of Advanced Manufacturing Technology, Faculty of Engineering, Pathumwan Institute of Technology, 833 Rama 1 Road, Wangmai, Pathumwan, Bangkok,10330,Thailand

[2]Department of Manufacturing Engineering, Faculty of Engineering, Pathumwan Institute of Technology, 833 Rama 1 Road, Wangmai, Pathumwan, Bangkok 10330, Thailand

E-mail : Kaonadod@gmail.com

Abstract. This research aimed to enhance strength of Nanofiber like Biopolymer as raw material for characterization which is popularly used in food and cosmetic industry and as medical material because it did not hazardously affect human body. This study employed Nano fiber characterization with Electrospinning Method by using gelatin/ CMC, common and inexpensive biopolymer with cellulose that could strengthen Nanofiber.The fabrication of fiber scaffold by electrospinning method produced nano-fiber by dissolving gelatin in organic solvent, 2,2,2-trifluoroethanol, and dissolving CMC in DI water. The design of experiment was surface responds and 3 factors were analyzed including 1) gelatin concentration varied from 8-12%, CMC concentration in range of 0.6-1.0% and 3) mixture percent of CMC in range of 10-30%. After the test of tension, the highest value was 13.99Mpaderived from gelatin with 10.64% concentration, CMC with 0.81%concentration and mixture of CMC with 17.79% concentration. After that, the work piece obtained from mixing CMC with concentration at 10, 20 and 30%, 10% Gelatin and 0.80%CMC as constant due to highest volume of product from such 3 values, was tested for decomposition compared with pure-scaffold gelatin. The findings revealed that decomposition rate of CMC mixture at 20% concentration was slowest and it could be found that the result of tension testing and decomposition testing was consistent. That is, 10% Gelatin, 0.8% CMC and 20% CMC mixture gave the best value in both experiments. Additionally, it was obvious that pure-scaffold gelatin took only 1 hour to totally decompose while CMC mixture at 10%, G91 that degraded entirely in 24 hours, CMC mixture at 30%,G73 degraded in 48 hours, CMC mixture at 20%,G82 degraded in 54 hours to decompose entirely with CMC mixture. It means that using CMC which is cellulose derivative as ingredient in scaffold gelatin could strengthen it better.

1. Introduction

Electrospinning is a simple but highly effective technique to use degradable synthetic polymer like PLA, PGA, PLGA, and PCL as well as natural polymer for biological and medical purposes because natural polymers can go along with human body better [1]. It is a popular technique to apply fibers such as cosmetic skin mask, military protective clothing,Nano sensor, Application in life science and tissue.Electrospinning's principle is dependence on electric potential difference. Biopolymer will be induced and become miniature fibers to supporting areas [2].

This study examined best tension of mixing gelatin and carboxymethylcellulose at different proportion with Electrospinning technique of artificial skin.Design of Experiment (DOE) method was used as an instrument to find proper level of factors affecting tension [3]. This is because natural polymer obtained from collagen found in skin, flesh and bone of animals is biodegradable, despite being hydrophilic polymer, but it cannot be mixed and characterized with electrospinning method. Therefore, this research used a solvent that could be dissolved with gelatin which was 2,2,2trifluoroethanol. Carboxymethylcellulose, CMC was employed to strengthen artificial because carboxymethylcellulose (CMC) is cellulose derivative. Normally, cellulose fiber with high alpha-cellulose or high-quality cellulose was required. Additionally, it is applied in various industries such as laundering, paint, glue, textile, paper, ceramic, food and medicines as CMC is colorless, odorless, and unharmful to humans and well soluble.

Furthermore, the study was conducted to change crosslinking to enhance strength of mechanical properties. Accordingly, crosslinking was employed with temperature use or Dehydrothermal treatment (DHT). DHT is a physical process by giving high heat in vacuum condition (Normally, temperature higher that 140 Celsius Degree is required) [4]. The entire process will remove water from fibers resulting in condensation and crossing of bonds which can strengthen fibers.

2. Experimental procedure

2.1 Materials
Powdered gelatin from Porcine Skin, 180 G Bloom, TypeB from Fluke Analytical and2,2,2-trifluoroethanol (TFE)(purity 99.0%) from Sigma-Aldrich were used. To prepare gelatin, gelatin was dissolved in TFE and stirring in room temperature for 6 hours with concentration of 6.64, 8.00, 10.00, 12.00, 13.36% and powdered CMC from Sigma-Aldrich was dissolved with distilled water with concentration 0.46, 0.60, 0.80, 1.00, 1.14 % and stirring in 70° C temperature for 30 minutes. After preparation, substances were characterized with electrospinning method by mixing CMC in gelatin in proportion of 3.18, 10.00, 20.00, 30.00, 36.82% and irritating the mixture in room temperature for 30 minutes.

2.2 DHT crosslinking
Scaffolds were crosslinked with DHT method by desiccating scaffold of gelatin and CMC with silica gel in vacuum box for 2 days. After that, the scaffold was inserted into vacuum oven in 140 C° temperatures for 48 hours.

2.3 Electrospinning
Gelatin/ CMC was prepared to be characterized with electrospinning method by adding the substance in 5-ml syringe with needle spinneret #20 or 0.9-mm. hold. 10 KV high voltage power supply was employed and substance flow rate was controlled with syringe pump at 0.8ml/h. The substance was injected on supporting foil with distance from needle tip to foil sheet was 13 centimeters [5].

2.4 Experimental design
Enhancing efficiency of tension of Scaffold with Central Composite Designs (CCD) of Response Surface Mythology (RSM) with Minitab 17 Program depended on 3 factors including gelatin concentration (A), CMC concentration (B) and ratio of mixing CMC in gelatin (C) [6]. Table 1 showedParameters for experimental design andthe general regression equation to predict the effects of factors on responses is defined as

$$Y=b_o+\sum^{3}_{i=1}b_iX_i+\sum^{3}_{i=1}b_{ii}X_i^2+\sum^{2}_{i=1}+\sum^{3}_{j>1}b_{ij}X_iX_j+\ldots\ldots \quad (1)$$

Where Y was the response variable (impact strength), $b_0$was the constant, $\sum$ was the residual (error) term, b_i was thelinear coefficients, $b_{ii}$was the quadratic coefficient, b_{ij} was the interaction coefficient and X_i was the dimensionlesscoded independent variable.

IOP Conf. Series: Materials Science and Engineering **635** (2019) 012002 doi:10.1088/1757-899X/635/1/012002

Table 1.Parameters for experimental design.

Factor	Variation		
	Low(-1)	Medium(0)	High(+1)
Gelatin concentration (%),(A)	8	10	12
CMC concentration (%),(B)	0.60	0.80	1.00
Ratio of mixing CMC in gelatin (%),(C)	10	20	30

2.5 Degradation of Gelatin-CMC scaffolds

Lysozyme from chicken egg white (Bio Basic Inc.) concentration was 31.2 u/ml (0.1 mg/ml PBS buffer). After that it incubated in a thermostatted oven at 37 °C for 54 hr. The sample incubated is controlled in buffer without enzyme. The sample started to remove and rinse with purified water to remove the enzymes, dried and weighed in 0.5, 1, 1.5, 24, 48 and 54 hours consecutively. We calculated the degree of enzymatic degradation from the following equation [7, 8]:

$$\text{Weight remain (\%)} = 100 - \left[\frac{(W_0 - W_f)}{W_0} \times 100\right] \tag{2}$$

Where: W_0 is initial weight of the scaffold
$W_f$is final weight of the scaffold

3. Results and Discussion

According to table 2, it indicated differentresult of tension from experiment of Response Surface Mythology (RSM) in each experiment

Table 2. Experimental design and results

Run	Variables (Real value)			Response
	Gelatin concentration (%), (A)	CMC concentration (%), (B)	Ratio of mixing CMC in gelatin (%) (C)	
1	8	0.6	10	7.787
2	12	0.6	10	9.824
3	8	1	10	4.357
4	12	1	10	11.384
5	8	0.6	30	7.606
6	12	0.6	30	8.493
7	8	1	30	5.701
8	12	1	30	11.533
9	6.64	1	20	1.352
10	13.36	1	20	8.112
11	10	0.46	20	8.331
12	10	1.14	20	7.883
13	10	1	3.18	13.201
14	10	1	36.82	13.142
15	10	0.8	20	13.702
16	10	0.8	20	13.552
17	10	0.8	20	13.775
18	10	0.8	20	13.651
19	10	0.8	20	13.676
20	10	0.8	20	13.624

3.1. Analysis of Variance (ANOVA)

Table 3 showed the result of analysis indicated that factors affecting experimental response included gelatin concentration (%), (A)provided thatP-Value was 0.001 and CMC Concentration (%), (B) with P-Value of 0.019 These two factors' p-value was less than 0.05 and P-Value ofLack-of-Fit was 0.761 which was more than 0.05 It means that the regression equation was highly accurate. R-sq(adj) was 99.86% indicating that the experiment was substantially equitable and it could be accepted. For Response Surface Mythology (RSM), the regression equation of this experiment was given also.

Table 3. Analysis of Variance

Source	DF	Adj SS	Adj MS	F-Value	P-Value
Model	9	253.738	28.193	6634.12	0.000
Linear	3	54.136	18.045	4246.24	0.000
A	1	53.973	53.973	12700.29	0.001
B	1	0.162	0.162	38.18	0.019
C	1	0.001	0.001	0.25	0.630
Square	3	185.447	61.816	14545.89	0.000
A*A	1	144.311	144.311	33957.77	0.001
B*B	1	55.995	55.995	13176.09	0.005
C*C	1	0.470	0.470	110.70	0.012
2-Way Interaction	3	14.155	4.718	1110.25	0.000
A*B	1	12.339	12.339	2903.40	0.000
A*C	1	0.688	0.688	161.83	0.001
B*C	1	1.128	1.128	265.51	0.022
Error	10	0.042	0.004		
Lack-of-Fit	5	0.014	0.003	0.51	0.761
Pure Error	5	0.028	0.006		
Total	19	253.781			

Regression Equation in Uncoded Units

$$\text{Result} = 13.6646 + 1.9880\,A - 0.1090\,B - 0.0088\,C - 3.1644\,A*A - 1.9712\,B*B - 0.1807\,C*C + 1.2419\,A*B - 0.2932\,A*C + 0.3756\,B*C$$

3.2 Analysis of response surface

Analyzing effects of 3 factors including gelatin concentration(%)(A), CMC concentration (%)(B) andCMC concentration (%), (B) on tension by drawing Three-dimensional plots according to Figure 1, the researcher found effects of tension from interaction between gelatin concentration(A) 10% ,and CMC concentration (B) 0.8%, However, impacts of tension from interaction between ratio of mixing CMC in gelatin(%),(C) for 10-30% and gelatin concentration (%),(A) or CMC concentration (%), (B) slightly affected tension.

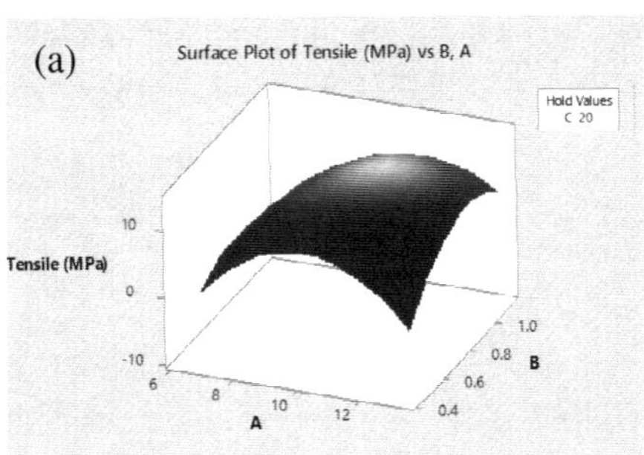

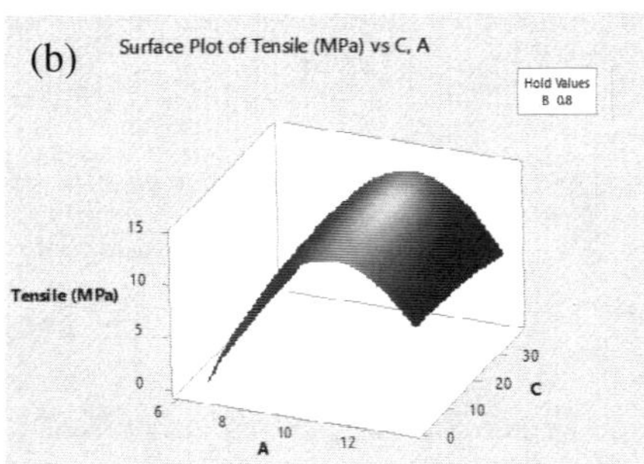

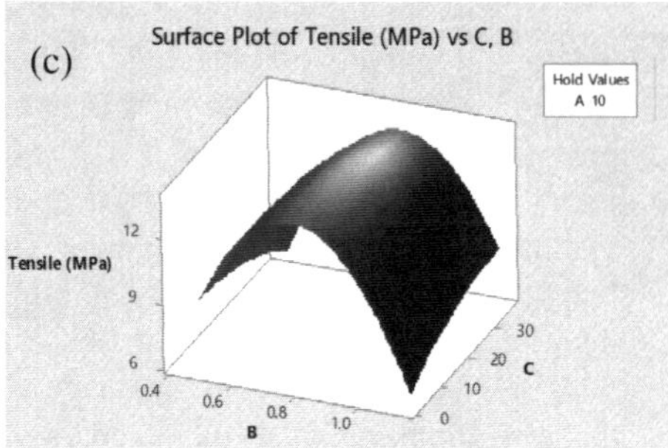

Figure 1 Impacts of tension from co-factors of (a) gelatinconcentration (%)(A)vs CMC concentration (%)(B), (b)gelatin concentration (%)(A) vs ratio of mixing CMC in gelatin (%)(C), (c)gelatin concentration (%) (A)vsratio of mixing CMC in gelatin (%)(C)

3.3 Experiment Analysis

Values of factors for setting to obtain highest response were 10.64% concentration of gelatin (A) 10.64%, 0.810% concentration of CMC (B), ratio of mixing CMC in gelatin (C) for17.79% and needed response value was 13.99 MPa according to Figure. Confidence Interval of mean was found that the possibility to obtain mean of response in this range was 95% and prediction interval was discovered that possibility to obtain response in this range was 95% also.

Multiple Response Prediction
Variable Setting
A 10.645
B 0.810
C 17.791

Response	Fit SE	Fit	95% CI	95% PI
Tensile (MPa)	13.9961	0.0262	(13.9377, 14.0545)	(13.8396, 14.1527)

3.4 Degradation of Gelatin-CMC scaffolds result

This study analyzed the result from Analysis of response surface with degradation method to make a comparison between Gelatin pure scaffolds and Gelatin-CMC scaffolds. Three best values were as follows in the table 4.

Table 4 Blending composition of gelatin/CMC

Samples	Concentration of Gelatin (A),	Concentration of CMC (B)	Ratio of mixing CMC in gelatin (C)
G100	10%	0.8%	0%
G91	10%	0.8%	10%
G82	10%	0.8%	20%
G73	10%	0.8%	30%

From the test, it was found that G100 which was Gelatin pure scaffolds could degrade in 1 hour when compared with G91 that degraded entirely in 24 hours, G73 degraded in 48 hours, G82 degraded in 54 hours with mixture of CMC as shown in figure 2.

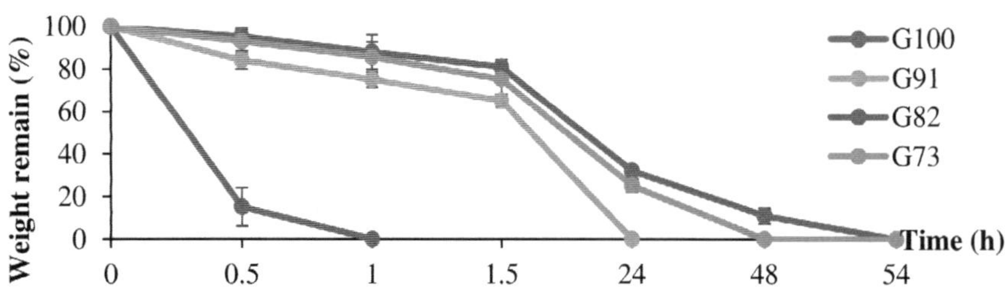

Figure 2 Result of Degradation of Gelatin-CMC scaffolds testing

4. Conclusions

According to the test result, maximum tension was 13.99Mpa originated from concentration of gelatin at 10.64%, concentration of CMC at 0.81% and mixture of CMC at 17.79%. The result of testing degradation between Gelatin-CMC scaffoldsincludingG100, G91, G82, G73 compared with pure scaffolds showed that G82 had slowest degradation rate as a result of best tension. It also was found that gelatin with pure scaffolds could degrade entirely in 1 hour while G91 that degraded entirely in 24 hours, G73 degraded in 48 hours, G82 degraded in 54 hours with mixture of CMC. It means that using CMC which was cellulose derivative as ingredient of scaffold gelatin could strengthen the scaffold gelatin substantially. So that adding CMC in appropriate ratio can help decaying especially G82. The result of decaying conforms to the result of tension from the analysis result of Response Surface Mythology (RSM).

5. Acknowledgments

This research conducted under the support of Department of Advanced Manufacturing Technology and Department of Manufacturing Engineering, Faculty of Engineering, Pathumwan Institute of Technology. There was gratefully acknowledged for provided facilities and experiment being required for this research.

6. References

[1] Z.-M. Huanga, Y.Z. Zhang, S.Ramakrishna, C.T. Lim 2004 Electrospinning and mechanical characterization of gelatin nanofibers, *Polymer* **45**, 5361–5368
[2] Y.Z. Zhang, J. Venugopal, Z.-M. Huang, C.T. Lim, S. Ramakrishna 2006 Crosslinking of the electrospun gelatin nanofibers, *Polymer* **47**, 2911–2917
[3] YuniKusumastuti, NurRofiqohEvianaPutri, and Ahmad RaihanDary2016Electrospinning optimization and characterization of Chitosan/Alginate/Polyvinylalcohol nanofibers*AIP Conference Proceedings* 1755,
[4] F.Wiwatwongwana, and S.Pattana, S. 2011. 'Modification gelatin scaffold with carboxymethylcellulose for dermal skin', *In Proceedings of the 5th International Conference on Bioinformatics and Biomedical Engineering,* 10-12 May, 2011, p.1-4. Wuhan, China. (Oral presentation)
[5] M. Pezeshki-Modaress, H. Mirzadeh, M. Zandi 2015 Gelatin–GAG electrospunnanofibrous scaffold for skin tissue engineering: Fabrication and modeling of process parameters, *Materials Science and Engineering C* **48**, 704–712
[6] S.M. Forghania, M.J. Ghazalia,n, A. Muchtara, A.R. Daudb, N.H.N. Yusoffc, C.H. Azharia 2013 Effects of plasma spray parameters on TiO2-coated mild steelusing design of experiment (DoE) approach*Ceramics International***39**3121–3127
[7] T.Sutjaritvorakul, F.Wiwatwongwana, P.Imsuwan, Anthony J.S. Whalley, S.Chutipaijit2018 Enzymatic degradation of modified gelatin and carboxymethyl cellulose scaffold*Material Today : Proceedings***5**14813-14817
[8] Y.S.Pek, M.Spector, I.V.Yannas, L.J.Gibson2004 Degradation of a collagen–chondroitin-6-sulfate matrix by collagenase and by chondroitinase, Biomaterials Volume **25**, Issue 3 473-482

Development of Natural Facial Mask for Skincare from Local Materials

Channarong Siri[1] and Wassanai Wattanutchariya*

Advanced Manufacturing Technology Research Centre (AMTech), Department of Industrial Engineering, Faculty of Engineering, Chiang Mai University, Chiang Mai 50200, Thailand.

*Corresponding author e-mail: wassanai@eng.cmu.ac.th, [1]channarong_s@cmu.ac.th

Abstract. This research aims to develop a natural facial mask from local materials. The latter are natural materials which could be found locally. In this study, we have selected local materials such as silk fibroin, mulberry, and chitosan. Silk fibroin is active ingredient in the prototype because of its high antioxidant activity. On the other hand, the mulberry extract is antioxidant and antibacterial activity in the facial mask. In the meantime, chitosan could be extracted from local materials such as squid pen and shrimp shells. Chitosan was chosen to be material of the facial mask to nourish good mechanical properties of the mask's structure. The research has developed the natural facial mask with the product design technique. Also, this study integrates consumer's requirements and quality function deployment (QFD) to transform consumer's needs into technical design specifications. The result show that the consumer's requirements which obtain high score are certification standard, ability to release substances, and skin adhesion. Therefore, consumer's preferences and technical specifications helped to achieve the prototype. The prototype was developed and evaluated according to consumer's satisfactions. The result of consumer satisfactions is natural products and skin adhesion. Nevertheless, consumers suggest further improvement in term of color, odor, thickness, and durability of the product.

1. Introduction

Nowadays, healthcare and beauty are becoming increasingly important in everyday life. Cosmetic products are widely used for skincare to enhance beautiful look on both facial appearance and body parts. Cosmetics market expands with a high growth rate. In addition, the herbal cosmetics market costs high value income because the ingredients are imported from abroad, causing consumers to spend high amount of expenses. The entrepreneurs are trying to expand their customer base by purchasing domestic raw materials instead of importing products from abroad. The ingredients are mostly from vegetables, fruits, and grains. In most cosmetics, there are ingredients chemically produced, which may have adverse effects on the consumers while safety is one of their most concerns. In the present, we bring materials from natural extracts used in cosmetics to replace the use of chemicals. So that, "Cosmeceuticals" is a term of combination between cosmetic and medicinal properties. These two equities share the same medicinal benefits and contain many natural active ingredients such as vitamins and antioxidants. Therefore, cosmeceuticals are used for skin rejuvenation. Otherwise, skin care products are created to maintain beauty, most of which are antioxidant activity, antimicrobial activity, wrinkles reduction, make skin radiant, helps skin smooth, add moisture to the skin, skin rejuvenation and old-skin exfoliation. Face mask is an alternative for consumers who need to take care of their facial appearance and gain facial rejuvenation. The facial mask sheet is very popular among people of all ages, both men and women. In addition, this beauty method is quite practical, available in every store, portable, and very refreshing. The mask offers many different recipes and ingredients, which conform to the purposes of consumers such as skin nourishment. In Thailand, there are plenty of ingredients with a variety of materials that can be developed as a cosmeceutical. Silk and mulberry are native plants and can be easily produced. There are great numbers of business of silk and mulberry. From the process of these businesses, it will leave some residues which seem to be useless. For example, silk from the cocoon of *Bombyx mori* consists of fibroin and sericin [1]. Furthermore, silk fibroin is a biomaterial which contains good antioxidant properties, biocompatibility, biodegradation, absorption properties, and good mechanical properties. Moreover, mulberry also has antioxidant and antibacterial properties due to its phenolic compounds, such as flavonoids and anthocyanins [2]. In addition, chitosan can be synthesized from squid pen and shrimp shell. Chitosan is a local material which has a good mechanical properties, antimicrobial activity, anti-inflammatory activity, biocompatibility, biodegradability, nontoxicity, and good absorption properties [3-4]. With all reasons stated, this research manipulates local materials including silk fibroin, mulberry, and chitosan to develop a prototype facial mask for skincare that is non-toxic and non-irritating.

The objective of this research is to synthesize, extract local materials and to analyse the properties to meet consumer's demands. Then, this research aims to develop a facial mask with QFD to meet the needs of consumers. Finally, this research shall evaluate the satisfaction to get a prototype that meets the requirements.

2. Materials and methods

This research has studied local materials to develop a facial mask for skin care by product design techniques. Afterwards, QFD was applied to transform the consumer's needs to the technical requirement. Thus, the prototype was produced and evaluated for the needs of consumers.

2.1. Local materials preparation

For the first process, raw cocoons were cut into small pieces and degummed for 30 minutes with a Na2CO3 solution. Then, filter out the solution. Next, the small pieces were boiled with DI water for 10 minutes and then filter out the water [5], repeat twice. After that, the degummed silk were dried in an oven drying at 37°C for 12 hours and dissolved in a solution of $CaCl_2$:CH_3CH_2OH:H_2O with mole ratio of 1 : 2 : 8 [6-8] at 70°C and applied to water bath at 70°C for 7 hours. Afterward, the solution was dialyzed in a cellulose tube in DI water for 3 days. Then, the silk fibroin was centrifuged at 2000 rpm for 10 minutes. Next, silk fibroin extract was frozen at -20°C and -80°C and then be processed lyophilization. The second process commenced by washing and drying mulberry leaves in an oven drying at 60°C for 48 hours. The dried leaves were crushed and extracted with water at 45°C for 3 hours and filtered by cheesecloth. The extract was centrifuged and filtered with filter paper. Next, the mulberry extract was evaporated with rotary evaporators at 45°C [2]. And finally, for the third process, squid pen chitosan was purchased from Taming Enterprises Co., Ltd, Thailand. Chitosan was soaked in lactic acid solution until the latter became homogeneous [9].

2.2. Consumer survey

The interviews were conducted to evaluate consumers' needs. The raw data was collected from consumers with a set of questionnaires concerning the facial mask product. Consumers' answers were converted to the consumers' needs and categorizing with the affinity diagram. Then, the latter was manipulated as a questionnaire online.

2.3. Quality function deployment

QFD was applied for product design and development. The QFD consists of four phases. This research was operated under the first phase which called "the House of Quality (HOQ)". Its main purpose is to convert customers' requirements into technical requirements. The HOQ consists of five major components: (A) customer requirement, (B) technical requirement, (C) relationship matrix, (D) Co-relationships and (E) Column weights as shown in figure 1. HOQ was employed to transform consumer requirements into technical requirements [10]. It was used to assess the needs of consumers and to improve the product [11]. HOQ assists the facial mask to meet consumers' needs by defining the relationship of consumer's demand with the technical specifications of the product. Information from online questionnaires was also utilized in the form of HOQ. An analysis of the common relationship of consumer's demand with the technical characteristics of the product shall reveal what features are important and which one needs to be developed to meet the needs of most consumers.

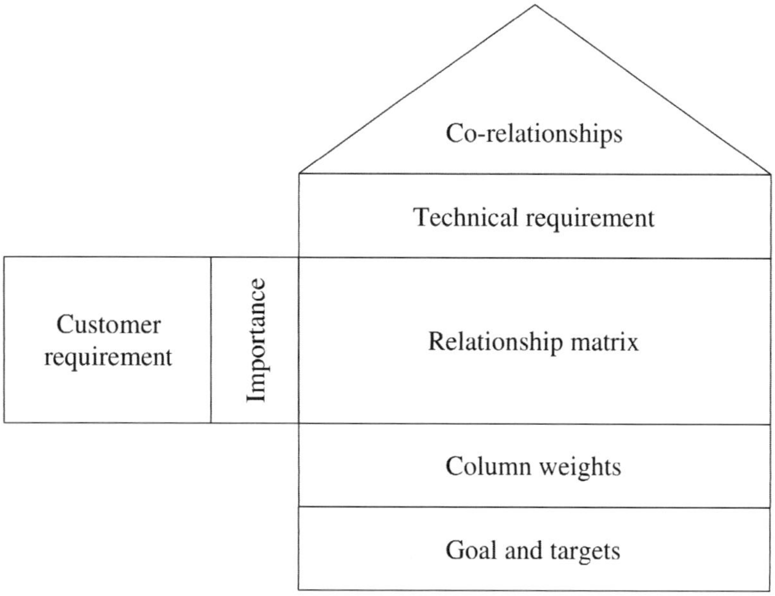

Figure 1. House of Quality (HOQ)

2.4. Develop prototype products

The prototype product characteristics was inquired by the questionnaires, by the results of HOQ before examining and by evaluation. The prototype was tested by cytotoxicity to ensure that the product is nontoxic.

2.5. Consumer Satisfaction Assessment

The prototype was evaluated by consumers to assess the level of satisfaction on it, by using sensory evaluation and by focusing on group questionnaires.

3. Results and discussion

3.1. Material characterization

The local materials were tested to verify their characterization. The silk fibroin extract was analyzed by Fourier transform infrared spectrometer (FTIR). The result reveals that the extract substance has a real fibroin. The mulberry extract was tested by DPPH assay, this test found that the compounds contain antioxidant activity.

3.2. Consumer survey

From an in-depth interview to gather information on consumer's needs, the result is adapted for facial mask production. The research divided consumer's needs into three groups, using the tree diagram as shown in figure 2.

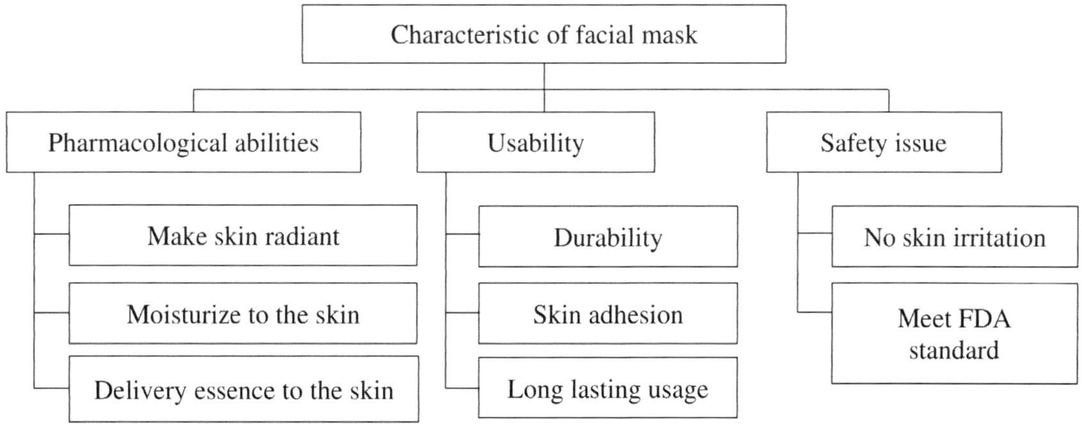

Figure 2. Tree diagram of consumer needs on facial mask.

3.3. House of Quality

HOQ is used to transform customer requirements into technical requirements. The research utilizes the HOQ to identify the most important technical requirements of the facial mask product. According to figure 4, the technical requirement with highest scores are the ability to release substance, skin adhesion, and certification standard. The results indicate that all three technical specifications should be used for product development.

3.4. Prototype of natural facial mask

The prototype of local material was developed according to the technical requirements and formed by lyophilization as shown in figure 3. The prototype was designed as an eye mask for testing. The size is 60×30 millimetres and has a thickness of 3 millimetres. The prototype from the local materials was tested by cytotoxicity and found to be non-toxic.

Figure 3. Prototype of natural facial mask.

3.5. Evaluation of consumer satisfaction

The result of the focus group assessment reveals that most consumers are satisfied with a product made from natural materials which contains a quality of good skin adhesion. The consumers affirm that they feel safe to try the product and the price is acceptable. However, consumers are also concerned about the colour, odour, thickness, and durability of the product. There are suggestions on how to improve the product such as the colour, it should be light brown or white to augment the friendly appearance to customer. Therefore, we have explained about the prototype that the colour is derived from the nature of mulberry, which is the main component of the product. In addition, the prototype odour is a natural extract which smells unpleasantly to the consumer. Hence, the process and technique should be improved to prevent further odour. The thickness of the prototype is too thick. Consumers suggest that the layer should be thinner to fit more precisely to the facial skin. For the durability of the prototype, it is recommended that the freeze-drying forming technique and proportion of mixture should improve the proportion of mixture, so that shall render prototype to be more durable and consistent with the need of consumers.

Symbol	Meaning	Score
	No relationship	0
△	Slight or possible	1
○	Moderate relationship	3
◉	Strong relationship	9

Symbol	Meaning
	No impact
++	Strong positive
+	Moderately positive
-	Moderately negative
--	Strong negative

Movement of Target Value
↑ Large the better
○ Target the best
↓ Smaller the better

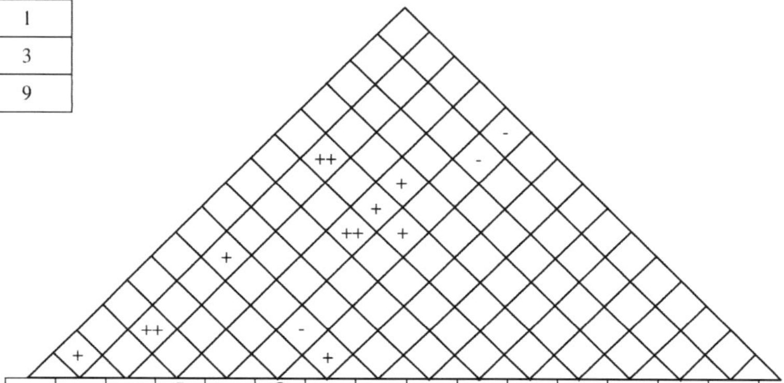

			Physical characteristics								Pharmacological characteristics				Mechanical properties	Safety	Cost
			↑	↑	↑	○	↓	○	↑	↑	↑	↑	↑	↑	↑	○	↓
Demand factors that affect product attributes		Importance score	Odor/smell	The amount Essence fit	Sensitive skin absorption	Ability to release substance	Sheet thickness	Face Fit	Adhesion to the surface well.	Long lasting usage	Skin lightening	Antioxidant	Add moisture	Acne treatment	Tensile strength	Certified standard	Cost of product
Physical abilities	No pungent smell	3.66	◉	△												△	
	Excess amount of Essence	3.69			△	△			○	○			△				○
	Essence volume is too low	3.38			△	△			○	○			△				○
Safety standards	Meet standard	4.51														◉	△
Pharmacological abilities	Skin Resurfacing	4.77		△		○					◉	◉		△			
	Moisturize the skin	4.74		△		○							◉				
Usability	Close the face	4.47					○	◉	◉	△						△	
	Not tear easily	3.69					○								◉	△	△
	Duration of use reasonable	4.11			◉	◉											

Target Value	Passed sensory evaluation	Passed sensory evaluation	Passed chromatography	Passed chromatography	Passed sensory evaluation	Passed sensory evaluation	Passed sensory evaluation	15-30 minutes	Tested by DPPH. assay	Tested by DPPH. assay	Passed sensory evaluation	Passed anti-acne bacteria test	Tested by tensile test	Regulated by FDA.	30Baht/sheets	Total score
Relative Technical Requirement	32.9	34.4	44.1	61.5	24.5	40.2	61.4	25.7	42.9	42.9	49.7	4.77	33.2	52.4	29.4	580
Absolute Technical Requirement	5.68	5.93	7.6	10.6	4.22	6.94	10.6	4.43	7.4	7.4	8.57	0.82	5.73	9.04	5.07	100

Figure 4. House of Quality (HOQ) for natural facial mask.

4. Conclusion

In this study, the active ingredient was extracted from local materials. In the test, silk fibroin and mulberry extracts contain antioxidant and antibacterial activity. The study of consumer's demands applies QFD techniques to develop the prototype of a natural facial mask. The technical requirements that consumers demand primarily are certification standard, ability to release substances, and skin adhesion. These are the technical requirements which led to the development of the prototype. The results of the prototype approach to sensory evaluation among consumers who tried facial mask, we found that the consumers were satisfied with the product at the moderate level. The consumers were satisfied with natural materials and skin adhesion. In addition, the results of the cytotoxicity assay affirmed that the product was initially standard in terms of non-toxicity to skin cells. However, consumers suggest further improvement in term of color, odor, thickness, and durability of the

IOP Conf. Series: Materials Science and Engineering **635** (2019) 012003 doi:10.1088/1757-899X/635/1/012003

prototype. From interviews and consumer suggest, the researchers shall develop and improve the prototype in the next generation of products to better meet the needs of consumers.

5. Acknowledgement
This research was support in financial sponsorship by the Thailand Research Fund under grant number RDG60T0106. The researchers also appreciate the support from Graduate School of Chiang Mai University.

References
[1] Altman G H, Diaz F, Jakuba C, Calabro T, Horan R L, Chen J, Lu H, Richmond J and Kaplan D L 2003 Silk-based biomaterials *Biomaterials* **24** 401-416
[2] Song-Hwan B and Hyung-Joo S 2007 Antioxidant activities of five different mulberry cultivars in Korea *LWT-Food Science and Technology* **40** 955-962
[3] Sionkowska, A and Płanecka A 2013 Preparation and characterization of silk fibroin/chitosan composite sponges for tissue engineering *Journal of Molecular Liquids* **178** 5-14
[4] Friedman A J, Phan J, Schairer D O, Champer J, Qin M, Pirouz A, Blecher-Paz K, Oren A, Liu P T, Modlin R L and Kim J 2013 Antimicrobial and anti-inflammatory activity of chitosan–alginate nanoparticles: A targeted therapy for cutaneous pathogens *Journal of investigative Dermatology* **133** 1231-1239
[5] Sah M K and Pramanik K 2010 Regenerated silk fibroin from B. mori silkcocoon for tissue engineering applications *International journal of environmental science and development* **1** 404
[6] Liu H, Li X, Zhou G, Fan H and Fan Y 2011 Electrospun sulfated silk fibroin nanofibrous scaffolds for vascular tissue engineering *Biomaterials* **32** 3784-3793
[7] Nguyen H T, Luong H T, Nguyen H D, Tran H A, Huynh K C and Vo T V 2017 Investigate the Effect of Thawing Process on the Self-Assembly of Silk Protein for Tissue Applications *BioMed research international* **2017** 1-16
[8] Yamada H, Nakao H, Takasu Y and Tsubouchi K 2001 Preparation of undegraded native molecular fibroin solution from silkworm cocoons *Materials Science and Engineering C* **14** 41-46
[9] Wattanutchariya W and Changkowchai W 2014 Characterization of porous scaffold from chitosan-gelatin/hydroxyapatite for bone grafting *In Proceedings of the International MultiConference of Engineers and Computer Scientists* (Hong Kong) vol 2 pp 12-14
[10] Lowe A, Ridgway K and Atkinson H 2000 QFD in new production technology evaluation *International Journal of Production Economics* **67** 103-112
[11] Wattanutchariya W and Royintarat T 2012 Implementation of quality function deployment and kansei engineering for GABA rice snack development *Environment and Natural Resources Journal* **10** 1-10

IOP Conf. Series: Materials Science and Engineering **635** (2019) 012004 doi:10.1088/1757-899X/635/1/012004

Bending strength and Biological properties of PLA-HA composites for femoral canine bone fixation plate

Saksorn Sadudeethanakul[1], Wassanai Wattanutchariya*, Wasawat Nakkiew[2], Anirut Chaijaruwanich[3], and Siwasit Pitjamit[4]

Advanced Manufacturing Technology Research Center (AMTech), Department of Industrial Engineering, Faculty of Engineering, Chiang Mai University, Chiang Mai 50200, Thailand

*Corresponding author's email address: wassanai@eng.cmu.ac.th, [1]saksorn_sadudee@cmu.ac.th, [2]wasawat@eng.cmu.ac.th, [3]anirut@eng.cmu.ac.th, [4]siwasit_p@cmu.ac.th

Abstract. This research is based on the purposes to find suitable hot compression conditions of composites materials, also to optimize mechanical and biological properties of the plate specimen. The type of composites materials under this investigation are bio-polymers (polylactic acid: PLA) and bio-ceramic (Hydroxyapatite: HA). In this research, the term "The design of the experiment (DOE)" was implemented to find the relation between forming factors including the ratio of HA ratio, molding temperature, and molding pressure. The result showed that the optimum conditions of the bending test were at 5%wt. of HA, 160 °C, 7 MPa of molding temperature and pressure respectively. Biodegradation rate of a specimen was 0.0175 g/day and the specimens acquired a good biocompatibility and they were non-toxic to living cells.

1. Introduction

The hind limb of canines is one of the areas which are interesting to veterinary orthopedic surgeons. In the surgical treatment of long bone diaphyseal fracture in canines, it is consistently attractive to use the bone plates [1,2] which are normally produced from metal such as stainless-steel grade 316L or titanium alloys. However, the strength of metal plate which is greater than the femoral bone may cause cracking around the surgical area. Furthermore, released ions from metal surface may lead to infection, inflammation, and swelling of the adjacent tissues [3]. According to the stated issues, this research shall find alternative materials for repairing tissues without toxicity. Thus, the compelling alternative materials in this research are bio-polymer: polylactic acid (PLA) and bio-ceramic: Hydroxyapatite (HA)

PLA is one of the most extensive thermoplastic aliphatic polyesters which is derived from a lactic acid monomer and the natural products of fermentation as sugar or corn. It becomes fusible at a temperature average 180 to 220 °C and at average 60 to 65 °C glass transition temperature (Tg). PLA has a good tensile strength, rigidity, and stiffness. Moreover, PLA can be degraded easily by a hydrolysis reaction between the ester group which occurs in the formation without injury and toxicity [4,5] and its properties are excellent for optimization with a biomedical application, for instance; internal plates or screw fixation of damaged or fractured bone [6].

HA is a fundamental inorganic component of bones and teeth because of its properties such as bone bonding ability and greater bioactivity to stimulate the formation of new cells surrounding damaged tissues. Therefore, HA can be applied in many fields; for example, coating on metallics implants and bone substitutes [7]. However, the disadvantage of HA is low biodegradability. Therefore, it is not suitable to use in bone scaffolds or cells promotion [8]. With references to the subject mentioned above, this research has attempted to combine HA with polymer composite to use in the biomedical application. As a polymer matrix composite has the advantage of being very adaptable, allowing the accommodation for its properties thus, this shall allow the accommodation to access its properties [6,9].

Factorial designs have been commonly used in experiments about several factors, which it is necessary to investigate a response in each effect of the factor. The 2^k design is basically the term which we use in the primary steps of experimental processes because it contributes to the fundamental number of runs from all experiments and complete factorial design could be studied in each k factor. Therefore, for optimal experiments, it is necessary to use these designs to verify this matter [10].

Then, this research aims to investigate and identify optimal forming conditions and degradation rates of PLA-HA composites. The fabrication method of PLA-HA compact specimens for canine's femoral bone plates is based on the 2^k factorial design with the center point. The optimal conditions for mechanical property was reported in terms of bending strength. Furthermore. Biodegradation was reported by calculating weight loss from the specimen in PBS solution and biocompatibility was tested with osteoblast cell lines. Finally, the actual determined bending, biodegradation and biocompatibility properties of the specimen from this research have benefits to apply for using in canine femoral in the future.

2. Materials and Methods

In this research, researcher manufactured PLA-HA composites based on the design of experiment, theory and fabricated all of specimen by hot compression molding.

2.1 Material Preparation

Polylactic acid (Molecular weight ≈ 60,000-80,000) was purchased from Natureworks®. Polymer pellets was reduced in particle size by Universal Milling machine with liquid nitrogen, then, we sieve the pellets to become powder and keep them in an oven under atmospheric temperature [11].

Hydroxyapatite powder (HA) was prepared from bovine bones. The first step is to reduce the bovine bones into small pieces by cutting or grinding method, then, use hydrogen peroxide to soak these bones to remove organic tissues. Afterwards, remove their collagen and organic substances with boiling water for a moment. Next, keep these bones to reduce moisture content in the hot air oven under ambient temperature. Finally, calcine all dry bones at 900°C for 3 hours and reduce particle size less than 20 μm [3] with a high-speed ball milling machine.

Add PLA and HA from fixing ratio (from table 1) into the ball mill grinder for 24 hours after, then, remove their moisture content in the vacuum oven at 80 °C for 12 hours.

2.2 Methods

All samples were fabricated by hot compression molding process. Taking 12.6 grams of composite powder to ensure fully gravity of die casting compression mold which is pre-heated average 80°C before forming a specimen. A specimen was fabricated under HA proportion 5-10 %wt. with the molding temperature of 140-160 °C and with molding pressure 3-7 MPa [12]. Also, the factorial design was applied with center-point and set up following table 1.

2.3 2^k Full factorial design

The experiment was designed for the various levels of composites forms according to 2^k factorial design with the center point. The factor in this experiment which regards to three factors contains HA proportion, temperature, and pressure following (table 1) repeat all of experiment twice in each condition then created a 2^k factorial table with software as Minitab18. The table showed 20 experiments include 16 factorial designs and 4 center points. The responses of all experiment were reported on bending strength, biodegradation, and biocompatibility

Table 1. Parameters for experimental design [12].

Factors	Levels		
	Low (-1)	Medium (0)	High (+1)
HA (%)	5	10	15
Molding Temperature (°C)	140	150	160
Molding Pressure (MPa)	3	5	7

2.4 Characterizations

All specimens were characterized on the mechanical property: bending strength. The specimen of optimum bending strength was tested to biological properties: biodegradation rates and biocompatibility. Minitab18 was used to consider a statically significant variability.

2.4.1 Bending strength

The bending strength of PLA-HA composites in each condition was measured by Universal Testing Machine (Instron 5566) with Three-point bending under the guidance of ASTM D790 at a loaded cell 100 N. Stress-strain curves were analyzed to find bending strength in each sample [13].

2.4.2 Biodegradation

The PLA-HA specimens were cut on cubic size 5 mm., into 36 pieces, then carried out these samples into the Phosphate Buffer Saline solution, pH ≈ 7.4, volume 1 ml. All samples were kept on disc gathered in the oven at 37 ºC for up to 1, 3, 7, 14, 30, and 50 days. At daily intervals, these samples were removed from the Phosphate Buffered Saline (PBS) solution and dried at room temperature, then calculated equation $[100(W_i\text{-}W_f)]/W_0$ to find the percentage of weight loss where Wi is the initial sample weight and W_f is the weight of dried residual [14].

2.4.3 Biocompatibility

PLA-HA samples were tested with osteoblast cell following on ASTM F895-11 and ASTM F1903-10. The percentage of cell viability was measured with the microplate reader at wavelength 570 nm. Finally, we calculated the following equation [OD of treated cell/OD of control cell] *100 and compared with osteoblast cells without Dulbecco's Modified Eagle Medium (Control sample).

2.4.4 *Statistical analysis*

2^k factorial design with center point was used to investigate the effect of hot compression molding and the statistical significance value between three factors in this experiment, which was identified by ANOVA. Statistically, signification was considered with P-value less than 0.05 at 95 confidential levels. Minitab18 software was employed for the statistical analysis.

3. Result and Discussion

All specimens were investigated to find a suitable condition of the mechanical property in term of bending strength, which we found at 5%wt. of HA, 160°C, 7 MPa. Biodegradation property of the specimen showed that it can be degraded in 0.0175 g/day or average 15 years. In addition, biocompatibility also showed cell viability at 88.97%, this can be concluded that this specimen is appropriate to use in femoral canine bone.

3.1 Bending strength

According to the experiment, we found the results of bending strength of three factors include; HA ratio, molding temperature, molding pressure to fabricate PLA-HA composites which were analyzed by minitab18 software. The results showed in table 2 and 3 that founded the highest bending strength was found at 5%wt. of HA ratio, temperature 160 °C and pressure 7 MPa. Thus, this highest bending value was nearly range with bending strength of cortical bone following in table 4. At higher HA ratio over 5%wt, the dispersal of HA particles reduced, which was the causes from higher viscosity than lower HA proportion in PLA matrix, and finally the bending strength of PLA-HA composites to get decreased [7]. In addition, the highest temperature appeared better distribution of the HA through PLA surface [15]. Researcher can assume that bending strength strongly depended on the effect of HA ratio and temperature because of its present p-value of both less than 0.005, R-sq and R-sq adjust are 90.07%, 82.85% respectively. Nevertheless, molding pressure of all experiment in each level of factor show the result which P-value reached more than 0.05. This indicates that it is not significantly different to bending strength test because of hot compression machine has limitation to be able do that.

Table 2. Bending strength of all condition of specimens tested

No.	HA (wt.%)	Temperature (°C)	Pressure (MPa)	Bending Strength (MPa)
1	5	140	3	73.092
2	15	140	3	41.670
3	5	160	3	63.711
4	15	160	3	58.701
5	5	140	7	48.857
6	15	140	7	51.627
7	5	160	7	70.336
8	15	160	7	49.591
9	5	140	3	64.889
10	15	140	3	47.915
11	5	160	3	61.627
12	15	160	3	55.432
13	5	140	7	41.001
14	15	140	7	55.766
15	5	160	7	75.275
16	15	160	7	42.400
17	10	150	5	56.292
18	10	150	5	50.889
19	10	150	5	46.252
20	10	150	5	47.084

Table 3. The estimated effects and coefficients bending strength from minitab18 software

Term	Coef	SE Coef	T	P
Constant	56.368	1.069	52.75	0.000
HA ratio	-5.980	1.069	-5.60	0.000
Temperature	3.266	1.069	3.06	0.011
Pressure	-2.011	1.069	-1.88	0.087
HA ratio*Temp	-2.123	1.069	-1.99	0.072
HA ratio*Pressure	1.470	1.069	1.38	0.196
Temp*Pressure	1.778	1.069	1.66	0.124
HA ratio*Temp*pressure	-6.772	1.069	-6.34	0.000
Ct Pt	-6.239	2.390	-2.61	0.024
R-Sq = 90.07%		**R-Sq (adj) = 82.85%**		

Table 4. The average bending strength of materials and cortical bone.

	Bending Strength (MPa)
Cortical Bone	61 [16]
Stainless Steel (316L)	645 [16]
Polylactic Acid	108 [17]
Hydroxyapatite	40 [18]

3.2 Biodegradation

Biodegradation of PLA-HA composite specimen at 5%wt. of HA ratio, molding temperature 160 °C and molding pressure 7 MPa. Figure 1. showed a relative between the percentage of degradation and the number of days to degrade that find its value from the graph. Predicted the percentage of degradation using least-square on Regression model to fit data to find the rate of an equation. The percentage of degradation Y in term of day of degradation X was explained by $Y = 0.0175X + 0.0376$ which means that a specimen can be degraded in 0.0175 g/day. We assumed that PLA-HA composite plate can be used appropriately in canines because of its bending property values nearly canines' bone and degrade in average 15 years.

3.3 Biocompatibility

PLA-HA sample was tested on cell viability with osteoblast cell for 24 hours and repeated in each condition 3 times to prove that all these composites samples are non-toxic. As a result, figure 2 showed PLA-HA samples have the percentage of cell viability at 88.97 (more than 80% when compared with a control sample) that means it is possible to be applied in the living body.

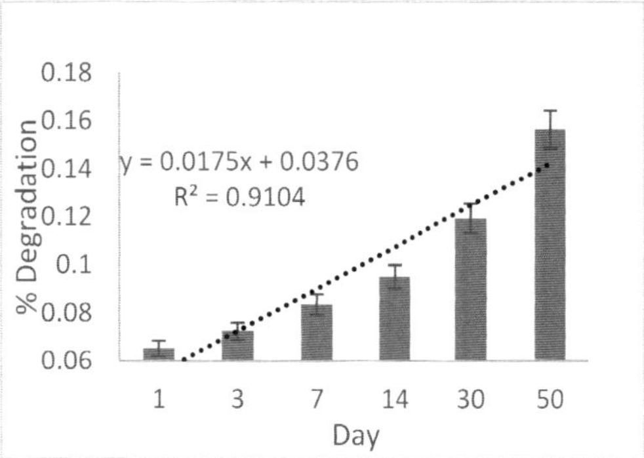

Figure 1. Relationship between percentage of degradation of PLA-HA composites at optimum condition

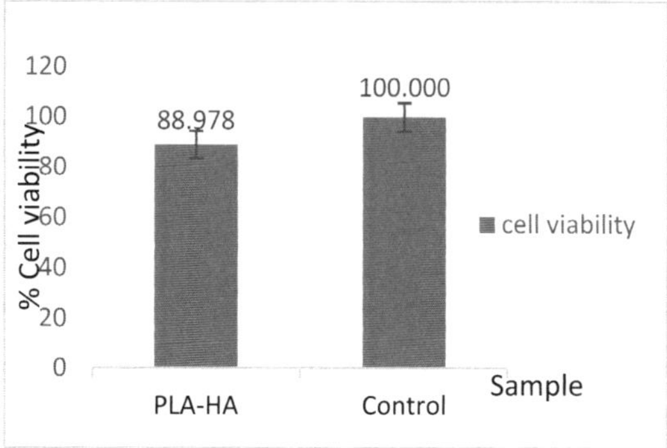

Figure 2. Comparison of percentage of cell viability between PLA-HA at optimum condition and control

4. Conclusion

In this study, researcher purpose is to fabricate a composites plate containing 2 kinds of biomaterial; polylactic acid and hydroxyapatite, by hot-compression molding method. All of experiment in this study, researcher employed 23 factorial design with the center point and 2 times repeated. The variable conditions that we observed were HA ratio (5, 10, 15 %wt.), molding temperature (140,150, 160°C) and pressure (3, 5, 7 MPa). Bending strength was investigated by a universal testing machine (UTM) and biodegradation rate was taken in PBS solution and Lysozyme for 1 3 7 14 and 50 days. The result showed that the optimum conditions of bending strength was found at the 5%wt. of HA, 160°C, 7 MPa. Moreover, biodegradation rate of specimen is 0.0175 g/day. In future works, in order to ensure the suitable femoral canine bone, biomechanical property including value comparison with finite element analysis shall be analysed. The investigation of biocompatibility shall be conducted to certify that composites plate are nontoxic and can be used in living canine's body without danger.

Acknowledgement

This research was supported in financial sponsorship by the National Research Council of Thailand (NRCT) under Project ID 270376. The researchers would like to thank Small Animal Hospital, Faculty of Veterinary Medicine and Graduate School, Chiang Mai University for their facility support.

Reference

[1] Pfeil, D. J. V., Déjardin, L. M., DeCamp, C. E., Meyer, E. G., Lansdowne, J. L., Weerts, R. J., & Haut, R. C. (2005). In vitro biomechanical comparison of a plate-rod combination–construct and an interlocking nail–construct for experimentally induced gap fractures in canine tibiae. American journal of veterinary research, 66(9), 1536-1543.

[2] Shahar, R., & Banks-Sills, L. (2002). Biomechanical analysis of the canine hind limb: calculation of forces during three-legged stance. The veterinary journal, 163(3), 240-250.

[3] Pitjamit, S., Sriprapha, P., & Nakkiew, W. (2016) Suitable forming condition of hydroxyapatite and bioactive glass composites for a bone fixation plate using Taguchi experimental design. Engineering and Applied Science Research, 43, 466-469.

[4] P Pawar, R., U Tekale, S., U Shisodia, S., T Totre, J., & J Domb, A. (2014). Biomedical applications of poly (lactic acid). Recent Patents on Regenerative Medicine, 4(1), 40-51.

[5] Wachirahuttapong, S., Thongpin, C., & Sombatsompop, N. (2016). Effect of PCL and compatibility contents on the morphology, crystallization and mechanical properties of PLA/PCL blends. Energy Procedia, 89, 198-206.

[6] Kasuga, T., Ota, Y., Nogami, M., & Abe, Y. (2000). Preparation and mechanical properties of polylactic acid composites containing hydroxyapatite fibers. Biomaterials, 22(1), 19-23.

[7] Shen, L., Yang, H., Ying, J., Qiao, F., & Peng, M. (2009). Preparation and mechanical properties of carbon fiber reinforced hydroxyapatite/polylactide biocomposites. Journal of Materials Science: Materials in Medicine, 20(11), 2259-2265.

[8] Wattanutchariya, W., Ruennareenard, J., & Suttakul, P. (2016). Appropriate forming conditions for hydroxyapatite-bioactive glass compact scaffold. Engineering Journal, 20(3), 123-133.
[9] Nainar, S. M., Begum, S., Ansari, M. N. M., & Anuar, H. (2014). Tensile Properties and Morphological Studies on HA/PLA Biocomposites for Tissue Engineering Scaffolds. International Journal of Engineering Research, 3(3), 186-189.
[10] Montgomery, D. C. (2017). Design and analysis of experiments. John wiley & sons.
[11] Kunioka, M., Ninomiya, F., & Funabashi, M. (2006). Biodegradation of poly (lactic acid) powders proposed as the reference test materials for the international standard of biodegradation evaluation methods. Polymer degradation and stability, 91(9), 1919-1928.
[12] Watcharaprapapong, P., Nakkiew, W., Wattanuchariya, W., & Pitjamit, S. (2018). Effect of forming conditions of poly-lactic acid/hydroxyapatite to tensile strength of canine bone fixation plate using full factorial experimental design. MATEC Web of Conferences (Vol. 192, p. 01049, 1-4.
[13] Standard test methods for flexural properties of unreinforced and reinforced plastics and electrical insulating materials. ASTM; D790.
[14] Wu, T. M., & Wu, C. Y. (2006). Biodegradable poly (lactic acid)/chitosan-modified montmorillonite nanocomposites: preparation and characterization. Polymer Degradation and Stability, 91(9), 2198-2204.
[15] Charles, L. F., Shaw, M. T., Olson, J. R., & Wei, M. (2010). Fabrication and mechanical properties of PLLA/PCL/HA composites via a biomimetic, dip coating, and hot compression procedure. Journal of Materials Science: Materials in Medicine, 21(6), 1845-1854.
[16] Kharazi, A. Z., Fathi, M. H., & Bahmany, F. (2010). Design of a textile composite bone plate using 3D-finite element method. Materials & Design, 31(3), 1468-1474.
[17] Hartmann, M. H. (1998). High molecular weight polylactic acid polymers. Biopolymers from renewable resources (pp. 367-411).
[18] Chen, Q., Zhu, C., & Thouas, G. A. (2012). Progress and challenges in biomaterials used for bone tissue engineering: bioactive glasses and elastomeric composites. Progress in Biomaterials, 1(1), 2.

IOP Conf. Series: Materials Science and Engineering **635** (2019) 012005 doi:10.1088/1757-899X/635/1/012005

Effect of silver nanoparticles with polyelectrolyte multilayer against Aggregatibacter actinomycetemcomitans

S wacharanad[1,*] and T kluaymai na ayudthaya[2]

[1]Department of preventive, Faculty of Dentistry, Naresuan university, Thailand
[2]Faculty of Dentistry, Naresuan university, Phitsanulok,Thailand

Corresponding Author Email: sirorat.w@gmail.com

Abstract. Silver nanoparticles capped with alginate (SNP-AL) were prepared by chemical reduction method. Three concentrations of SNP-AL were produced by varied alginate into 0.1, 1, and 5 mM. UV-Vis spectroscopy was used to characterize their specific wavelength at 400 nm. Subsequently, 10 bilayers thin film were established between PDADMAC and SNP-AL via polyelectrolyte multilayer technique. These were characterized by SEM and AFM. Their toxicity to anaerobe bacteria *Aggregatibacter actinomycetemcomitans* (Aa) a common cause of periodontal disease, was tested. Findings showed that these silver nano-chips could release SNP and reduce the Aa colony by 90%.

1. Introduction

Silver nanoparticles (SNP) are increasingly being applied in the fields of medical and dentistry treatment because the unique properties of these particles make them especially effective in fighting bacterial infections [1],[2],[3]. In addition, they can be useful in treating patients who receive antibiotics regularly, and may be at risk of developing drug resistance [2],[3],[4]. They are also useful for treating patients who have been found to be allergic to some antibiotics such as Penicillin, etc [5],[6]. Thus, SNP offers a new alternative for inhibiting or destroying bacteria, especially for anaerobic bacteria, which are a significant cause of diseases such as periodontal disease. *Aggregatibacter actinomycetemcomitans* (Aa) is one of several bacteria causing periodontal disease [7],[8]. The toxins released from this bacteria can destroy the gum, causing dissolution of bone around the tooth, gum recession, tooth rocking, and eventual tooth loss[8],[9],[10]. In general treatment, after having cleaned the tooth, the dentist will wash the gum with an antibiotic drug or put the periochip, which is a film containing an antibiotic drug, into the gum to inhibit bacterial growth [11],[12],[13]. Therefore, the researcher is interested in developing this film by using SNP with PDADMAC to coat in several layers with polyelectrolyte multilayer technique as it is the technique that can coat several materials on several kinds of surface easily and not costly [14],[15]. Moreover, the features of multilayer coating and the porosity between layers can also control the release of SNP from the film [16].

Thus, this research aims to develop several layers of film between PDADMAC and SNP, in three concentrations, for use in testing their efficacy in inhibiting the growth of Aa over a 24 hour period.

2. Materials and methods

2.1 Chemicals

Poly(diallyldimethylammoniumchloride) (PDADMAC, medium molecular weight, 20% by wt in water, typical Mw=200,000 – 350,000), Alginic acid (Al, Typical Mw = 10,000 – 600,000), poly(sodium 4-styrenesulphonate) (PSS, typical Mw = 70,000) and sodium borohydride were purchased from Sigma-Aldrich, (Thailand) Co., Ltd. Silver nitrate was purchased from VWR, Thailand. Sodium chloride was purchased from Labscan Asia Co., Ltd., Thailand. All chemicals and solvents were used as received without any further purification. Doubly distilled water was used in all experiments.

2.2 Silver nanoparticles preparation

Chemical reduction was used to prepare SNP capped with alginate. 1 mM of silver nitrate was mixed with AL varying concentration from 0.1, 1 and 5 mM. then added 10 mM of sodium borohydried in the previous solution and stirred 15 min. the solution was change into yellow color. They were left overnight in room temperature. After that they were analyzed via UV-Vis spectroscopy.

2.3 Multilayer thin films preparation

Ten millimolar PDADMAC were prepared and then added to 0.1M NaCl in PDADMAC solution. A quartz slide was cleaned with piranha solution (H_2SO_4:H_2O_2 (70:30) and then coated with four layers of primer (PDADMAC/PSS). The multilayer thin film between PDADMAC and SNP-AL was prepared by dipping the quartz slide in PDADMAC for 2 min, followed by three rinses in distilled water. The slide was then dipped in SNP-AL 1:0.1 for 2 min, followed by three rinses with distilled water. This step achieved the formation of one bilayer. Dipping was continued in PDADMAC and SNP-AL 1:0.1 until ten bilayers of PDADMAC/SNP-AL 1:0.1 had formed. The preparation of PDADMAC and SNP-AL 1:1 and PDADMAC and SNP-AL 1:5 thin film followed the same method [17]. These SNP-AL films were analyzed their surface by scanning electron microscopy and measured their thickness by atomic force microscopy.

2.4 Antibacterial property against Aa

The antibacterial activity against Aa was tested using the standard method. The SNP films were exposed to 20 µl of Aa in brain heart infusion broth (2 ml). After 24 h incubation at 37 ◦C, 5% CO_2 condition, the bacteria/broth mixture was diluted five times. Then 50 µl of diluted bacteria was placed onto brain heart infusion agar using the spread plate method. After 24 h incubation the bacteria were counted. The result was corrected by the dilution factor to give the number of colony forming units (CFU) per millilitre. The percentage of bacterial reduction was then calculated and compared to blank condition.

3. Results and discussion

3.1 UV-Vis spectroscopy

The SNP-AL with the concentration of alginate 0.1, 1, and 5 mM will be prepared through chemical reduction. The three substances have the concentration in yellow color as shown in Figure 1. When measuring the light absorbance with UV-VIS spectroscopy, the three substances have the peak value of light absorbance at 400 nm which is the specific absorbance value of SNP [18]. This can confirm the competence of SNP coated with alginate.

3.2 multilayers of PDADMAC/SNP-AL

When 20 layers of PDADMAC-SNP are coated on quartz slide, multicolored film will be gained as shown in Figure 2. For PDADMAC/SNP-AL 1:0.1, the color will be dark gray which is the feature indicating the film with much thickness as SNP-AL 1:0.1 has low negative charging value surrounding particles resulting in when being coated on PADAMAC, the particles will arrange in the short distance as there is low repulsive force between the charge. This causes the film in each layer contains the

IOP Conf. Series: Materials Science and Engineering **635** (2019) 012005 doi:10.1088/1757-899X/635/1/012005

particles of SNP-AL densely. At 20 levels, it will be in metallic gray color corresponding with the features of the film PDADMMAC/SNP-AL 1:0.1 [18],[19]. When being analyzed with SEM, it is found that the features of particles are large and spread densely on the film as shown in Figure 3. On the contrary, for PDADMAC/SNP-AL 1:1 and 1:5 mM with the increasing concentration of alginate causes the charge density surrounding SNP to increase. Consequently, the repulsive force between the particles us high. The SNP particles will spread resulting in when being coated as multilayer film, SNP will spread over PDADMAC layer non-densely. Multilayer film to be gained will be thin and in yellow color [19]. When analyzing it with SEM, the particles will be seen as smaller as shown in Figure 3D. The thickness of the film was analyzed by atomic force microscopy and the results are compiled in Table 1. An average thickness is can be measured when considering the height of the film with respect to the substrate.

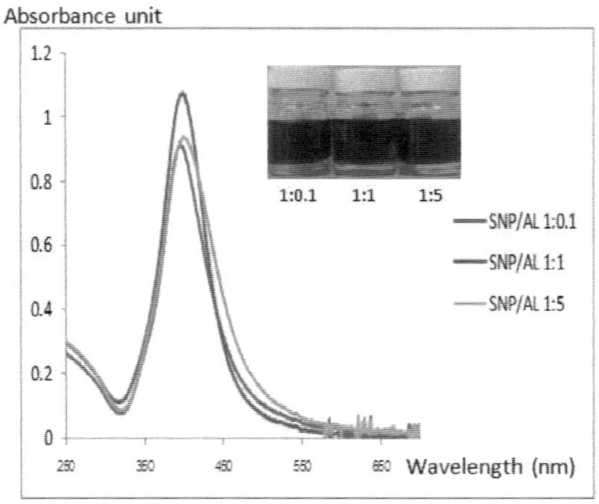

Figure 1. The UV–vis absorbance spectra of SNP-AL solution in a various concentration of alginate.

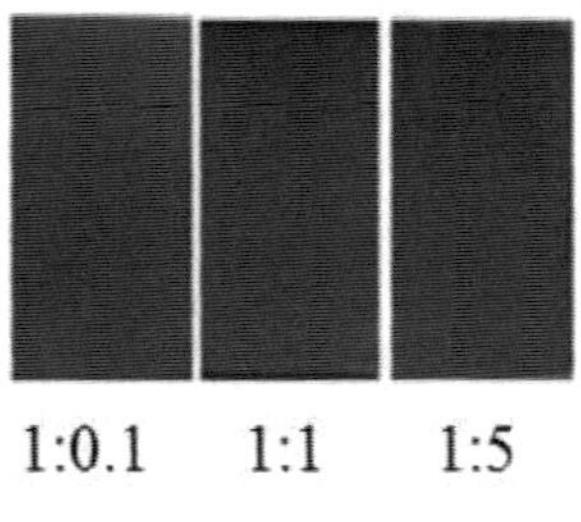

Figure 2. Ten bilayer of PDADMAC/SNP-AL in a various conditons were coated on glass slide.

Table 1. The thickness of SNP-AL films were showed in this table

samples	Thickness(μm) Mean average ± SD
Blank	0
PDADMAC/SNP-AL 1:0.1	1.28± 0.16
PDADMAC/SNP-AL 1:1	0.52± 0.09
PDADMAC/SNP-AL 1:5	0.75±0.06

3.3 Antibacterial effect (ATCC29523)

For testing the three kinds of film on Aa, it is found that when the film is exposed to it for 24 hours, the increase in amount of colony of Aa can reduce compared to the condition without the film. As shown in Table 2, it can be seen that the Aa without the SNP film will have the amount of 2.02 x 10^8

CFU. However, when the film is inserted for 24 hours, the amount of colony of Aa will reduce. PDADMAC/SNP-AL 1:0.1 causes colony to remain 1.45 x 10^7 CFU, PDADMAC/SNP-AL 1:1 remains 1.5 x 10^7 CFU and PDADMAC/SNP-AL 1: 5 remains 1.75 x 10^7CFU. When calculating in % of reduction, the amount is around 92.80, 92.57, and 91.34 %, respectively. It can be seen that three kinds of multilayer film cannot totally destroy Aa, only reduce the amount of colony. According to such result, it can be seen that SNP can spread from the film to Aa.

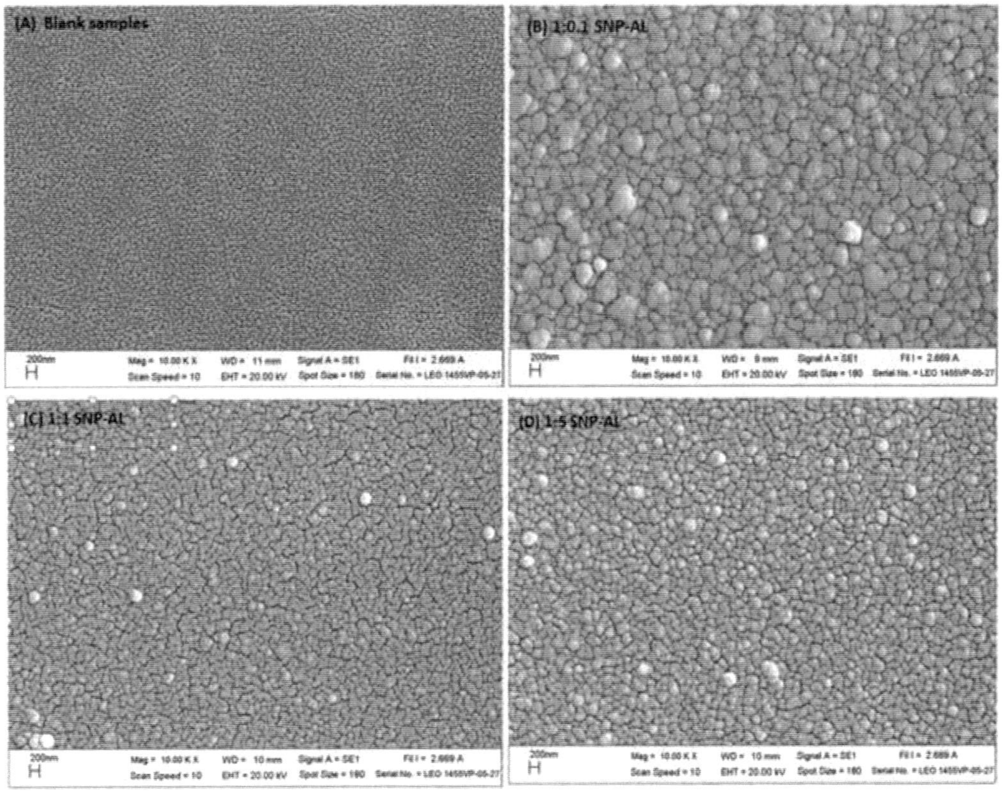

Figure 3. SEM image showing the surface of blank condition (A), PDADMAC/SNP-AL 1:0.1 film (B), PDADMAC/SNP-AL 1:1 film (C) and PDADMAC/SNP-AL 1:5 film (D).

According to the theory, the film PDADMAC/SNP-AL 1:0.1 can release SNP more easily than other kinds of film as the charge density surrounding the particles is less causing the adhesive force between molecules is less [18],[19]. When considering with this experimental result, it is found to follow the aforementioned principle. However, the difference of result on Aa of each kind of film is very little. The mechanism of SNP influencing the destruction of bacteria is unclear. However, there are three mechanisms which can be explained. The first mechanism is that the very tiny size of SNP in nanoparticle level causes it to penetrate the cell wall of bacteria to destroy the systems such as respiratory system [20],[21],[22]. The second mechanism is that SNP can penetrate into the bacteria and destroy the compound of sulfur and phosphorus such as DNA [20],[21],[22]. The last mechanism is that it can release silver ion which has the properties to kill the bacteria. Its positive charge will seize the cell wall of bacteria causing the cell wall to change in the structure by building electron-dense granules to destroy the cell . Moreover, silver ion can also reacts with thiol group which is the component of enzyme causing the enzyme not to react and the cell will die eventually [20],[21],[22]. From these mechanisms, it can be explained that SNP-AL can destroy Aa.

Table 2. Antibacterial activity of multilayer thin film

samples	Colony of Aa (CFU) 24hr	% of reduction
Blank	2.02×10^8	0
PDADMAC/SNP-AL 1:0.1	1.45×10^7	92.80
PDADMAC/SNP-AL 1:1	1.5×10^7	92.57
PDADMAC/SNP-AL 1:5	1.75×10^7	91.34

4. Conclusion

Silver nano-chips are built through the method of polyelectrolyte multilayer technique in order to control the release of SNP. When testing the competence of film in anting Aa, it is found that the number of Aa can be reduced for 90 % Therefore, it is the chance in developing silver nano-chips to be more effective. This can be done by increasing the numbers of layers of the film, increase the amount of SNP, and change the kind of coating substance on the nanoparticles, etc. For future study, silver nano-chips can be used to test with other antibacterial agents which are the causes of periodontal disease and then should be tested in humans to treat periodontal disease instead of local antibiotics.

5. Acknowledgements

This work was funded by Faculty of dentistry, Naresuan university, Thailand and supported by Mrs. Suttimas Yuakyong , Faculty of Dentistry, Naresuan University

6. References

[1] Ouay B L, Stellacci F 2015Antibacterial activity of silver nanoparticles: A surface science insight *Nano Today*. **10** 339

[2] Guzman M, Dille J, Godet S 2012 Synthesis and antibacterial activity of silver nanoparticles against gram-positive and gram-negative bacteria *Nanomed nanotechnol*. **8** 37

[3] Saxena A,Tripathi R M , Zafar F, Singh P 2012 Green synthesis of silver nanoparticles using aqueous solution of Ficus benghalensis leaf extract and characterization of their antibacterial activity *Mater. Lett.*. **67** 91

[4] Rai M, Yadav A and Gade A 2009 Silver nanoparticles as a new generation of antimicrobials *Biotechnol Adv*. **27** 76

[5] Macy E and Contreras R 2014 Health care use and serious infection prevalence associated with penicillin ''allergy'' in hospitalized patients: A cohort study *J. Allergy Clin. Immunol.* **133** 790

[6] Mendes C, Marin M E, Quiñones F, Sifuentes-Osornio J, Siller C C, Castanheira M, Zoccoli C M, López H, Súcari A, Rossi F, Angulo G B, Segura A J A, Starling C, Mimicaand I and Felmingham D 2003 Antibacterial Resistance of Community-Acquired Respiratory Tract Pathogens Recovered from Patients in Latin America: Results from the PROTEKT Surveillance Study (1999–2000). *Braz Jl Infect Dis.* **7** 44

[7] Fine D H, Markowitz K, Furgang D, Fairlie K, Ferrandiz J, Nasri C, McKiernan M and Gunsolley J 2007 Aggregatibacter actinomycetemcomitans and Its Relationship to Initiation of Localized Aggressive Periodontitis: Longitudinal Cohort Study of Initially Healthy Adolescents. *J. clin. microbiol.* **45** 3859

[8] Li Y, Messina C, Bendaoudl M, Fine D H, Schreiner H and Tsiagbe V K 2010 Adaptive immune response in osteoclastic bone resorption induced by orally administered Aggregatibacter actinomycetemcomitans in a rat model of periodontal disease *MOL Oral Microbiol.* **25** 275

[9] Cochran D L 2008 Inflammation and Bone Loss in Periodontal Disease *J PERIODONTO.* **79** 1569

[10] Pihlstrom B L, Michalowicz B S and Johnson N W 2005 Periodontal diseases *Seminar.* **366** 1809

[11] Quirynen M, Teughels W, Soele M D and Steenberghe D V 2002 Topical antiseptics and antibiotics in the initial therapy of chronic adult periodontitis: microbiological aspects. *PERIODONTOL 2000.* **28** 72

[12] Heasman P A, Heasman L, Stacey F, McCracken G I 2001 Local delivery of chlorhexidine gluconate (PerioChipTM) in periodontal maintenance patients *J. clin. Periodontol.* **28** 90

[13] Drisko C H 2001 Nonsurgical periodontal therapy *PERIODONTOL 2000.* **25** 77

[14] Etienne O, Picart C, Taddei C, Keller P, Hubsch E, Schaaf P, Voege J C, Haikel Y, Ogier J A and Egles C 2006 Polyelectrolyte multilayer film coating and stability at the surfaces of oral prosthesis base polymers: An in vitro and in vivo study *J. DenT. Res.* **85** 44

[15] Harnet J C, Guen E L, Ball V, Tenenbaum H, Ogier J, Haikel Y and Vodouhe C 2009 Antibacterial protection of suture material by chlorhexidine-functionalized polyelectrolyte multilayer films *J.Mater Sci.Mater. Med.* **20** 185

[16] Berg M C, Zhai L, Cohen R E and Rubner M F 2006 Controlled Drug Release from Porous Polyelectrolyte Multilayers *Biomacromolecules.* **7** 357

[17] Kittitheeranun P, Sanchavanakit N, Dubas S T and Sajomsang W 2010 Loading of curcumin in polyelectrolyte multilayers *Langmuir.* **26** 6869

[18] Wacharanad S and Dubas S T 2013 Tunable silver leaching from polyelectrolyte multilayers thin films *Advanced Materials Research.***701** 145

[19] Dubas S T, Wacharanad S and Potiyaraj P 2011 Tunning of the antimicrobial activity of surgical sutures coated with silver nanoparticles *Colloids Surf. A: Physicochem. Eng. Aspects.* **380** 25

[20] Prabhu S and Poulose E K 2012 Silver nanoparticles: mechanism of antimicrobial action, synthesis, medical applications, and toxicity effects *International Nano Letters.* **2** 1

[21] You C, Han C, Wang X, Zheng Y, Li Q, Hu X, Sun H 2012 The progress of silver nanoparticles in the antibacterial mechanism, clinical application and cytotoxicity *Mol Biol Rep.* **39** 9193

[22] Marambio-Jones C and Hoek E M V 2010 A review of the antibacterial effects of silver nanomaterials and potential implications for human health and the environment *J. Nanoparticle Res.***12** 1531

IOP Conf. Series: Materials Science and Engineering **635** (2019) 012006 doi:10.1088/1757-899X/635/1/012006

Effects of Si concentrations on microstructure and mechanical properties of as-cast Co–Cr–Mo alloys

P. Tunthawiroon[1*] and A. Chiba[2]

[1]Department of Industrial Engineering, Faculty of Engineering, King Mongkut's Institute of Technology Ladkrabang, Bangkok 10520, Thailand.
[2]Institute for Material Research, Tohoku University, Sendai 980-8577, Japan

*Email: phacharaphon.tu@kmitl.ac.th

Abstract. This research investigates the effects of different Si concentrations on the microstructure and mechanical properties of Co–Cr–Mo–xSi alloys. The Si concentration (x) was 0.1, 0.5, 1.0, 3.0 and 5.0 wt%. The mechanical properties of the experimental heat-treated Co–Cr–Mo–xSi alloys was determined by hardness and tensile tests. The microstructure was analyzed using SEM. The microstructure analysis revealed that Si concentration played a role in the formation of the intermetallic phase of Co–Cr–Mo–xSi alloys. The presence of the intermetallic phase significantly decreased the ductility of the experimental alloys. However, the hardness was improved and was increased directly proportional to the fraction of the intermetallic phase precipitating in the Co–Cr–Mo–xSi alloys.

1. Introduction

Co-based alloys are extensively used as biomedical materials owing to their good biocompatibility and mechanical properties such as high tensile strength and high resistance to abrasive wear [1–3]. Recently, the Co–Cr–Mo alloys are the prospective candidate materials for replacing the currently used hot-working tool steels [4,5]. This is attributed to the excellent oxidation and corrosion resistance at high temperature [6]. Co–Cr–Mo alloys are divided into two groups, cast and wrought alloys. The cast Co–Cr–Mo alloys, referred to the ASTM F75, are the most common alloys for dentistry and artificial joints. The wrought or thermo-mechanically processed Co–Cr–Mo alloys (ASTM F799, F1537) are used as implant devices which require the superior strength [7]. The chemical compositions of Co–Cr–Mo alloys mainly consists of 27-30wt% of Cr and 5-7wt% of Mo and balance with Co. The microstructure of Co–Cr–Mo alloys possibly composed of γ-fcc and ε-hcp phase depending on the temperature [8]. The intermetallic phase, σ, phase is the common intermetallic compound precipitating these alloys. The Mo and Cr are added to enhance the solid solution strengthening as well as the corrosion resistance. The trace elements of N and C are also added to improve the mechanical properties of Co–Cr–Mo alloys due to the presence of fine Cr_2N and carbide phases. However, the fatigue and corrosion resistance would be deteriorated when the precipitates occupy at the grain boundaries of alloys.

Recently, the previous researches indicated that the addition of Si in Co–Cr alloys could stabilize the ε-hcp phase which is the harder phase compared to γ-fcc [9,10]. Moreover, the alloying Si suppressed the σ phase formation and decreased the melting point of Co–Cr alloys. Besides, the oxidation resistance of Co–Cr–Mo and Co–Cr–W alloys was simultaneously improved by the small amounts of Si [6,11]. The previous studies on the oxidation behavior of Co–Cr–Mo–Si alloys have indicated that Si effectively promoted the formation of a continuous Cr_2O_3 oxide film [6]. In addition, the internal oxidation of SiO_2 improved the bonding strength of Cr_2O_3 and matrix as well as the inhibition of further oxidation process [6]. The addition of Si wold protect Co–Cr–Mo alloys from the oxidation or corrosion attack. In other words, the reliability of alloys is increased. However, there has been no related research on the effects of Si on the microstructure and mechanical properties of Co–Cr–Mo alloys. Therefore, the aim of the present work is to investigate the effects of Si on the microstructure and mechanical properties of Co–Cr–Mo alloys.

2. Experimental details

2.1. Materials

The as-cast Co–Cr–Mo–xSi alloys were prepared using high-frequency vacuum induction furnace in an Ar atmosphere, where x is the Si concentration between 0.1, 0.5, 1.0, 3.0, and 5.0 wt%. The molten alloys were cast to be the rectangular ingots with the dimension of 20 × 30 × 150 mm^3 in a steel mold. The nominal compositions of the as-cast Co–Cr–Mo–xSi alloys are tabulated in Table 1. Hereafter, the alloys are named as 0.1Si, 0.5Si, 1.0Si, 3.0Si and 5.0Si in accordance with the concentrations of Si used in this study. Prior to analysis, the ingots were subjected to isothermally heat treatment at 1200 °C for 24 h, followed by rapid cooling in the furnace using Ar gas.

Table 1. The nominal compositions of the experimental Co–Cr–Mo–xSi alloys

Alloy compositions (wt%)	Co	Cr	Mo	Si	Mn	C	N
Co–Cr–Mo–0.1Si	Bal.	27.5	6.10	0.156	0.614	0.056	0.135
Co–Cr–Mo–0.5Si	Bal.	28.1	6.15	0.466	0.521	0.054	0.148
Co–Cr–Mo–1.0Si	Bal.	28.2	6.17	0.966	0.593	0.054	0.128
Co–Cr–Mo–3.0Si	Bal.	27.1	5.89	3.167	0.75	0.067	0.101
Co–Cr–Mo–5.0Si	Bal.	27.6	6.03	4.445	0.55	0.034	0.111

2.2. Microstructural observation

In the microstructure analysis, the heat-treated ingot was sliced into round-shaped thin-sheet specimens with the dimension of 15 × 2 mm (Ø× t) using a wire electrical discharge machine (EDM). The specimens were ground off by abrasive paper and polished with 1-μm and 0.3-μm alumina suspensions (AP-D suspension, Struers). A 0.04-μm colloidal silica suspension (OP-S suspension, struers) was used for final polishing. The polished specimens were then cleaned with ethanol in an ultrasonic cleaner and air-dried. The microstructure of experimental Co–Cr–Mo–xSi alloys was investigated using a scanning electron microscope (SEM; S-3400N, Hitachi).

2.3. Mechanical testing

The mechanical properties of the experimental heat-treated Co–Cr–Mo–xSi alloys were characterized by hardness and tensile tests at room temperature. Vickers micro-hardness tester (HMV-2 Series, Shimadzu) with an applied load of 9.8 N and an indentation duration of 15 s was used for the hardness tests. The measurement was carried out five times for each specimen to obtain the reproducibility. The tensile tests and sample preparations were conducted according to the ISO 22674 standard: international standard for metallic materials for fixed and removable restorations and appliances. The tensile experiments were carried out using a uniaxial tensile testing machine (Autograph AGS-100kNX, Shimadzu) equipped with the extensometer system. The specimens were individually strained in triplicate until failure at a speed of 0.1 $mm\text{-}min^{-1}$ (corresponding to a nominal strain rate of approximately 1.5×10^{-4} s^{-1}). The morphology of the fracture surface of the broken tensile specimens was observed using a scanning electron microscope (SEM; S-3400N, Hitachi).

3. Results and discussion

3.1. Microstructure analysis

Figure 1(a-e), respectively show back-scattered electron BSE-SEM images of the experimental 0.1Si, 0.5Si, 1.0Si, 3.0Si and 5.0Si alloys. The microstructure of experimental alloys shows a dendritic structure with the interdendritic intermetallic phase (bright phase), which uniformly distributed over the area. For 3.0Si and 5.0Si alloys, the interdendritic network of the intermetallic phase was clearly observed. The area fractions of the intermetallic phase precipitated in the 0.1Si, 0.5Si, 1.0Si, 3.0Si and 5.0Si alloys measured by the image processing software were 0.75%, 1.73%, 1.3%, 10.16% and 26.64%, respectively. The image processing analysis of intermetallic phase revealed that the elevated Si concentrations contributed to the increase in the fraction of intermetallic phase. This result is consistent to Çelik and Kaplan [12], reported in which the increasing fraction of intermetallic phases was due to the formation of various kinds of precipitates in high alloying Co-based alloys.

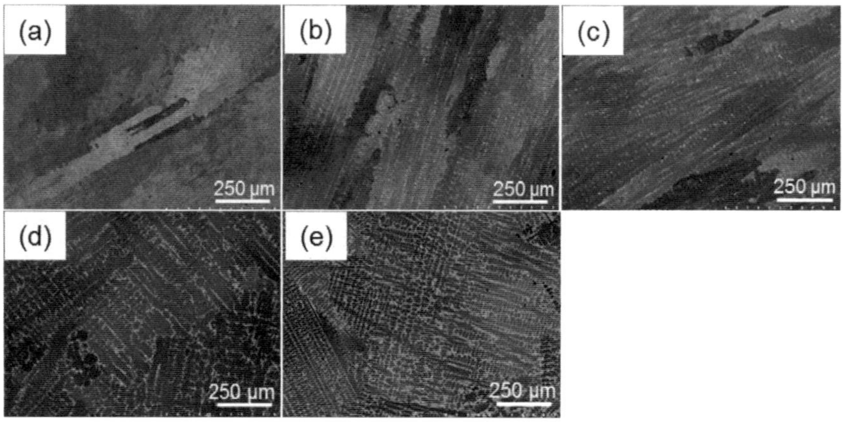

Figure 1. BSE-SEM images of the experimental (a) 0.1Si, (b) 0.5Si, (c) 1.0Si, (d) 3.0Si and (e) 5.0Si alloys.

3.2. Mechanical properties

Figure 2 illustrates the hardness and the standard deviation of the experimental 0.1Si, 0.5Si, 1.0Si, 3.0Si and 5.0Si alloys. The hardness of the experimental alloys increased with an increase in Si concentrations. The average hardness of the 0.1Si, 0.5Si, 1.0Si, 3.0Si and 5.0Si alloys was 287, 292, 295, 356 and 515 HV1, respectively. An increase in hardness is attributable to the presence of intermetallic phase which is naturally harder and stiffer than the surrounding matrix. However, the amount of Si below 1.0 wt% had a slight effect on the hardness evolution of Co–Cr–Mo alloys. The hardness of Co–Cr–Mo alloys was significantly elevated when the concentration of Si was greater than 1.0 wt%.

Figure 3 shows the representative stress-strain curves of the experimental alloys conducted at the room temperature. The tensile properties and hardness results along with the standard deviations of the experimental 0.1Si, 0.5Si, 1.0Si, 3.0Si and 5.0Si alloys were summarized in Table 2. The effects of Si on tensile properties is similar to the hardness analysis in that low content of Si in Co–Cr–Mo alloys (below 1.0 wt%) has a slight effect on the 0.2% proof stress, tensile strength and elongation of the Co–Cr–Mo alloys. The average 0.2% proof stress, tensile strength, and elongation to failure of 0.1Si, 0.5Si, and 1.0Si slightly increased. However, a high amount of Si over 3.0wt% significantly reduced tensile strength and elongation to failure of the alloys as seen in the stress-strain curves of 3.0Si and 5.0Si alloys. The stress-strain curves of 3.0Si and 5.0Si alloys show the linear elastic behavior with the extremely low elongation to failure (3.5 and 0.91%) indicating the ductility of the alloys was reduced. The diminished properties is attributed to the fraction of intermetallic phase precipitating in the Co–Cr–Mo alloys. The intermetallic phases which are brittle in nature deteriorates the ductility of alloys resulting in a significantly decrease in elongation to failure [13].

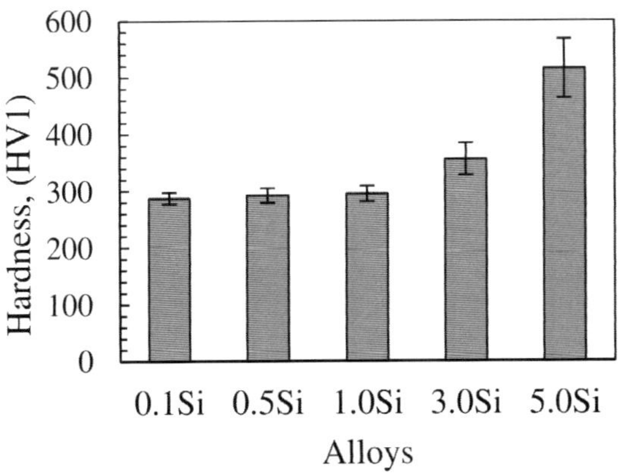

Figure 2. Hardness of the experimental 0.1Si, 0.5Si, 1.0Si, 3.0Si and 5.0Si alloys.

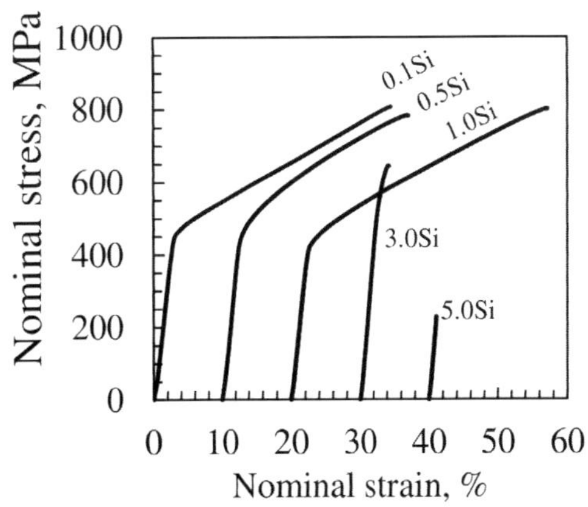

Figure 3. Stress-strain curves of the experimental alloys conducted at the room temperature

Table 2. Vicker hardness and tensile properties of the experimental alloys

Alloys	0.1Si	0.5Si	1.0Si	3.0Si	5.0Si
Hardness (HV1)	287±10.3	292±12.0	295±14.0	356±28.0	515±51.0
0.2% proof stress (MPa)	445.0±22.1	473.4±14.6	445.6±6.8	529.6±22.1	N/A.
Tensile strength (MPa)	778.6±23.4	773.7±7.4	796.3±9.6	592±65	194.1±26.5
Elongation to failure (%)	31.8±2.5	23.2±2.8	35.6±1.2	3.55±0.7	0.91±0.13

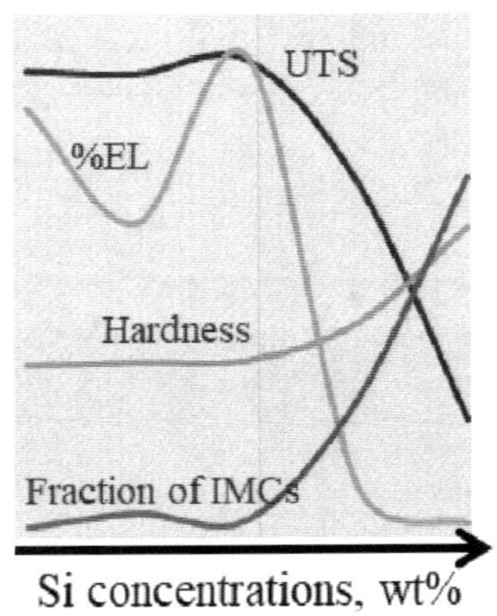

Figure 4. Relationship between tensile strength, elongation to failure, hardness and the area fraction of intermetallic phase of Co–Cr–Mo as a function of Si concentrations.

Based on this finding, the relationship between the mechanical properties and the fraction of intermetallic phase of Co–Cr–Mo–xSi alloys could be summarized as shown in Fig. 4. The Co–Cr–Mo alloys with low Si concentrations between 0.1 and 1.0wt% pointed showed that the tensile strength, elongation to failure, hardness and the area fraction of intermetallic phase were almost constant. Inversely, alloys with high Si concentrations (3.0-5.0wt%), exhibited the extremely low ductility due to the brittleness of the intermetallic phase. Thus, to suppress the formation of large intermetallic phase, the Si concentration in Co–Cr–Mo alloys should be of between 0.1 and 1.0wt%.

The fracture surface of tensile specimens induced by the tensile test was investigated as shown in Fig. 5. An intergranular quasi-cleavage fracture was observed in all alloys. For the alloys with low Si concentrations, crack initiation was possibly induced by the isolated particles of the intermetallic phase as indicated by the white arrow in Fig. 5(g-i). On the other hand, for 3.0Si and 5.0Si alloys, flat fracture surface together with the intergranular quasi-cleavage fracture was observed. The flat fracture observed in the 3.0 and 5.0Si alloys probably induced by the network of intermetallic phase as discussed in the microstructure analysis.

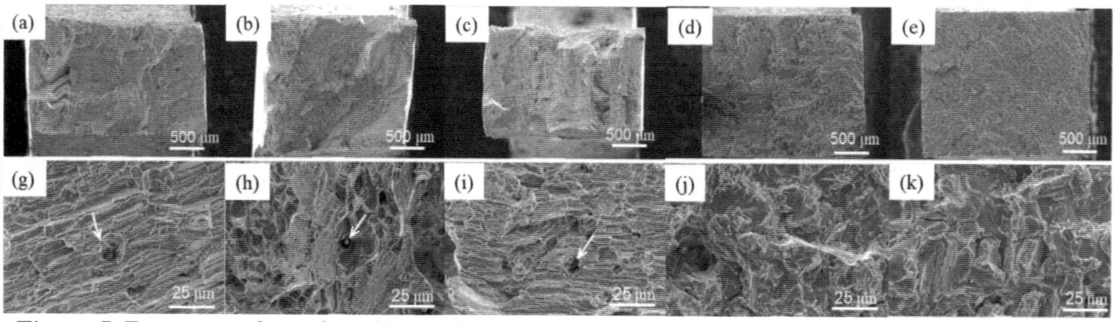

Figure 5. Fracture surface of tensile specimens (a) 0.1Si, (b) 0.5Si, (c) 1.0Si, (d) 3.0Si and (e) 5.0Si alloys and corresponding high magnification images (g) 0.1Si, (h) 0.5Si, (i) 1.0Si, (j) 3.0Si and (k) 5.0Si.

4. Conclusions

This research investigated the effects of different Si concentrations (0.1, 0.5, 1.0, 3.0 and 5.0wt%) on the microstructure and mechanical properties of Co–Cr–Mo–xSi alloys. The microstructure analysis revealed that Si concentration played a role in the formation of the intermetallic phase of Co–Cr–Mo–xSi alloys. The increasing of Si concentrations promoted the formation of the larger fraction of the intermetallic phase in Co–Cr–Mo–xSi alloys. The presence of the intermetallic phase significantly decreased the ductility of the experimental alloys. However, the hardness was elevated and was increased proportionally to the fraction of the intermetallic phase. To limit the formation of intermetallic phase, thus, the Si concentrations in Co–Cr–Mo alloys should be of between 0.1 and 1.0wt%.

5. Acknowledgements

The authors would like to thank Mr. Kouji Kuramoto and Mr. Kimio Wako of IMR, Tohoku University, for contributing to the alloys fabrication. The authors also like to thank Mr. Kamtorn Sukpimai for assistance in tensile tests and Assoc.Prof.Dr. Kannachai Kanlayasiri for precious discussion and support.

6. References

[1] Mineta S, Namba S, Yoneda T, Ueda K and Narushima T 2012 *Metall. Mater. Trans.* A **43** 3351–8
[2] Li Y, Yamashita Y, Tang N, Liu B, Kurosu S, Matsumoto H, Koizumi Y and Chiba A 2012 *Mater. Chem. Phys.* **135** 849–54
[3] Chen Y, Li Y, Kurosu S, Yamanaka K, Tang N and Chiba A 2014 *Wear.* **319** 200–10
[4] Tunthawiroon P, Li Y, Koizumi Y and Chiba A 2017 Mater. Sci. Eng. A **703** 27-36
[5] Tang N, Li Y, Koizumi Y, Kurosu S and Chiba A *Corros. Sci.* **75** 262–8
[6] Tunthawiroon P, Li Y, Tang N, Koizumi Y and Chiba A 2015 *Corros. Sci.* **95** 88–99
[7] Niinomi M 2002 *Metall. Mater. Trans.* A **33** 477–86
[8] Kurosu S, Matsumoto H and Chiba A 2010 *Metall. Mater. Trans.* A **41** 2613–25
[9] Yamanaka K, Mori M, Kuramoto K and Chiba A 2014 *Mater. Des.* **55** 987–98

IOP Conf. Series: Materials Science and Engineering **635** (2019) 012006 doi:10.1088/1757-899X/635/1/012006

[10] Yamanaka K, Mori M, Torita Y and Chiba A 2018 *Mater. Sci. Eng.* C **92** 112–20
[11] Yamanaka K, Mori M and Chiba A 2015 *Corros. Sci.* **94** 411–9
[12] Çelik H and Kaplan M 2004 *Wear.* **257** 606–11
[13] Yoda K, Takaichi A, Nomura N, Tsutsumi Y and Doi H 2012 *Acta Biomater.* **8** 2856–62

IOP Conf. Series: Materials Science and Engineering **635** (2019) 012007 doi:10.1088/1757-899X/635/1/012007

Characterization of intermetallic phase in as-cast Si-doped Co–Cr–Mo alloys

P. Tunthawiroon[1*] and A. Chiba[2]

[1]Department of Industrial Engineering, Faculty of Engineering, King Mongkut's Institute of Technology Ladkrabang, Bangkok 10520, Thailand.
[2]Institute for Material Research, Tohoku University, Sendai 980-8577, Japan

*Email: phacharaphon.tu@kmitl.ac.th

Abstract. This research aims to investigate the effects of Si concentrations on the microstructure and formation of the intermetallic phase of Co–Cr–Mo–xSi alloys. The Si concentration (x) was varied between 0.1, 0.5, 1.0, 3.0 and 5.0 wt%. The microstructure and intermetallic phase of the solution heat-treated Co–Cr–Mo alloys were characterized using SEM-EDX, XRD and TEM. The microstructure analysis revealed that Si concentration played a role in the evolution of intermetallic phase precipitating in Co–Cr–Mo alloys. Two types of intermetallic phased were found in the Si-doped Co–Cr–Mo alloys. The σ phase was identified in the alloys containing Si between 0.1-1.0 wt% while the increase in Si (3.0-5.0Si) contributed to the formation of the χ phase. The calculated lattice parameters of the intermetallic phase precipitated in the 3.0Si and 5.0Si alloys correlated well with those of $Co_5Cr_3Si_2$ phase.

1. Introduction

Cobalt-chromium-molybdenum (Co–Cr–Mo) alloys are extensively used in the manufacture of medical implants owing to their outstanding combination of mechanical properties and corrosion resistance as well as biocompatibility [1-4]. Such medical implants are expected to have a long lifetime and physically reliability. However, the as cast Co–Cr–Mo alloys usually contain the casting defects, for example, coarse dendritic structure, microporosity and cored matrix [5–7] which caused the early failure of medical devices. So, the deformation processes are applied to improve the mechanical properties. It has been reported that the appropriate hot rolling condition could produce an extremely high yield strength (1100-1400 MPa) with no reduction in ductility [8,9]. Furthermore, alloying element and heat treatment which are much more economical could be an effective method to enhance the mechanical properties of Co–Cr–Mo alloys.

In the Co–Cr–Mo alloys, the small amount of the σ phase, which is the common intermetallic phase, could improve the tensile strength. In addition, C is added to form the precipitate carbide, such as $M_{23}C_6$ and M_6C which could simultaneously limit the formation of σ phase and enhance the tensile properties of Co-based alloys [10,11]. However, the carbide phases induce the surface fatigue when the alloys are used in the abrasive applications [12]. Therefore, to prevent the carbide precipitation, the carbon content in the Co-based alloys should keep below 0.1wt% [9].

Recently, Yamanaka et al. [13] suggested that the addition of Si could suppress the formation of the σ phase in Co–Cr alloys. It is well-known that the alloying elements would possibly change the precipitation behavior in Co–Cr–Mo alloys owing to the thermodynamic activity changes [14].

IOP Conf. Series: Materials Science and Engineering **635** (2019) 012007 doi:10.1088/1757-899X/635/1/012007

However, information on the effects of Si on the microstructure and precipitation behavior of Co–Cr–Mo alloys has not systematically investigated so far. Thus, the aim of present work is to investigate the microstructure and the evolution of the intermetallic phase of Si-doped Co–Cr–Mo alloys.

2. Experimental

2.1. Materials

The as-cast Si-doped Co–Cr–Mo alloys were prepared using high-frequency vacuum induction furnace in an Ar atmosphere. The Si concentrations were added in the Co–Cr–Mo alloys between 0.1, 0.5, 1.0, 3.0, and 5.0 wt%. The molten alloys were cast to be the rectangular ingots with the dimension of 20 × 30 × 150 mm^3 in a steel mold. The nominal compositions of the as-cast Co–Cr–Mo–xSi alloys are tabulated in Table 1. Hereafter, the alloys are named as 0.1Si, 0.5Si, 1.0Si, 3.0Si and 5.0Si in accordance with the concentrations of Si used in this study. Prior to analysis, the ingots were subjected to solution heat treatment at 1200°C for 24 h, followed by rapid cooling in the furnace using Ar gas.

Table 1. The nominal compositions of the experimental Co–Cr–Mo–xSi alloys

Alloy compositions (wt%)	Co	Cr	Mo	Si	Mn	C	N
Co–Cr–Mo–0.1Si	Bal.	27.5	6.10	0.156	0.614	0.056	0.135
Co–Cr–Mo–0.5Si	Bal.	28.1	6.15	0.466	0.521	0.054	0.148
Co–Cr–Mo–1.0Si	Bal.	28.2	6.17	0.966	0.593	0.054	0.128
Co–Cr–Mo–3.0Si	Bal.	27.1	5.89	3.167	0.75	0.067	0.101
Co–Cr–Mo–5.0Si	Bal.	27.6	6.03	4.445	0.55	0.034	0.111

2.2. Microstructure and phase analysis

For the microstructure analysis, the heat-treated ingot was sliced into thin-sheet specimens using a wire electrical discharge machine (EDM). The specimens were ground off by abrasive paper and polished with 1-μm and 0.3-μm alumina suspensions (AP-D suspension, Struers). A 0.04-μm colloidal silica suspension was used for final polishing (OP-S suspension, Struers). The polished specimens were then cleaned with ethanol in an ultrasonic cleaner and air-dried. The microstructure of experimental Si-doped Co–Cr–Mo alloys was investigated using an electron backscatter diffraction (EBSD) under a field-emission scanning electron microscope (FEI, XL30S-FEG) operated at 20kV and a scanning electron microscope (SEM; S-3400N, Hitachi) coupled with an energy-dispersive spectroscopy (EDS). X-ray diffraction (XRD) patterns were acquired with a diffractometer (X'Pert MPD, PANalytical) using a Cu Kα (0.1547 nm) radiation source. The samples for transmission electron microscope (TEM) observation were prepared by mechanical polishing and ion slicer (EM-09100IS, JEOL). The detailed structure was observed on a TEM (EM-002B, Topcon) operated at 200 kV.

3. Results

3.1. EBSD and SEM analysis

Figure 1(a-c) illustrates the EBSD mapping analyses of the experimental 0.1Si, 0.5Si, 1.0Si, 3.0Si and 5.0Si alloys, including inverse pole figure (IPF) maps, phase maps and kernel average misorientation (KAM) maps. The EBSD analysis revealed that the experimental alloys have a coarse microstructure, consisting of large-columnar and small-equiaxed grains. The grain size varied from a few hundred micrometers to a few millimeters. The IPF maps in Fig. 1(a) show the maps with random colors indicating the variation in the crystallographic orientation of the individual grains. The EBSD phase maps (Fig. 1(b)) show that the microstructure of experimental alloys exhibited a nearly single phase of γ- phase or face-centered cubic (fcc) structure. Since, the homogenizing process was carried out at 1200 °C following by rapid cooling, the γ-phase which is stable at high temperature was remained.

The KAM maps in Fig. 1(c) show the scattering of local distortion as indicated by the high different contrast of color within grains and at grain boundaries. This distortion might attribute to the remaining of local strain which would develop during the solidification and/or rapidly cooling stage [2].

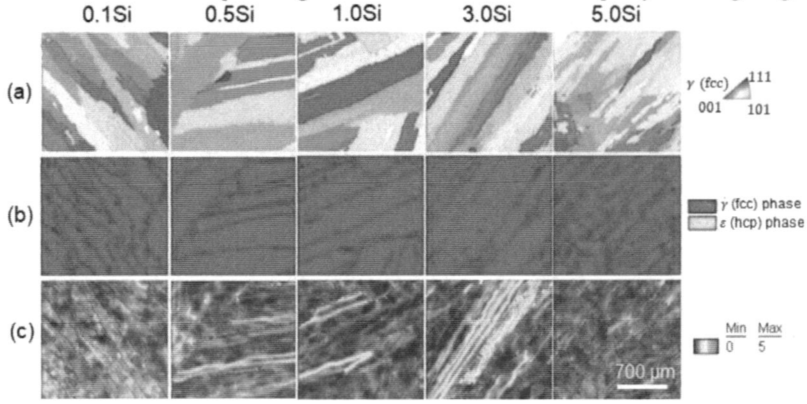

Figure 1. EBSD map of 0.1Si, 0.5Si, 1.0Si, 3.0Si and 5.0Si alloys (a) IPF map, (b) phase map, and (c) KAM maps.

Figure 2 shows the SEM images of the 0.1Si, 0.5Si, 1.0Si, 3.0Si and 5.0Si alloys and the corresponding EDS spectra acquired from the matrix and intermetallic phase. The EDS spectra of 0.1Si, 0.5Si and 1.0Si alloys revealed that the intensities of Cr and Mo in the intermetallic phase were higher than that of Co. On the other hand, the EDS spectra of 3.0Si and 5.0Si alloys showed the high concentration of Si as displayed in Fig 2(d) and (e). Quantitative EDS analysis of the intermetallic phase was summarized as tabulated in Fig. 3. The quantitative EDS analysis indicated that some Si dissolved in the intermetallic phase and the amount of Si in the intermetallic phase increased with increasing of Si in the initial compositions. Inversely, the amount of Cr and Mo are slightly dropped in the intermetallic phase. This results implied that the concentrations of Si affected on the evolution of the intermetallic phase precipitating in the Co–Cr–Mo–xSi alloys.

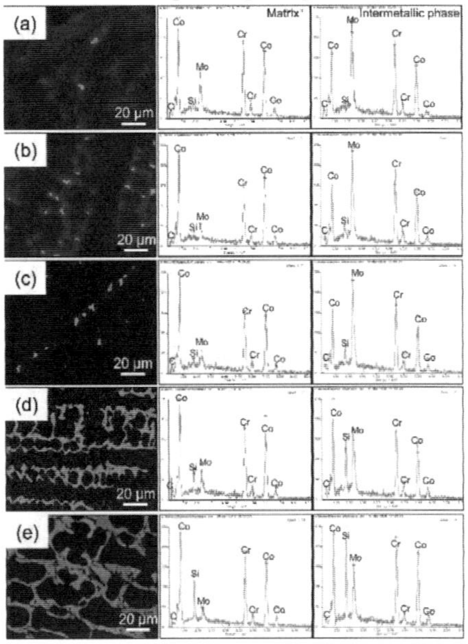

Figure 2. SEM images of the experimental as-cast 0.1Si, 0.5Si, 1.0Si, 3.0Si and 5.0Si alloys and the corresponding EDS spectra acquired from the matrix and intermetallic phase.

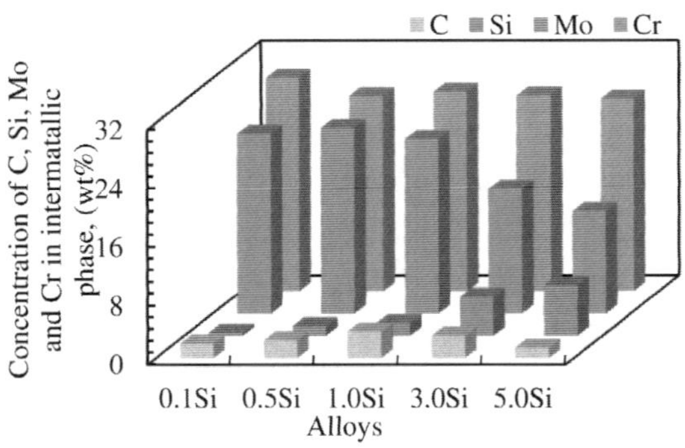

Figure 3. Quantitative EDS analysis of the intermetallic phase precipitating in Co–Cr–Mo alloys

3.2. XRD and TEM analysis

Figure 4 shows the XRD patterns of the experimental as-cast 0.1Si, 0.5Si, 1.0Si, 3.0Si and 5.0Si alloys. The XRD analysis revealed that the microstructure of alloys exhibited a dominant γ-phase, which was consistent with the EBSD phase map analysis. The weak peak of ε-phase or hcp structure was also detected in this analysis. The peak at $2\theta \approx 47.0°$ was identified as the σ phase. In addition, the peaks at $2\theta \approx$ 48.7° and 53.4° which were found in the XRD patterns of 3.0Si and 5.0Si alloys might be the σ and/or $Co_5Cr_3Si_2$ intermetallic phase.

Figure 5 shows the TEM bright-field images of the experimental 0.1Si, 0.5Si, 1.0Si, 3.0Si and 5.0Si alloys and the corresponding selected-area diffraction (SAD) patterns acquired from the intermetallic phase as indicated by an arrow in each TEM bright-field image. The SAD analyses revealed that the σ phase (space group $P4_2/mnm$) precipitated in the 0.1Si, 0.5Si and 1.0Si alloy. However, the SAD patterns of 3.0Si and 5.0Si alloys disclosed the formation of $Co_5Cr_3Si_2$ intermetallic phase and its lattice constant was calculated to be 0.8708 nm. According to the EDX, XRD and TEM analysis, the intermetallic phase precipitating in the 3.0Si and 5.0Si alloys could be confirmed to be a ternary intermetallic phase of $Co_5Cr_3Si_2$ (space group $I\bar{4}3m$) [15,16].

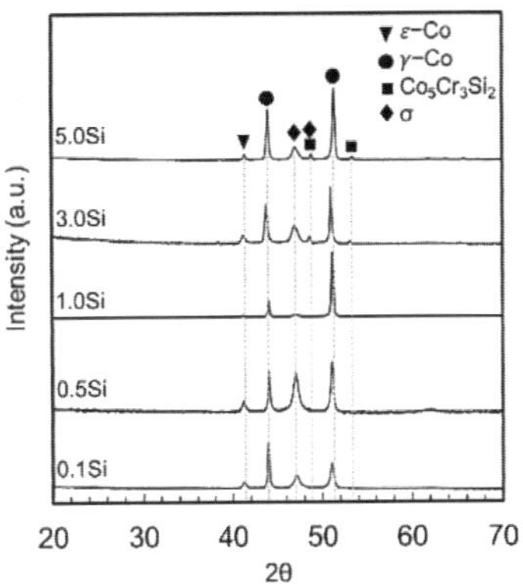

Figure 4. XRD patterns of the experimental as-cast 0.1Si, 0.5Si, 1.0Si, 3.0Si and 5.0Si alloys.

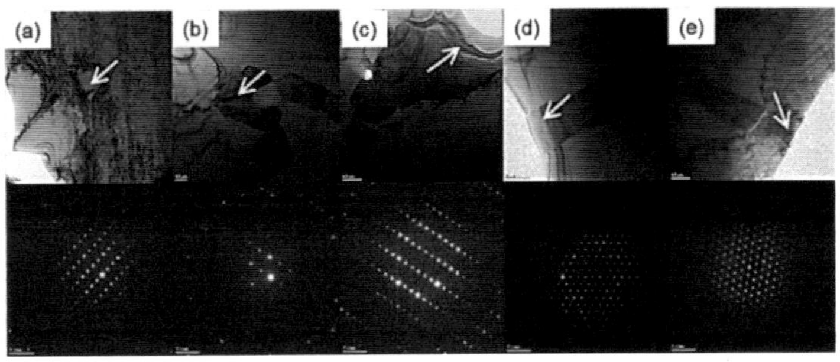

Figure 5. TEM bright-field images of the experimental 0.1Si, 0.5Si, 1.0Si, 3.0Si and 5.0Si alloys and the corresponding selected-area diffraction (SAD) patterns taken from the intermetallic phase.

4. Discussion

The intermetallic phases precipitating in Co–Cr–Mo–xSi alloys have been investigated in this study. The results revealed that an increase in Si concentrations has a significant influence on the evolution of intermetallic phase in the Co–Cr–Mo alloys. Two types of intermetallic phases, σ and $Co_5Cr_3Si_2$, were found in the experimental heat-treated Co–Cr–Mo–xSi alloys. Based on a study on the formation of intermetallic phase in Co-based alloys [17], a number of the intermetallic phases could possibly precipitate in the Co–Cr–Mo alloys. For instance, the σ and carbide $M_{23}X_6$ (M refer to the metallic elements and X refers to carbon C) were basically found in equilibrium with γ-fcc [18]. An increase in C promoted the formation of carbide phase and suppressed the σ phase in the Co–Cr–Mo system alloys. In addition, the precipitation of the other intermetallic phases in Co-based alloys such as η, χ, R and π phase has been found depending on the heat-treated temperature, cooling processes and alloying elements [2,17,19,20]. However, in this study, no carbide phase was observed in the experimental alloys due to the low concentrations of C added in initial compositions [20]

To clarify the impact of Si on the formation of the σ phase in Co–Cr–Mo alloys, the fraction of σ phase is plotted as a function of Si concentration using a thermodynamic calculation software, ThermocalC, as shown in Fig. 6. At 1200°C, corresponding the heat-treated temperature in this experiment, the fraction of the σ phase greatly decreases with increasing of Si concentrations and becomes near zero when the Si reaches 3.0 wt%. The reduction of the σ phase was consistent with the

XRD and TEM analysis in Fig. 4 and 5 in that the σ phase was found in the 0.1Si, 0.5Si and 1.0Si alloys and almost disappeared when Si concentrations increased to 3.0 and 5.0 wt%. From the EDX analysis of in intermetallic phase of 0.1Si, 0.5Si and 1.0Si alloys, the amount of Co, Cr, Mo and Si was about 48 wt%, 35 wt%, 15 wt%, and 1 wt%, respectively, which closed to chemical compositions of the σ phase, reported in the previous studies [3,21].

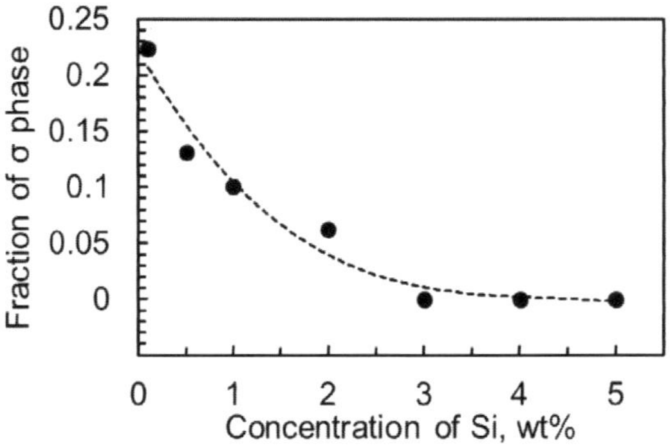

Figure 6. Fraction of σ phase as a function of Si concentration calculated using a thermodynamic calculation software, ThermocalC.

On the other hand, the $Co_5Cr_3Si_2$ ternary intermetallic phase formed in the alloys with Si concentrations over 1.0 wt%. However, information for the formation of the $Co_5Cr_3Si_2$ phase in the biomedical Co–Cr–Mo alloys is limited so far. Gupta [15] reported that the $Co_5Cr_3Si_2$, namely χ, could be found to be in equilibrium with the γ and σ phase in the Co–Cr–Si system. Since the maximum solubility of Si in σ phase was limited at 17 at% at 1175 °C, the extension of Si would lead to the formation of χ phase. In the present study, the amount of Si in the intermetallic phase of the 3.0Si and 5.0Si alloys phase was found to be 5.6 wt% (11.51 at%) and 7.8 wt% (15.37 at%), respectively. Thus, the χ phase should not be formed in the experimental 3.0Si and 5.0Si alloys because the concentrations of Si was below the maximum solubility limit in σ phase. Narushima et al. [17] and Mineta et al. [22] similarly reported that the additions of 1.0 wt% Si significantly increase the heat-treatment temperature and time for complete precipitate dissolution of σ phase in Co–Cr–Mo alloys. Therefore, the formation of $Co_5Cr_3Si_2$ phase in this study was possibly attributed to the insufficient heat-treated temperature and time. In contrast, Nová et al. [23] reported that the silicide (Cr_3Si) phase was observed in Si alloyed Co–Cr–Mo when the concentrations of Si in the initial compositions was reached 3.0 wt%. However, the lattice parameters, a, of Cr_3Si phase is 0.4564 nm (space group ($Pm\bar{3}m$)) which is different from the calculated lattice parameters of $Co_5Cr_3Si_2$ phase in this study. Thus, the intermetallic phase precipitated in the 3.0Si and 5.0Si alloys is χ-phase.

5. Conclusions

In the present study, the microstructure of Co–Cr–Mo alloys with different Si contents was investigated. Alloying Si in Co–Cr–Mo alloys has no effect on the crystallographic orientations and phase presented. The two types of intermetallic compounds were found in the Si-doped Co–Cr–Mo alloys. The $\boldsymbol{\sigma}$ phase was identified in the alloys containing Si between 0.1 and 1.0 wt% while the increase in Si from 3.0 to 5.0 wt% contributed to the formation of the $\boldsymbol{\chi}$ phase. From the TEM analysis, a lattice parameter of the intermetallic phase precipitated in the 3.0Si and 5.0Si alloys was correlated well with those of $Co_5Cr_3Si_2$ phase.

Acknowledgements
The authors would like to thank Mr. Isamu Yoshii, Mr. Kouji Kuramoto and Mr. Kimio Wako of IMR, Tohoku University, for contributing to the alloy fabrication. A special thank you also goes to Mr. Shun Ito of IMR, Tohoku University, for valuable help in the TEM observation.

References

[1] Valero-Vidal C, Casabn-Julin L, Herraiz-Cardona I and Igual-Muñoz A 2013 Mater. Sci. Eng. C 33 4667–76
[2] Yamanaka K, Mori M and Chiba A 2015 J. Mech. Behav. Biomed. Mater. 50 268–76
[3] Tunthawiroon P, Li Y, Koizumi Y and Chiba A 2017 Mater. Sci. Eng. A 703
[4] Ghalme S G, Mankar A and Bhalerao Y 2016 Int. J. Mater. Sci. Eng. 4 113–25
[5] Giacchi J V., Morando C N, Fornaro O and Palacio H a. 2011 Mater. Charact. 62 53–61
[6] Giacchi J V, Fornaro O and Palacio H 2012 Mater. Charact. 68 49–57
[7] Lee S, Takahashi E, Nomura N and Chiba A 2005 Mater. Trans. 46 1790–3
[8] Mori M, Yamanaka K, Sato S, Wagatsuma K and Chiba A 2012 Metall. Mater. Trans. A Phys. Metall. Mater. Sci. 43 3108–19
[9] Yamanaka K, Mori M and Chiba A 2012 Metall. Mater. Trans. A Phys. Metall. Mater. Sci. 43 5243–57
[10] Chen Y, Li Y, Kurosu S, Yamanaka K, Tang N and Chiba A 2014 Wear 319 200–10
[11] Yamanaka K, Mori M and Chiba A 2014 J. Mech. Behav. Biomed. Mater. 29 417–26
[12] Chiba A, Kumagai K, Nomura N and Miyakawa S 2007 Acta Mater. 55 1309–18
[13] Yamanaka K, Mori M, Kuramoto K and Chiba A 2014 J. Mater. 55 987–98
[14] Mineta S, Namba S, Yoneda T, Ueda K and Narushima T 2010 Adv. Mater. Res. 89–91 377–82
[15] Gupta K P 2010 J. Phase Equilibria Diffus. 31 300–7
[16] Podrez-Radziszewska M, Haimann K, Dudziński W and Morawska-Sołtysik M 2010 Arch. Foundry Eng. 10 51–6
[17] Narushima T, Mineta S, Kurihara Y and Ueda K 2013 65 489–504
[18] Kurosu S, Matsumoto H and Chiba A 2010 Metall. Mater. Trans. A 41 2613–25
[19] Mori M, Sato N, Yamanaka K, Yoshida K, Kuramoto K and Chiba A 2016 J. Mech. Behav. Biomed. Mater. 64 187–98
[20] Kurosu S, Nomura N and Chiba A 2007 Adv. Mater. Res. 26–28 777–80
[21] Chen Y, Li Y, Kurosu S, Yamanaka K, Tang N, Koizumi Y and Chiba A 2014 Wear 310 51–62
[22] Mineta S, Namba S, Yoneda T, Ueda K and Narushima T 2011 Metall. Mater. Trans. A 42 1941–9
[23] Nová K, Novák P and Dvorský D 2017 Mater. Tehnol. 51 443–7

Microstructure and Mechanical Properties of Recycled Al-7%Si-1%Fe Cast Alloy with Part Addition of Beverage Cans

J Luijan[1], C klaynil[1], N Mookam[2], P Muangnoy[1]

[1]Department of Production Engineering Technology, Faculty of Industry and Technology, Rajamangala University of Technology Rattanakosin, Prachuap Khiri Khan Province, 77110, Thailand.
[2]Department of Industrial Engineering Technology, Faculty of Industry and Technology, Rajamangala University of Technology Rattanakosin, Prachuap Khiri Khan Province, 77110, Thailand.

E-mail: jinkamon.lui@rmutr.ac.th

Abstract. The effect of un-modified addition of beverage cans material on the microstructure and mechanical properties of Al-7%Si-1%Fe alloy were investigated. Beverage can material included lids and bodies which were added into the liquid metal at 5 and 10 wt.% respectively. The microstructure of the beverage can material was found to consist of $Al(Mn,Fe)_3$ and Mg_2Si phase, with a larger phase fraction of $Al(Mn,Fe)_3$ in the bodies compared to lids. The X-ray diffraction revealed that the lids and bodies contained Al, Si, $Al(Mn,Fe)_3$ and Mg_2Si phases. An Al_3Mg_2 phase unique to the lids was also found. The microstructure of un-modified Al-7%Si- 1%Fe reference alloy consisted of acicular silicon and Al_5FeSi (β phase) in α-Al matrix. Addition of can lids at 5 and 10 wt.% decreased the β phase. Addition of can bodies at 5 and 10 wt.% also decreased the β phase, formed an $Al_{15}(MnFe)_3Si_2$ (α phase) and also formed a Mg_2Si phase especially with 10 wt.% addition. The addition of both lids and bodies modified the β phase similarly to the bodies only samples. A fibrous silicon morphology was found for all conditions. Finally, the ultimate tensile strength increased by 20-30% while the elongation increased with beverage can addition compared to the un-modified Al-7%Si-1%Fe reference cast alloy.

1. Introduction

Nowadays, aluminum alloys are widely used in component manufacture for example in the automotive industry such as wheels, pistons etc. Furthermore, household and food industries also use aluminum alloys for wires/cables, beverage cans, food packages etc. This is because of several desired properties which include, good casting ability, improved mechanical properties through precipitation hardening [1], excellent electrical conductivity, beautiful surface, high corrosion resistance and recycling process.

Aluminum-silicon alloy casting grade (A319, A356 and A380) is a preference material in automotive part production in Thailand which is also known as thick aluminum. However, when this material is brought back to recycle, there is often iron contamination in molten metal. It may come from several sources such as poor casting process, using iron molds in casting or being recycled with other grade of aluminum alloy. This contaminated iron can cause intermetallic compound platelet and flake-like β -A_5FeSi (β phase) in microstructure which influences quality in casting and a decrease in its mechanical properties especially the low cooling rate [2]. The high Fe content (1-1.5 wt.%) in aluminum casting is needed for the die cast process (high cooling rate) providing high surface quality and eases release from the mold. The short solidification time reduces the β phase formation and thus improves the mechanical properties. At present, a process for β-phase reduction will use manganese, lithium, potassium and chromium [3-8] as additives to form α-$Al_{15}(Mn,Cr,Fe)_3Si_2$ (α phase) which look like Chinese script to improve mechanical properties [7-8].

Wrought grade aluminum is commonly used in producing food packaging and building components. One of the most popular shapes of aluminum is a thin sheet which is used in beverage can production. Aluminum alloy used for making lids of beverage cans is grade 5182 which consists of manganese 0.2-0.5 wt.% and magnesium 4-5 wt.%, while the bodies of beverage cans use aluminum grade 3004 which has manganese at 1-1.5 wt.% and magnesium 0.8-1.3 wt.% . For the recycled beverage cans process, all cans are compressed as rods before being brought to the melt and added with various elements to correct the chemical ratio of the molt.

Even though recycling beverage cans for Al-7%Si-1%Fe alloy (A356) has been approved, the process may be improved. Furthermore, in the field of research and industry, there is no research which brings a part of beverage can to use as additive for recycled Al-7%Si-1%Fe alloys. Therefore, it is important to find an appropriate ratio of manganese from the recycled parts of beverage cans. This research will focus on studying the effect of adding beverage cans to microstructure and mechanical properties of Al-7%Si-1%Fe alloy in the recycle process.

2. Experimental procedure

The beverage cans were decoded by heat treatment for 30 min. at 400 °C. After that, the lids and bodies were separated. The lids and bodies were melted separately in a silicon carbide crucible by resistant electric furnace. The specimens of lids and bodies beverage cans were grinded with alumina powder and silica colloid. All specimens were etched using the Keller's reagent and the microstructure was examined by optical microscope. The phase identity of intermetallic compound was determined by X-ray diffraction. In the β phase modification efficiency, the Al-7% Si-1% Fe alloy was used to test the efficiency of modification. First the alloy was melted in a silicon carbide crucible at 800°C. When the alloy started to melt, flux was added to cover the surface. After that, a part of beverage can was added into the molt at 5 and 10 wt.% with strontium to fix weight constant at the same level for each sample. The molt was held for 30 minutes before purging with argon gas for 5 minutes and then poured at 720°C into a thin walled stainless steel and steel mold according to ASTM B108-03a. The microstructure was examined by optical microscope. Tensile specimens were treated for 30 min. at 520°C. The mechanical properties were tested with universal testing machine.

3. Result and Discussion

A microstructure inspection of remolded beverage can lids show that the microstructure consists of the $Al(Mn,Fe)_3$, Mg_2Si and Al_3Mg_2 phases shown in Fig 1. A microstructure inspection of beverage can bodies show that the microstructure consisted of $Al(Mn,Fe)_3$ and Mg_2Si phases, shown in Fig 2. The analysis result of remolded beverage can by X-ray diffraction technique is shown in Fig 3, the alloy is formed of Al, Si, Mg_2Si, $Al(Fe,Mn)_3$ and Al_3Mg_2.

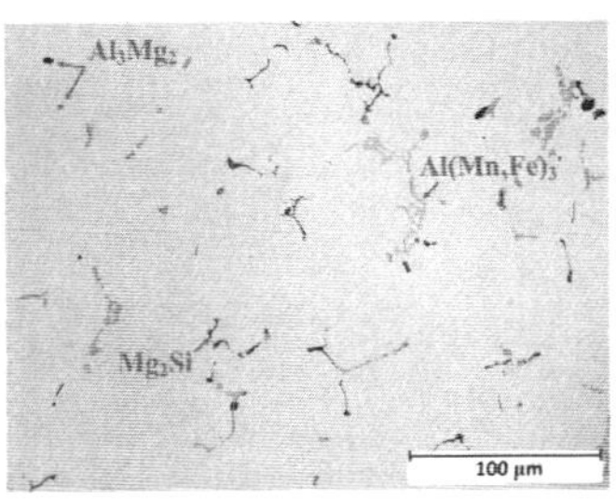

Figure 1. Microstructure of remolded beverage can lids

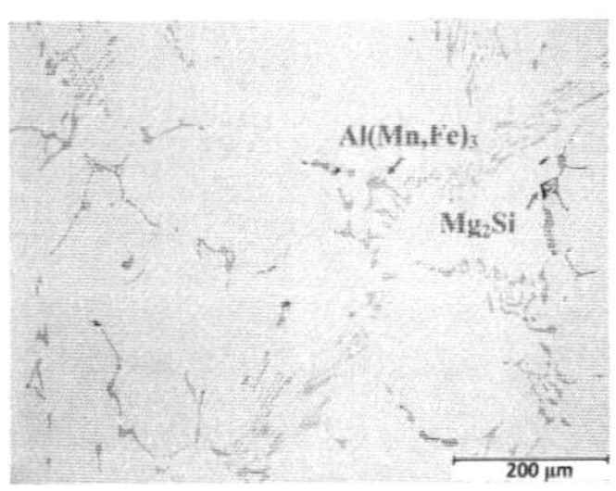

Figure 2. Microstructure of remolded beverage can bodies

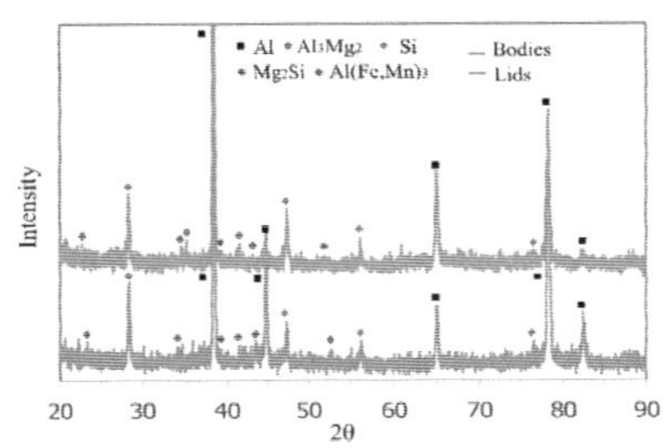

Figure 3. X-ray diffraction of remolded lids and bodies of beverage cans

IOP Conf. Series: Materials Science and Engineering **635** (2019) 012008 doi:10.1088/1757-899X/635/1/012008

3.1 *Microstructure of Al-7%Si-1%Fe*

An investigation of the microstructure of Al-7% Si-1% Fe alloy found that the microstructure is combined with eutectic silicon phase which has an acicular morphology and the β phase which look like fibers with length of 100-150 μm widely spread in the area of α-Al as shown in Fig. 4.

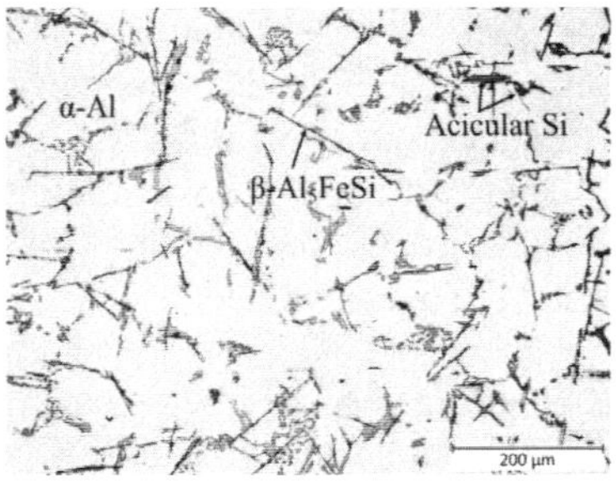

Figure 4. Microstructure of Al-7% Si-1% Fe alloy

3.2 *Microstructure of Al-7%Si-1%Fe after adding lids of beverage can*

The result of improving the alloy of Al-7%Si-1%Fe by adding lids 5 wt.% show that the β phase is still present, but has a decrease in size and length (100 μm). From the result, it has demonstrated that although adding lids at 5 wt.% cannot modified the β phase to α phase, it can reduce size and length of the β phase as shown in Fig5(a). The result from 10wt.% indicates further reduction of the β phase as shown in Fig 5(b).

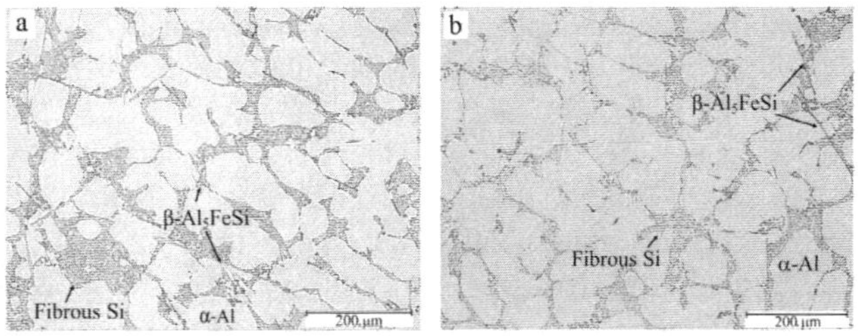

Figure 5. Microstructure of Al-7% Si-1% Fe after adding lids of beverage cans at (a) 5wt%, (b) 10wt.%

3.3 *Microstructure of Al-7%Si-1%Fe after adding body of beverage can*

The result of improving Al-7% Si-1% Fe by adding bodies of beverage cans at 5wt.% found that the size and quantity of the β phase decreased; and there was also a phase of Mg_2Si (Fig 6) spread around an arm of dendrite as shown in Fig 6(a). However, adding bodies of can at 10wt.% resulted in the α phase and Mg_2Si phase in microstructure. The forming of Mg_2Si which appear around an arm of dendrites indicated that phase eutectic silicon nearby the Mg_2Si phase had decreased due to the binding of silicon and magnesium as shown in Fig 6(a-b).

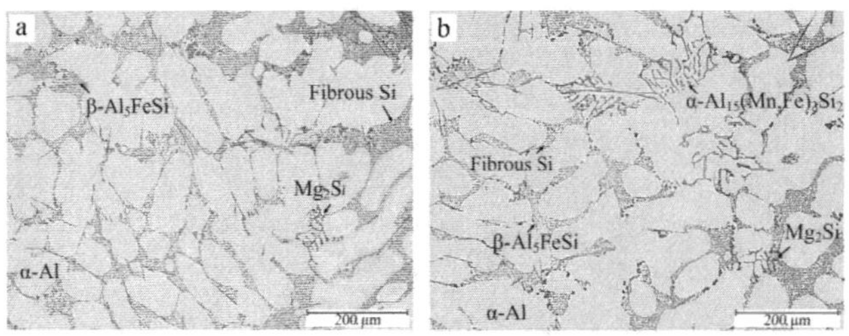

Figure 6. Microstructure of Al-7% Si-1% Fe after adding bodies of beverage can at (a) 5wt%, (b) 10wt.%

3.4 Microstructure of Al-7%Si-1%Fe after adding both lid and body of beverage can

The result of improving Al-7% Si-1% Fe with bodies and lids of beverage cans at 5 and 10 wt.% produced a decrease in the β phase and increased the amount the α phase. The quantity of the α phase increased with addition level as shown in Fig 7 (a-b).

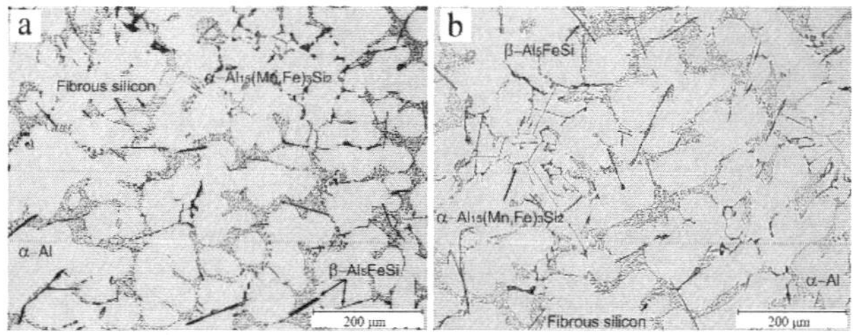

Figure 7. Microstructure of Al-7% Si-1% Fe after adding bodies and lids of beverage can at (a) 5wt%, (b) 10wt.%

3.5 Mechanical properties

All specimens were treated to decrease the stress from solidification and to improve distribution of chemical composition. The ultimate tensile strength of Al-7%Si-1%Fe was 98 MPa. The addition with 5 and 10 wt.% of lids increased this to 116 and 134 MPa while the addition with 5, and 10 wt.% of bodies was 112 and 137 MPa. Addition of both lids and bodies with 5 and 10 wt.% resulted in tensile strength of 135 and 152 MPa as shown in Fig.8.

Fig. 9 shows the elongations of casting with lids, bodies and both lids and bodies additions. The Al-7% Si-1% Fe had 3.6% while the addition of lids with 5 and 10 wt.% was 4.3, 5.6%, respectively. The addition with bodies and both lids and bodies was very effective to increase the elongation compared to lids alone.

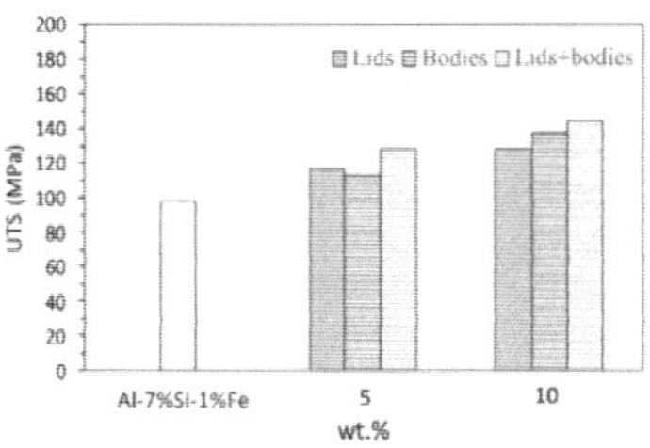

Figure 8. Ultimate tensile strength of Al-7%Si-1%Fe and Al-7%Si-1%Fe with added lids, bodies and both lids and bodies.

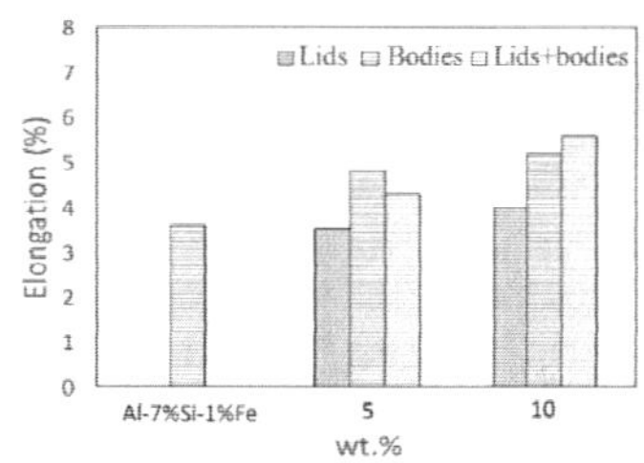

Figure 9. Elongations of Al-7%Si-1%Fe and Al-7%Si-1%Fe with added lids, bodies and both lids and bodies.

The addition of lids and bodies of beverage cans into molten Al-7%Si-1%Fe alloy have an effect to transition the β phase to α phase. Moreover, the higher level of phase $Al(Fe,Mn)_3$ in microstructure of can bodies when compared with can lids (Fig 1) gave a better the β phase adjustment. When adding lids and bodies of cans which have phase $Al(Fe,Mn)_3$ during solidification process, α-Al phase formation will happen at 577°C. This causes other elements such as Si, Fe, Mn, Si, Mg and Sr to be rejected into the melt [5,9], because α-Al has a low dissolution ability. When concentrations of Si, Mn and Fe are high in liquid phase, there will be a combination and formation of Mg_2Si and α phases while phase eutectic silicon can transform its morphology from acicular to be fibrous in every experiment. This is the result of adding Al-10%Sr at 0.04wt.% in every experiment. However, in all cases a phase in lids and bodies of cans have an effect to modify of the β phase.

The remold of the Al-7%Si was contaminated of the Fe element and formed into β phase. The result shown that the part beverage addition made modified the β phase in to α phase which improve the mechanical properties and decrease porosity of casting. For casting part industry can use the Al-7%Si-1%Fe cast alloy installed the usage of the primary Al-7%Si alloy can reduced. This is the cost saving due to Al-7%Si is more expensive than the recycle Al-7%Si-1%Fe alloy. Moreover, the bodies and lids parts can used to compensate the Al-Mn or Al-Cr alloy which is the import product.

4. Conclusions

The microstructure of the lid consists the $Al(Mn,Fe)_3$, Al_3Mg_2 and Mg_2Si phase while the body consists $Al(Mn,Fe)_3$ and Mg_2Si. Both lid and body of beverage cans were used to modify the β phase into α phase. The efficacy of modification from the β phase depended on the amount of $Al(Mn,Fe)_3$ phase in the part of beverage cans and the percentage addition to the molt. The mechanical properties of Al-7%Si-1%Fe cast alloys can be improved with beverage can lid, body and mixed lid and body additions.

References

[1] Lumley R 2011 *Fundamentals of aluminium metallurgy, production, processing and applications* (Woodhead publishing limited, Great Abington)
[2] Davis J.R 1993 Aluminum and Aluminum Alloys(ASM International, Materials Park,) pp 199-230
[3] Ashtari P, Tezuka H and Sato T 2004 The 9th International conference on aluminum alloy (Australia) Modification of Fe-containing intermetallic compounds of Al-Si-Cu-Fe cast alloy using Sr, Li and Mn additions pp 951-955
[4] Ashtari P, Tezuka H and Sato T 2004 Influence of Li addition on intermetallic compound morphologies in Al–Si–Cu–Fe cast alloys Scripta Materialia vol 51 pp 43-46
[5] Ashtari P, Tezuka H and Sato T 2005 Modification of Fe-containing intermetallic compounds by K addition to Fe-rich AA319 aluminum alloys Scripta Materialia vol 53 pp 937-942
[6] Eidhed W 2008 Modification of β-Al5FeSi Compound in Recycled Al-Si Cast Alloys by Using Sr, Mg and Cr Addition Journal of Materials Science and Technology vol 24 pp 45-47
[7] Sreeja Kumari S S, Pillai R M and Pai B C 2008 A study on the structural, age hardening and mechanical characteristics of Mn and Ca added Al–7Si–0.3Mg–0.6Fe alloy Journal of Alloys and Compounds vol 453 pp 167–173
[8] Sreeja Kumari S S, Pillai R M, Rajan T P D and Pai B C 2007 Effects of individual and combined additions of Be, Mn, Ca and Sr on the solidification behaviour, structure and mechanical properties of Al–7Si–0.3 Mg–0.8Fe alloy Materials Science and Engineering A vol 460 pp 561–573
[9] Li Z Limodin N, Tandjaoui A, Quaegebeur P, Osmond P and Balloy D 2017 Influence of Sr, Fe and Mn content and casting process on the microstructures and mechanical properties of

Acknowledgments

The authors are grateful to Rajamangala University of Technology Rattanakosin for the financial support under contract number 270604.

Wettability and printability of SAC305-xTiO_2 Pb-free solder paste on Cu substrate

K. Sukpimai, W. Suwannakrue, K. Kanlayasiri

Department of Industrial Engineering, Faculty of Engineering, King Mongkut's Institute of Technology Ladkrabang, Bangkok 10520, Thailand

E-mail: kkkannac@kmitl.ac.th

Abstract. This paper was aimed to study on effects of TiO_2 nanoparticles on wettability and printability of SAC305-xTiO_2 Pb-free solder on Cu substrate. TiO_2 nanoparticles at various concentrations (0.00, 0.05, 0.10, 0.50, and 1.00 wt%) were introduced to SAC305 solder paste by mechanical mixing. Wettability of the solder on Cu substrate was measured in terms of contact angle while printability was calculated from the ratio of printed volume on Cu substrate to ideal volume of solder on the substrate. Results showed that the introduction of TiO_2 nanoparticles decreases the contact angle and 0.33 wt% TiO_2 is the optimal concentration providing the lowest contact angle within the experimental range. Moreover, the addition of TiO_2 nanoparticles reduced the printability of solder paste on Cu substrate. The higher amount of nanoparticles gave the lower printability. Simultaneous optimization on wettability and printability was also performed to find the optimal TiO_2 concentration.

1. Introduction

In recent decades, environmental problems due to electronics wastes are the major global concern. Many countries issue environmental regulations to control the use of toxic materials in electronics products such as Restriction of Hazardous Substances (RoHS) directive of European Union. This directive has a key impact on electronics equipment manufacturers because it prohibits the use of some hazardous materials in electronics equipment, especially lead (Pb) which is the main component in various customary solders [1-2]. The legislation of RoHS forces the development of Pb-free solders to replace the traditional Pb-bearing solders.

Many Pb-free solders have been developed and commercialized to the market such as Sn-Cu, Sn-Ag, Sn-Ag-Cu (SAC) alloys. Among these new solders, SAC305 (Sn-3.0Ag-0.5Cu) alloy has been widely used in electronics industries because of its good mechanical and electrical properties [3]. However, various properties of Pb-free solders are inferior to the long-established Pb-bearing solders [4-5]. Therefore, research on Pb-free solders is still intensified to improve their properties to level to those of traditional Pb-bearing solders. With the advancement of nanotechnology, many nanoparticles have been introduced to Pb-free solders in order to enhance their properties [6-9]. In case of surface mount technology (SMT), although wettability and printability of solder paste are among the critical characteristics for gaining reliable solder joint, only few articles related to these properties have been reported in the literature [10-15].

This paper investigated effects of TiO_2 nanoparticles on wettability and printability of SAC305-xTiO_2 Pb-free solder paste on Cu substrate. Simultaneous optimization on wettability and printability was also performed to find optimal concentration of the nanoparticles. Desirability function was used as the optimization tool in this research.

2. Experimental Setup

In this research, SAC305 solder paste was used as the base solder and TiO_2 nanoparticles were used as the dispersed phase. TiO_2 nanoparticles were mechanically mixed for 30 min with the solder paste for the uniform mixture. A concentration of TiO_2 nanoparticles at 0.00, 0.05, 0.10, 0.50, and 1.00 wt% was employed in this experiment. Pure copper sheet with a thickness of 0.35 mm was used as the substrate for wettability test. The wettability test was performed using reflow soldering at 260 °C for 30 s according to IPC/JEDEC: J-STD-020D-01 standard. Wettability of SAC305-xTiO_2 on Cu substrate was determined in terms of contact angle. The contact angle (θ) was calculated from equation (1) where H and R is height and radius of the solder joint, respectively [16].

$$\theta = sin^{-1}\left(\frac{2}{\frac{R}{H}+\frac{H}{R}}\right) \tag{1}$$

Printability of SAC305-xTiO_2 was tested based on JIS Z3198-3:2003 standard and it was computed using equation (2) where V is the printed volume on the substrate and V_o is the ideal printed volume of solder on the substrate [17]. MINITAB software was used for optimization of TiO_2 concentration via desirability function.

$$\%P = \frac{V}{V_o} \times 100 \tag{2}$$

3. Result and Discussion

3.1 Wettability

Effect of TiO_2 nanoparticles on contact angle of SAC305 solder paste on Cu substrate was illustrated in Figure 1. The introduction of TiO_2 nanoparticles decreased the contact angle of virgin SAC305 until 0.50 wt% TiO_2 and the contact angle was increased again at 1.00 wt%. The decline of contact angle at the beginning was due to the fact that nanoparticles could lower the surface tension of molten solder [18]. However, simultaneously, adding the nanoparticles resulted in increasing of viscosity of the molten solder. Due to these two concurrent phenomena, the contact angle of the solder was therefore decreased to a certain point before upsurge again. Among the five concentrations of TiO_2 in this experiment, the concentration of TiO_2 nanoparticles at 0.50 wt% gave the lowest contact angle of 9.27°. Therefore, adding the nanoparticles at this concentration provided the best wettability of SAC305 solder paste on Cu substrate. In fact, the degree of wettability can be classified into five levels based on the contact angle as presented in Table 1.

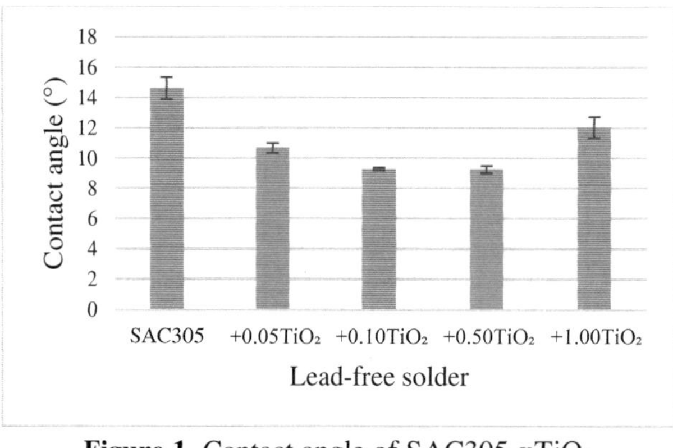

Figure 1. Contact angle of SAC305-xTiO_2.

IOP Conf. Series: Materials Science and Engineering **635** (2019) 012009 doi:10.1088/1757-899X/635/1/012009

Table 1. Classification of wettability [19, 20].

Degree of wettability	Contact angle
Excellent	$0^o < \theta < 30^o$
Good	$30^o < \theta < 40^o$
Fair	$30^o < \theta < 40^o$
Poor	$50^o < \theta < 70^o$
Extremely poor	$\theta > 70^o$

From the results, 0.50 wt% TiO_2 was the best concentration among the investigated values, however it was difficult to determine the optimal concentration of TiO_2 nanoparticles in the experimental range. Desirability function was therefore used to find the optimal concentration. The objective function was to minimize the contact angle of the solder on Cu substrate and the optimization result showed that 0.33 wt% TiO_2 nanoparticles provide the minimum contact angle at about 6.18^o as presented in Figure 2. Confirmation tests were performed 3 times at 0.33 wt% TiO_2 and the measured contact angle was $6.61^o \pm 0.51^o$.

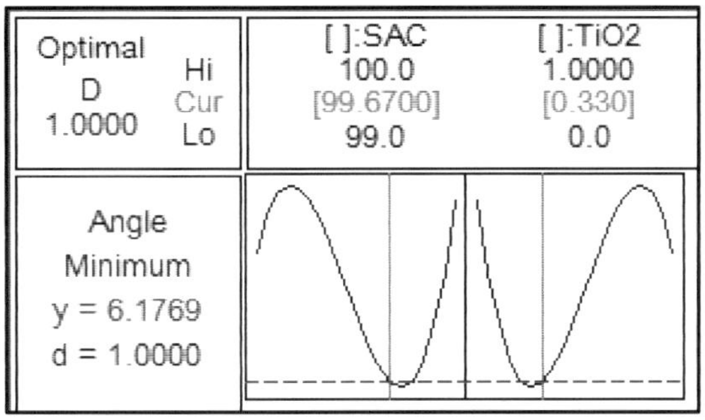

Figure 2. Optimization on contact angle.

3.2 Printability

High printability is always desired in solder printing process for SMT. However, in this experiment, it was obvious that printability of SAC305-xTiO_2 decreases with the increase of TiO_2 nanoparticles as shown in Figure 3. This was due to the fact that the increase of TiO_2 nanoparticles provided the higher viscosity of solder paste. With the higher viscosity, the solder paste was more stuck to the stencil's wall when printed resulting in the lower printability. In practice, the acceptable printability is no less than 80% [21]. As a result, SAC305-1.00TiO_2 was not acceptable for solder printing process for SMT while SAC305-0.50TiO_2 marginally met the criterion and the best solder paste for printing was the virgin SAC305.

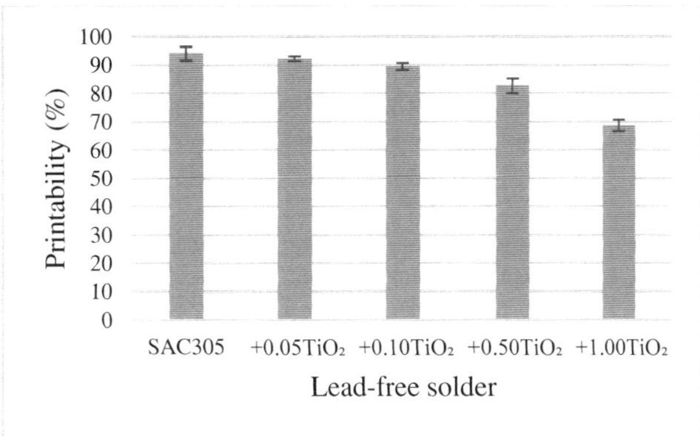

Figure. 3 Printability of SAC305-xTiO_2.

3.3 Optimal TiO_2 Concentration for Wettability and Printability

Since the optimal concentration of TiO_2 of the wettability and printability was not at the same value, in this section, simultaneous optimization on wettability and printability was therefore performed to find the optimal concentration of TiO_2 nanoparticles. Desirability function was used again to obtain the simultaneous optimal concentration. The objective function was to minimize the contact angle and maximize the printability. Using the desirability function, the optimal concentration for both wettability and printability was at about 0.04 wt% TiO_2 providing contact angle and printability of about 11.61° and 92.04%, respectively, as illustrated in Figure 4. However, the optimal solution was normally dependent on the selected importance of each response as well as the desired lower and upper limits of each response. Confirmation tests were then performed three times to verify the results and the measured contact angle and printability was 11.05° ± 0.65° and 89.88 ± 5.74 %, respectively.

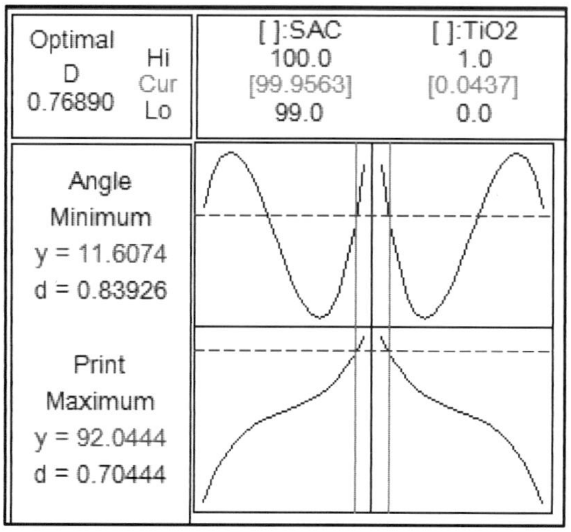

Figure 4. Simultaneous optimization on wettability and printability.

4. Conclusion

This paper investigated effects of TiO_2 nanoparticles on wettability and printability of SAC305-xTiO_2 on Cu substrate. It was found that the addition of 0.33 wt% TiO_2 nanoparticles provides the highest wettability and the printability of SAC305-xTiO_2 decreases with the increase of TiO_2. However, using simultaneous optimization, 0.04 wt% TiO_2 was the optimal concentration for both contact angle and printability. The developed solder paste could be used as lead-free nano-composite solder paste in electronics industries. However, in order to be used as composite solder, other properties such as creep and fatigue resistance should be further investigated for this solder.

Acknowledgements

The authors would like to thank King Mongkut's Institute of Technology Ladkrabang and the National Research Council of Thailand for the financial sponsorship of this project.

References

[1] D. Shangguan 2005 *Lead-Free Solder Interconnect Reliability* (Materials Park Ohio: ASM International)

[2] K._Kanlayasiri and R. Kongchayasukawat 2018 Property alterations of Sn-0.6Cu-0.05Ni-Ge lead-free solder by Ag, Bi, In and Sb addition *Trans. Nonferrous Met. Soc. China* **28** no. 6 pp 1166-1175

[3] M. Abtew and G. Selvaduray 2000 Engineering Lead-free Solders in Microelectronic *Mater. Sci. Eng.* **27** pp 95–106

[4] H. Conrad, Z. Guo, Y. Fahmy, and D. Yang 1999 Influence of microstructure size on the plastic deformation kinetics, fatigue crack growth rate, and low-cycle fatigue of solder joints *J. Electron. Mater* **28** no. 9 pp 1062–1070

[5] H. Mavoori and S. Jin 1998 New creep-resistant low melting point solders with ultrafine oxide dispersions *J. Electron. Mater* **27** no. 11 pp 1216–1222

[6] J. L. Marshall, J. Calderon, J. Sees, G. Lucey, and J. S. Hwang 1991 Composite Solders *IEEE Trans. Components, Hybrids, Manuf. Technol.* **14** no. 4 pp 698–702

[7] H. S. Betrabet, S. M. McGee, and J. K. McKinlay 1991 Processing dispersion-strengthened Sn-Pb solders to achieve microstructural refinement and stability *Scr. Metall. Mater* **25** no. 10 pp 2323–2328

[8] A. A. El-Daly, A. Fawzy, S. F. Mansour, and M. J. Younis 2013 Thermal analysis and mechanical properties of Sn-1.0Ag-0.5Cu solder alloy after modification with SiC nano-sized particles *J. Mater. Sci. Mater. Electron.* **24** no. 8 pp 2976–2988

[9] J. Shen and Y. C. Chan 2009 Research advances in nano-composite solders *Microelectron. Reliab* **49** no. 3 pp 223–234

[10] R. R. Lathrop 1997 Solder paste print qualification using laser triangulation *IEEE Trans. components, Packag. Manuf. Technol. Part C. Manuf.* **20** no. 3 pp 174–182

[11] C. C.H., W. W., L. J.C.C., S. F., and L. S.W.R. 2007 Characterization and comparison of five SAC-based solder pastes for Pb-free reflow soldering *Proceedings of HDP07*

[12] A. Araźna, G. Kozioł, W. Stęplewski, and J. Borecki 2008 Assessment of lead-free solder pastes with different type of powder for fine-pitch application *31st Int. Spring Semin. Electron. Technol. Reliab. Life-time Predict. ISSE 2008* pp 579–584

[13] B. An and Y. Wu 2010 Evaluating the printability of solder paste from paste roll characteristics *11th Int. Conf. on Electron. Packaging Technol. & High Density Packaging* pp 1186–1189

[14] K. Li, Y. Lei, J. Lin, B. Liu, H. Bai, and J. Qin 2013 Development of Sn-Bi systems lead-free solder paste *14th Int. Conf. Electron. Packag. Technol. ICEPT 2013* no. 3132006 pp 160–162

[15] K. Kanlayasiri and N. Meesathien 2018 Effects of Zinc Oxide Nanoparticles on Properties of SAC0307 Lead-Free Solder Paste *Adv. Mater. Sci. Eng.* **2018** Article ID 3750742 https://doi.org/10.1155/2018/3750742

[16] G. Humpston and D. M. Jacobson 2004 *Principles of Soldering* (United States of America: ASM International)

[17] E. H. Amalu, N. N. Ekere and S. Mallik 2011 Evaluation of rheological properties of lead-free solder pastes and their relationship with transfer efficiency during printing process *Mater. Des.* **32**. pp 3189-3197

[18] G. Chen, H. Peng, V. V. Silberschmidt, Y. C. Chan, and C. Liu. 2016 Performance of Sn–3.0Ag–0.5Cu composite solder with TiC reinforcement: Physical properties, solderability and microstructural evolution under isothermal ageing *J. of Alloys Compd.* **685** pp 680-689

[19] L. Zang, Z. Yuan, H. Xu, B. Xu 2011 Wetting process and interfacial characteristic of Sn-3.0Ag-0.5Cu on different substrates at temperatures ranging from 503 K to 673 K *Appl. Surf. Sci.* **257** pp 4877–4884

[20] A. Wedi, D. Baither, G. Schmitz 2011 Contact angle and reactive wetting in the SnPb/Cu system, *Scripta Mater.* **64** pp 689–692

[21] L. Bruno and T. Johnson 2014 Printing of Solder Paste – A Quality Assurance Methodology *IPC APEX EXPO 2014* (Las Vegas: United States)

A study on determination of optimum gear ratios of a worm - helical gearbox

Vu Ngoc Pi[1], Tran Thi Hong[2], Le Xuan Hung[1], Luu Anh Tung[1], Nguyen Khac Tuan[3]

[1]Mechanical Engineering Faculty, Thai Nguyen University of Technology, Thai Nguyen city, Vietnam
[2] Nguyen Tat Thanh University, Ho Chi Minh city, Vietnam
[3] Automotive and Power Machinery Faculty, Thai Nguyen University of Technology, Thai Nguyen city, Vietnam

vungocpi@tnut.edu.vn

Abstract. This paper introduces a new study on the optimum calculation of the gear ratios of a worm-helical gearbox. For finding the gear ratios, the cross section area of the gearbox was chosen as the objective function of the optimization problem. Moreover, the influence of the input parameters including the total gearbox ratio, the allowable contact stress and the wheel face width coefficient of the helical gear set, the output torque was investigated. For evaluation of the influence of these parameters on the optimum gear ratios, an "experiment" was designed and a computer program was built for performing the "experiment". In addition, models for determining the optimum gear ratios of the gearbox were proposed. Based on those models, the gear ratios can achieve high accuracy with simplification.

1. Introduction

To date, there have been many study to determine the gear ratios of worm gearboxes. For a two-stage worm gearbox, for getting the reasonable housing structure of the gearbox, the gear ratios can be determined as follow ([1], [2]):

$$u_1 \approx u_2 \approx \left(u_h\right)^{1/2} \tag{1}$$

Also, the optimum gear ratio can be determined from tabulated form which was found based the practical data [3]. In addition, it is reported that the optimum gear ratio of the second stage of a two-stage worm gearbox is $u_2 \approx 30.97$ when the objective is the gearbox structure reasonable [4].

For worm –helical gearboxes, the gear ratio of the helical gear set is calculated by the following equation [1]:

$$u_2 = \left(0.03 \div 0.06\right) u_h \tag{2}$$

In fact, the above formula was found from an actual experience and it is quite simple. Nevertheless, the use of it may lead to the gearbox structure unreasonable. For ensuring the effectiveness of oil lubrication for both stages of the gearbox, in [5] presented a graph for determining the gear ratio of the worm gear set from the total ratio of the gearbox (Figure 1). From the figure it is clear that, the determination of the gear ratio of the worm gear u_1 is very complicated. Moreover, for determining u_1, the user must select the coefficient c (with c = 1.3 to 1.6) so it is very complicated and we can not reach to the optimum values. To exclude these points, the optimum gear ratios of the gearbox with the same objective as in [5] can be calculated by the following model [6]:

$$u_2 = 6.86 \cdot \psi_{ba2}^{1/2} \tag{3}$$

Where, ψ_{ba2} is the coefficient of wheel face width ($\psi_{ba2} = 0.3 \ldots 0.4$).

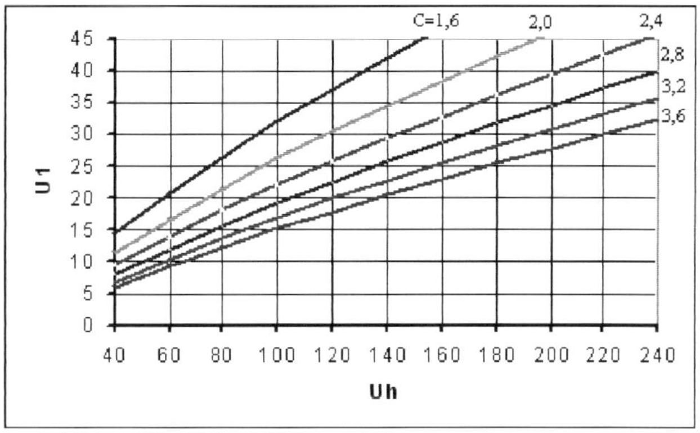

Figure 1. Transmission ratio of worm gear unit versus the total transmission ratio [5**]**.

Also for worm –helical gearboxes, for getting the minimum gearbox length it was reported that the optimum gear ratio of the helical gear unit u_2 was the allowable maximum gear ratio of it ($[u_{2\max}] = 8 \ldots 10$) [7].

From the above analysis, although there have been several studies on for determining the optimum gear ratios of the worm gearboxes, there is still lack of a study in which the objective is the minimum acreage of the cross section. This paper introduces a study for calculation of the optimum gear ratios of the worm-helical gearboxes with this objective.

2. Optimization problem

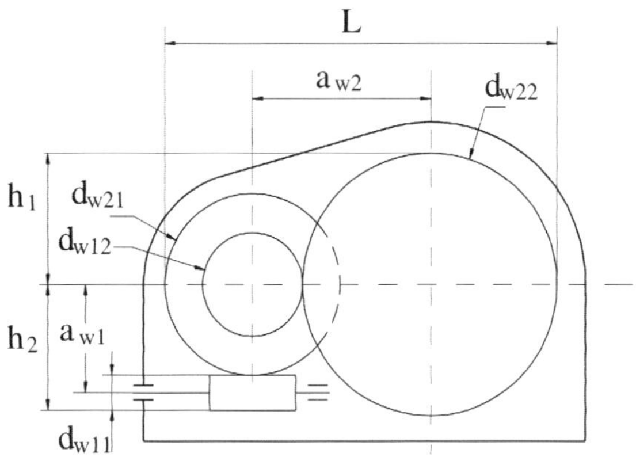

Figure 2. Calculation schema

The cross section area of the gearbox is calculated by the following equation (Figure 2):

$$A = h \cdot L \tag{4}$$

In which,

$$h = \max h_1 + \max h_2 \tag{5}$$

Where, h_1 and h_2 are determined as (see Figure 2):

$$h_1 = \max\left(d_{w21}/2; d_{w22}/2\right) \tag{6}$$

$$h_2 = \max\left(d_{w11}/2 + a_{w1}; d_{w22}/2\right) \tag{7}$$

$$L = d_{w21}/2 + a_{w2} + d_{w22}/2 \tag{8}$$

In the above equations, a_{w1} and a_{w2} are the center distances of the first and the second stages; d_{w11} and d_{w21} are the pitch diameters of the worm and the worm gear; d_{w22} is the pitch diameter of the driven gear of the helical gear unit. With the worm gear set, we have [8]:

$$d_{w21} = m \cdot z_2 = 2 \cdot a_{w1} \cdot z_2 / \left(z_2 + q\right) \tag{9}$$

$$d_{w11} = m \cdot q = 2 \cdot a_{w1} \cdot q / \left(z_2 + q\right) \tag{10}$$

In which, q is the worm pitch diameter coefficient and q can be determined as [8]:

$$q = k_q \cdot z_2 \tag{11}$$

In which, k_q is the coefficient for calculation of the worm pitch diameter coefficient: $k_q = 0.25 \ldots 0.3$ [8].

For the helical gear set, the driven diameter can be determined as [8]:

$$d_{w22} = 2 \cdot a_{w2} \cdot u_2 / \left(u_2 + 1\right) \tag{12}$$

From Equations (4) to (12), the following equation is given:

$$A = f\left(a_{w1}, a_{w2}, k_q, u_2\right) \tag{13}$$

Thus, the optimization problem is defined as:

$$\text{minimize } A = L \cdot h \tag{14}$$

With the following constraints

$$8 \le u_1 \le 80 \tag{15}$$

$$1 \le u_2 \le 9$$

From equations (13), (14) and (15), it is revealed that to solve the optimization problem it is required to determine the center distances of the worm gear set a_{w1} and the helical gear set a_{w2}.

2.1. Determining the center distance of the worm gear set

The center distance of the worm gear set (mm) can be determined by [8]:

$$a_{w1} = \left(z_2 + q\right)\left(\left[170/\left(z_2 \cdot [\sigma_H]\right)\right]^2 \cdot T_{21} \cdot K_H / q\right)^{1/3} \tag{16}$$

Substituting (11) into (16) we have:

$$a_{w1} = \left(1 + k_q\right)\left(\left(170/[\sigma_H]\right)^2 \cdot T_{21} \cdot K_H / k_q\right)^{1/3} \tag{17}$$

$$a_{w1} = 60.168 \cdot \left(K_H \cdot [T_{21}] / [\sigma_H]^2\right)^{1/3} \tag{18}$$

Where, T_{21} is the torque on the wheel (Nmm); With the worm-helical gearbox, T_{21} is calculated by:

$$T_{21} = T_{out} / \left(\eta_g \cdot \eta_b^3 \cdot u_2\right) \tag{19}$$

In which, T_{out} is the output torque (Nmm); η_{hg} is the helical gear transmission efficiency ($\eta_{hg} = 0.96\ldots0.98$ [8]); η_b is the transmission efficiency of a pair of rolling bearing ($\eta_b = 0.99\ldots0.995$ [8]). Choosing $\eta_{hg} = 0.97$, $\eta_b = 0.992$ and substituting them into (19) gives:

$$T_{21} = 1.056 \cdot T_{out} / u_2 \quad (20)$$

K_H is the load factor; $K_H = 1.1\ldots1.3$ [8]; Choosing $K_H = 1.2$ we have

$$a_{w1} = 33.2084 \cdot (1 + k_q) \left(\frac{T_{out}}{k_q \cdot u_2 \cdot [\sigma_H]^2} \right)^{1/3} \quad (21)$$

Wherein, $[\sigma_H]$ is the allowable contact stress (N/mm^2); $[\sigma_H]$ depends on the wheel material. If the wheel material is tinless bronze or soft grey iron, from the tabulated data in [8], the following regression equation was found (with the determination coefficient $R^2 = 0.9906$) for determination of the allowable contact stress:

$$[\sigma_H] = 5.0515 \cdot v_{sl}^2 - 49.742 \cdot v_{sl} + 189.9 \quad (22)$$

In which, v_{sl} is the slip velocity; v_{sl} can be calculated as [8]:

$$v_{sl} = 0.0088 \cdot \left(P_1 \cdot u \cdot n_1^2 \right)^{1/3} \quad (23)$$

If the wheel material is tin bronze, $[\sigma_H]$ is determined by [8]:

$$[\sigma_H] = K_{HL} \cdot v_{sl} \cdot [\sigma_{H0}] \quad (24)$$

In which, $[\sigma_{H0}]$ is the allowable contact stress when the stress change cycle is 10^7:

$$[\sigma_{H0}] = (0.7\ldots0.9) \cdot \sigma_t \quad (25)$$

Where, σ_t is the tensile stress (N/mm2); $\sigma_t = 260$ if $v_{sl} = 5\ldots8$; $\sigma_t = 230$ if $v_{sl} = 8\ldots12$ and $\sigma_t = 285$ if $v_{sl} = 8\ldots25$.

K_{HL} is the service life ratio; K_{HL} is determined by:

$$K_{HL} = \left(10^7 / N_{HE} \right)^{1/8} \quad (26)$$

In which N_{HE} is the equivalent loading cycle number for the worm-wheel teeth for the whole lifetime of the gearing:

$$N_{HE} = 60 \cdot n_2 \cdot t_\Sigma \quad (27)$$

Where, t_Σ is t the service lifetime of the gearing (h); n_2 is rotational frequencies of the wheel (rpm);

2.2. Determining the center distance of the helical gear set

The center distance of the second stage a_{w2} is calculated by [8]:

$$a_{w2} = K_a \cdot (u_2 + 1) \cdot \left(T_{12} \cdot k_{H\beta} / \left([\sigma_H]^2 \cdot u_2 \cdot \psi_{ba2} \right) \right)^{1/3} \quad (28)$$

Similarly, for the second stage we have

$$T_{out} = T_{12} \cdot \eta_{hg} \cdot \eta_{be}^2 \cdot u_2 \quad (29)$$

Choosing $\eta_{hg} = 0.97$ and $\eta_{be} = 0.992$ as in section 2.1 gives

$$T_{12} = 1.0476 \cdot T_{out} / u_2 \quad (30)$$

Substituting (30), k_a=43 and $k_{H\beta} = 1.1$ (as in section 2.1) into (28) gets:

IOP Conf. Series: Materials Science and Engineering **635** (2019) 012010 doi:10.1088/1757-899X/635/1/012010

$$a_{w2} = 45.0814 \cdot (u_2 + 1) \cdot \left(T_{out} / \left([\sigma_H]^2 \cdot u_2^2 \cdot \psi_{ba2}\right)\right)^{1/3} \quad (31)$$

2.3. Experimental work

To inspect the effect of the input parameters on the optimum gear ratios, an "experiment" was designed and performed. The experiment was designed with a 2-level full factorial design. Table 1presents the input parameters which were selected for the investigation. Consequently, the design was organised with $2^5 = 32$ number of experiments. To perform the experiment, a computer program was built based on equations (14) and (15). Table 2 shows different levels of the input parameters and the results of the output of the computer program (the optimum gear ratio of the worm gear set u_1).

Table 1. Input parameters.

Factor	Code	Unit	Low	High
Total gearbox ratio	u_g	-	50	300
Coefficient for calculation q	k_q	-	0.25	0.3
Coefficient of wheel face width of helical gear set ψ_{ba2}	X_{ba2}	-	0.35	0.4
Allowable contact stress of helical gear set $[\sigma_H]$	AS	MPa	350	410
Output torque	T_{out}	Nmm	10^5	10^6

3. Optimization results and discussions

For imagining the influence of the input parameters on the response and for evaluating the relative strength of the effect, a graph of the main effect of each parameter is plotted in Figure 3. From the Figure, it is found that the optimum gear ratio of the worm gear set u_1 increases significantly with the increase of the total gearbox ratio u_g. Also, it is effected by coefficient k_q, the allowable contact stress of the helical gear set AS, the coefficient of wheel face width of the helical gear set ψ_{ba2} and the output torque T_{out}.

Table 2. Experimental plans and output response.

StdOrder	RunOrder	CenterPt	Blocks	u_g	Xba1	Xba2	AS (MPa)	Tout (Nm)	u_1
28	1	1	1	30	0.35	0.35	410	1000	9.53
14	2	1	1	30	0.3	0.4	410	100	8.66
13	3	1	1	5	0.3	0.4	410	100	2.62
32	4	1	1	30	0.35	0.4	410	1000	9.11
30	5	1	1	30	0.3	0.4	410	1000	8.66
6	6	1	1	30	0.3	0.4	350	100	8.66
3	7	1	1	5	0.35	0.35	350	100	2.88
27	8	1	1	5	0.35	0.35	410	1000	2.88
...									
20	31	1	1	30	0.35	0.35	350	1000	9.53
8	32	1	1	30	0.35	0.4	350	100	9.11

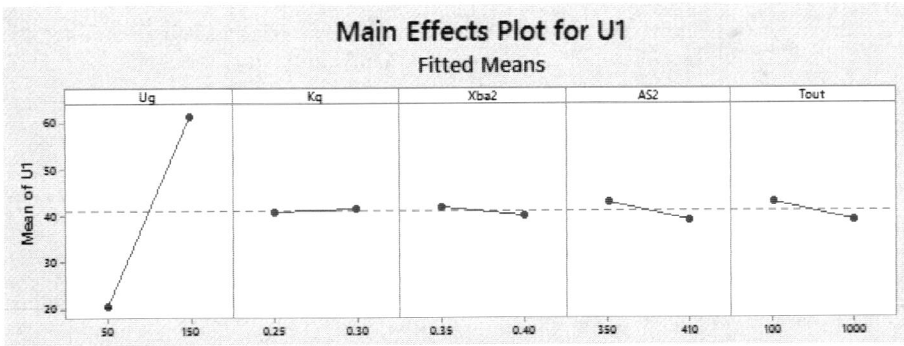

Figure 3. Main effects plot for optimum gear ratio of the worm gear set.

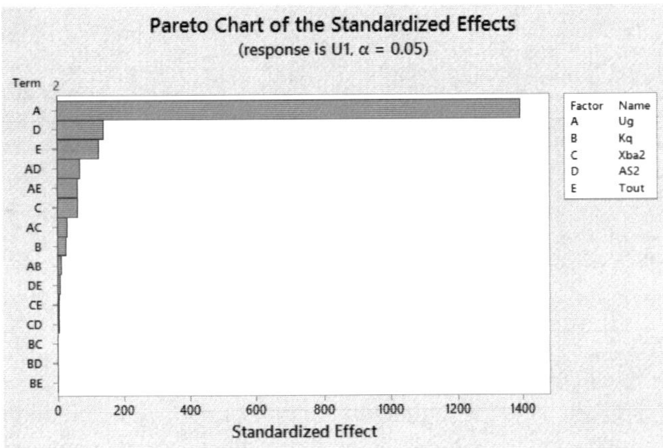

Figure 4. Pareto Chart of the Standardized Effects.

Figure 4 presents the Pareto chart of the standardized effects from the largest effect to the smallest effect. It can be seen from this graph that the bars that represent parameters including the total gearbox ratio (factor A), the coefficient k_q (factor B), the coefficient of wheel face width of the helical gear set (factors C), the allowable contact stress of the helical gear set (factor D), the output torque (factor E), and the interactions between them (AD, AE, AC, AB...) cross the reference line. Consequently, these factors are statistically significant at the 0.05 level with the response model.

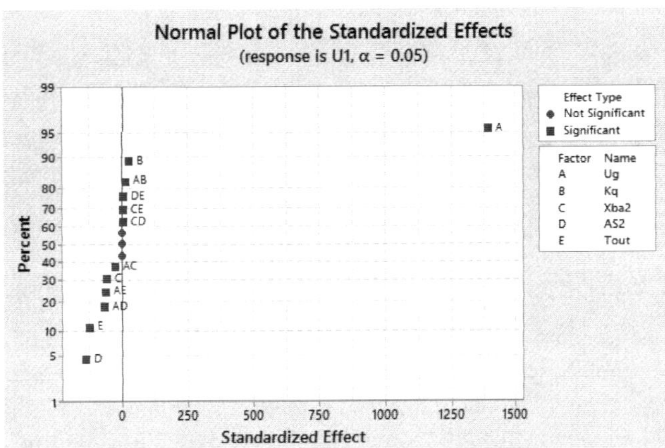

Figure 5. Normal Plot for u_1.

Coded Coefficients

Term	Effect	Coef	SE Coef	T-Value	P-Value	VIF
Constant		41.0717	0.0147	2792.00	0.000	
Ug	41.0717	20.5359	0.0147	1396.00	0.000	1.00
Kq	0.7229	0.3615	0.0147	24.57	0.000	1.00
Xba2	-1.7402	-0.8701	0.0147	-59.15	0.000	1.00
AS	-4.0468	-2.0234	0.0147	-137.55	0.000	1.00
Tout	-3.6712	-1.8356	0.0147	-124.78	0.000	1.00
Ug*Kq	0.3615	0.1807	0.0147	12.29	0.000	1.00
Ug*Xba2	-0.8701	-0.4351	0.0147	-29.57	0.000	1.00
Ug*AS	-2.0234	-1.0117	0.0147	-68.77	0.000	1.00
Ug*Tout	-1.8356	-0.9178	0.0147	-62.39	0.000	1.00
Kq*Xba2	0.0328	0.0164	0.0147	1.12	**0.281**	1.00
Kq*AS	0.0312	0.0156	0.0147	1.06	**0.305**	1.00
Kq*Tout	-0.0025	-0.0013	0.0147	-0.09	**0.933**	1.00
Xba2*AS	0.0685	0.0343	0.0147	2.33	0.033	1.00
Xba2*Tout	0.0940	0.0470	0.0147	3.19	0.006	1.00
AS*Tout	0.1741	0.0870	0.0147	5.92	0.000	1.00

Model Summary

S	R-sq	R-sq(adj)	R-sq(pred)
0.0832151	100.00%	100.00%	100.00%

Figure 6. Estimated Effects and Coefficients for u_1.

As the Pareto chart cannot show which effects increase or decrease the response, the Normal Plot of the standardized effects (Figure 5) is used for that. From Figure 5, it was found that the total gearbox ratio is the most significant factor for the optimum gear ratio of the worm gear set. Furthermore, the total gearbox ratio (factor A), the coefficient k_q (factor B) and the interaction AB have a positive standardized effect. If they increase, the optimum gear ratio of the worm gear set increases. Also, the coefficient of wheel face width of the helical gear set, the allowable contact stress of the helical gear set, the output torque and the interactions AC, AD and AE have a negative standardized effect. When they increase the optimum gear ratio of the worm gear set decreases.

Figure 6 describes the estimated effects and coefficients for u_1. It is found from the figure that parameters which have a significant effect on a response have P-values lower than 0.05 are the total gearbox ratio u_g, the coefficient k_q, the allowable contact stress of the helical gear set AS (or $[\sigma_H]$, the coefficient of wheel face width of the helical gear set X_{ba2} (or ψ_{ba2}), the output torque T_{out} and their interactions (except $k_q \cdot X_{ba2}$, $k_q \cdot AS$ and $k_q \cdot T_{out}$). Therefore, the relation between the optimum gear ratio of the worm gear set and the significant effect factors can be described as follows:

$$\begin{aligned} u_1 = {} & 13.58 + 0.78021 \cdot u_g - 17.7 \cdot k_q - 26.87 \cdot \psi_{ba2} - 0.02639 \cdot [\sigma_H] - 0.003985 \cdot T_{out} + \\ & + 0.1446 \cdot u_g \cdot k_q - -0.348 \cdot u_g \cdot \psi_{ba2} - 0.000674 \cdot u_g \cdot [\sigma_H] - 0.000041 \cdot u_g \cdot T_{out} + \\ & + 26.3 \cdot k_q \cdot \psi_{ba2} + 0.0208 \cdot k_q \cdot [\sigma_{H2}] - -0.00011 \cdot k_q \cdot T_{out} + 0.457 \cdot \psi_{ba2} \cdot [\sigma_H] + \\ & + 0.00418 \cdot \psi_{ba2} \cdot T_{out} + 0.000006 \cdot [\sigma_H] \cdot T_{out} \end{aligned} \tag{32}$$

From Figure 6, the high values of adj-R2 and pred-R2 indicate that the estimated model (Equation (32)) fit the experimental data very well. Equation (32) is used to calculate the gear ratio of the first stage u_1. After having u_1, the gear ratio of the helical gear set is determined by $u_2 = u_g / u_1$.

4. Conclusions

A study on determination of the optimum gear ratios of a worm helical gearbox in order to get the minimum cross-sectional acreage of the gearbox was carried out. In the study, the effect of the input factors including the total gearbox ratio, the allowable contact stress and the wheel face width coefficient of the helical gear set, the output torque was analysed. Also, a model for calculation of the optimum gear ratio of the worm gear set for getting the minimum cross-sectional acreage of the gearbox was proposed. The gear ratios can be calculated simply by using the explicit model.

Acknowledgements

The work described in this paper was supported by Thai Nguyen University of Technology for a scientific project.

References

[1] N T Hiep and N V Lam 1999 *Design of machine elements* (Hanoi: Educational Republishing House) p 380
[2] Chernavsky S A et al. 1984 *Design of mechanical transmissions: Manual for high technical schools* (Moscow: Mashinostroenie) p 560
[3] G Milou G. Dobre et al. 1996 *Optimal Design of Two Step Gear Units, regarding the Main Parameters* (VDI Berichte No 1230) p 227
[4] V N Pi and V Q Dac 2007 Calculation of total transmission ratio of two step worm reducers for the best reasonable gearbox housing structure (in Vietnamese), *J. of Science and Technology - Thai Nguyen University* **Vol. 1 (41**) pp 65-69
[5] T Chat 1993 Some problems of kinematics calculation of transmission mechanics system (in Vietnamese) *Proc. of the National Conference on Engineering Mechanics* **Vol. 2** (Hanoi) pp. 7-12.
[6] V N Pi and V Q Dac 2005 Optimal calculation of total transmission ratio of worm- helical gear reducers (in Vietnamese) *J. of Science and Technology - Thai Nguyen University, Vol. 1,* **No. 4 (36)** pp. 70-73.
[7] V N Pi and V Q Dac, Optimal calculation of partial transmission ratios of worm-helical gear reducers for minimal gearbox length *J. of Science & Technology - Technical Universities* **No. 61,** 2007, pp. 73-77.
[8] T Chat and L V Uyen 2007 *Design and calculation of Mechanical Transmissions Systems Vol. 1* (Hanoi: Educational Republishing House)

IOP Conf. Series: Materials Science and Engineering **635** (2019) 012011 doi:10.1088/1757-899X/635/1/012011

Determining optimal partial transmission ratios of mechanical driven systems using a V-belt drive and a two-stage helical reducer

Vu Ngoc Pi[1], Tran Thi Hong[2], Le Xuan Hung[1]

[1]Mechanical Engineering Faculty, Thai Nguyen University of Technology, Thai Nguyen city, Vietnam
[2] Nguyen Tat Thanh University, Ho Chi Minh city, Vietnam

vungocpi@tnut.edu.vn

Abstract. This article introduces a study on the optimal calculation of partial transmission ratios of mechanical drive systems using a V-belt drive and a two-stage helical reducer. In the study, the dimension of the cross section of the system was chosen as the objective function of the optimization problem. Also, the design equation for pitting resistance of a gear set was investigated and equations on moment equilibrium condition of a mechanic system including a V-belt drive and a two-stage helical reducer and their regular resistance condition were analyses. From the results of the study, models for determining the optimal partial ratios of the V-belt drive and two stages of the reducer were proposed. Using these models, the partial ratios can be calculated accurately and simply.

1. Introduction

Until now, in optimization design of a mechanical driven system, many studies have concentrated in optimum determination of the gear ratios of the reducer in the system. The reason of that is the gear ratios strongly affect the size, the dimension, the mass as well as the cost of the reducer.

The gear ratios have been determined for two-stage reducers [1, 2, 3 and 4], three-stage reducers [4, 5, 6, 7 and 8], and four-stage reducers [4, 8, 9 and 10]. Also, the gear ratios have been found by different methods. They were found by graph method [1, 2 and 4], by "practical method" [2] or by model method [5, 6, 7, 8, 9 and 10].

In the graph method, gear ratios were determined graphically. For example, for three-stage helical reducers, gear ratios of the first stage and the second stage can be found from the graph in Figure 1 [1]. In the "practical method", gear ratios were found from practical data. For instance, based on the data from reducer factories, it was found that the weight of two-stage reducer is minimum when the ratio of the center distance of the second stage to that of the first stage is from 1.4 to 1.6 [2]. After that, the optimum gear ratios were introduced in the tabulated form. In model method, models for calculation of gear ratios have been proposed for numerous objectives. The objective can be the minimal volume of gears [4], the minimal cross section of the reducer [5], the minimal mass of gears [6, 9 and 10], the minimal reducer length [7] and the minimal reducer mass [8].

Newly, the gear ratios of mechanical systems which use a V-belt drive [11 and 12] or a chain drive [13] and a two-stage reducer. In this studies, the objective was the minimum height of the system including the reducer and the V-belt drive of the chain drive.

This paper presents a study for optimum determining partial transmission ratios and for driven systems using a V-belt drive and a two-stage helical reducer with the objective is the minimum system cross-sectional dimension.

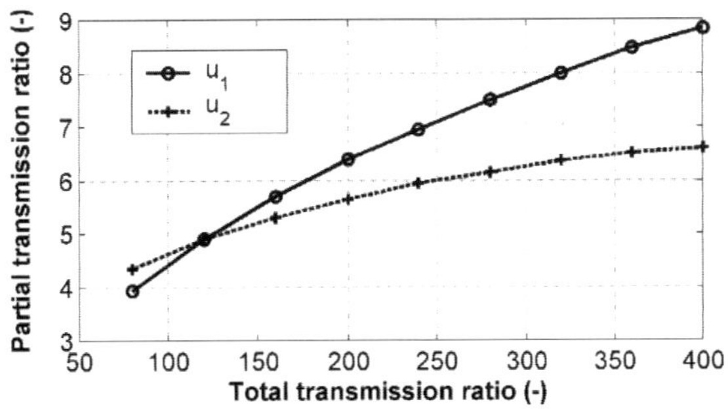

Fig. 1: Transmission ratio of step 1 and 2 versus the total transmission ratio [1]

2. Optimization problem for determining optimum partial transmission ratios

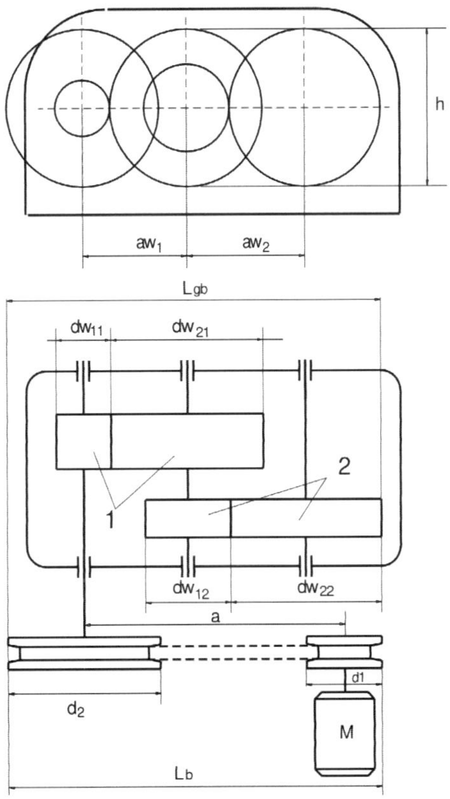

Figure. 2: Calculation schema

The objective of the optimization problem is to determine the optimum transmission ratios in order to get the minimum acreage of cross section of a mechanical driven system using a V-belt drive and a two-stage helical reducer. From Figure 2, the acreage of cross section of the system can be determined as:

$$A = L_{max} \cdot h_{max} \tag{1}$$

Where:

$$L_{max} = \max\left(L_{gb}, L_b\right) \quad (2)$$

$$h_{max} = \max\left(d_2, d_{w21}, d_{w22}\right) \quad (3)$$

In which, L_{gb} and L_b are determined as (see Figure 2):

$$L_{gb} = d_2/2 + a_{w1} + a_{w2} + d_{w22}/2 \quad (4)$$

$$L_b = d_1/2 + a + d_2/2 \quad (5)$$

In the above equations, a_{w1} and a_{w2} are the center distance of the first and the second step, respectively; d_2, d_{w21} and d_{w22} is the diameter of the driven pulley of the V-belt drive, the driver gear of the first and the second step, respectively.
For the V-belt drive, the diameter of the drive pulley can be calculated by:

$$d_1 = d_2/\left[u_b \cdot (1-\varepsilon)\right] \quad (6)$$

Substituting (6) into (5) gives

$$L_b = d_2/\left[u_b \cdot (1-\varepsilon)\right] + a + d_2/2 \quad (7)$$

For the first and the second-stage of the reducer, the driven diameters can be determined as [14]:

$$d_{w21} = 2 \cdot a_{w1} \cdot u_1/\left(u_1+1\right) \quad (8)$$

$$d_{w22} = 2 \cdot a_{w2} \cdot u_2/\left(u_2+1\right) \quad (9)$$

From (2), (3), (4), (7), (8) and (9) the following equations are given:

$$L_{max} = f\left(d_2, a_{w1}, a_{w2}, u_b, u_1, u_2\right) \quad (10)$$

$$h_{max} = f\left(d_2, a_{w1}, a_{w2}, u_1, u_2\right) \quad (11)$$

Thus, the optimization problem is defined as minimize

$$A = L_{max} \cdot h_{max} \quad (12)$$

With the following constraints

$$\begin{aligned} 1 &\le u_b \le 6 \\ 1 &\le u_1 \le 9 \\ 1 &\le u_2 \le 9 \end{aligned} \quad (13)$$

From (10), (11) and (12), it is clear that for solving the optimization problem it is necessary to calculate the driven pulley diameters d_2, the center distance of the first stage a_{w1}, and the center distance of the second stage a_{w2}.

2.1. Determining the driven pulley diameter d_2

For a V-belt drive, from tabulated data for determining allowable power [14], the following regression model for calculation of driver diameter d_1 (with the determination coefficient R^2=0.9156) was found:

$$d_1 = 269.7721 \cdot [P_1]^{0.7042} / v^{0.5067} \quad (14)$$

Theoretically, the peripheral velocity of the belt can determined as follow:

$$v = \pi \cdot d_1 \cdot n_1 / 60000 \tag{15}$$

From (14) and (15) the diameter of the driving pulley can be determined by the following equation:

$$d_1 = 1093.8 \cdot [P_1]^{0.7923} / n_1^{0.6369} \tag{16}$$

Also, the diameter of driven pulley of a V-belt drive is calculated by [14]:

$$d_2 = u_b \cdot d_1 \cdot (1-\varepsilon) \tag{17}$$

Substituting (16) into (17) gives

$$d_2 = 1093.8 \cdot u_b \cdot (1-\varepsilon) \cdot [p_1]^{0.7923} / n_1^{0.6369} \tag{18}$$

Where, ε is slippage coefficient; $\varepsilon = 0.01 \ldots 0.02$ [14]; u_b is the transmission ratio of the V-belt drive; $[P_1]$ is the allowable power of the drive (kW); $[P_1]$ is calculated by the following equation:

$$[P_1] = n_1 \cdot [T_1] / (9.55 \cdot 10^6) \tag{19}$$

Choosing $\varepsilon = 0.015$ and substituting its and (19) into (18) gives:

$$d_2 = 0.0032 \cdot u_b \cdot n_1^{0.1554} \cdot [T_1]^{0.7923} \tag{20}$$

Where, $[T_1]$ is the permissible torque on the driving shaft which can be calculated from permissible torque on the output shaft $[T_{out}]$ by:

$$[T_1] = [T_{out}] / (u_t \cdot \eta_t) \tag{21}$$

In the equation (21), u_t is the total transmission ratio of the system; η_t is the total efficiency of the system:

$$\eta_t = \eta_d \cdot \eta_{br}^2 \eta_o^3 \tag{22}$$

In which, η_d is V-belt efficiency (η_d is from 0.956 to 0.96[2]); η_{br} is helical gear transmission efficiency (η_{br} is from 0.96 to 0.98 [14]); η_o is transmission efficiency of a pair of rolling bearing (η_o is from 0.99 to 0.995 [14]). Choosing $\eta_d = 0.955$, $\eta_{br} = 0.97$ and $\eta_o = 0.992$ [14] and substituting (22) into (21) gives

$$T_1 = 1.14 \cdot T_{out} / u_t \tag{23}$$

2.2. Determining the center distance of the first and the second stage

The center distance of the first stage of the reducer can be calculated by [14]:

$$a_{w1} = K_a \cdot (u_1 + 1) \cdot \sqrt[3]{T_{11} \cdot k_{H\beta} / \left([\sigma_H]^2 \cdot u_1 \cdot \psi_{ba1}\right)} \tag{24}$$

Where,

- $K_{H\beta}$ is contact load ratio for pitting resistance; for the first stage of a two-stage helical reducer $k_{H\beta} = 1.02 \div 1.28$ [14]. Therefore, we can chose $k_{H\beta} = 1{,}1$;
- $[\sigma_H]$ is allowable contact stress (MPa); In practice, $[\sigma_H] = 350 \ldots 410$ (MPa) and we can chose $[\sigma_H] = 380$ (MPa);
- k_a is material coefficient; As the gear material is steel, k_a =43 [14];
- ψ_{ba} is coefficient of wheel face width; for the for the first stage of a two-stage helical reducer $\psi_{ba1} = 0.3$.

Substituting the above values into (24) gets:

$$a_{w1} \approx 1.2639 \cdot (u_1 + 1) \cdot \sqrt[3]{T_{11} / u_1} \tag{25}$$

Calculating in the same way and with the note that the coefficient of wheel face width for the second stage $\psi_{ba2} = 0.35$, the center distance of the second stage is calculated by:

$$a_{w2} \approx 1.2006 \cdot (u_2 + 1) \cdot \sqrt[3]{T_{12} / u_2} \tag{26}$$

3. Results and discussions

Based on Equations (12) and (13), a computer program was performed to determine the optimum partial transmi0ssion ratios for getting the minimum dimension of the cross section of the system. The following data was used in the optimization program: $u_t = 20\ldots50$; $T_{out} = 10^6\ldots10^7$.

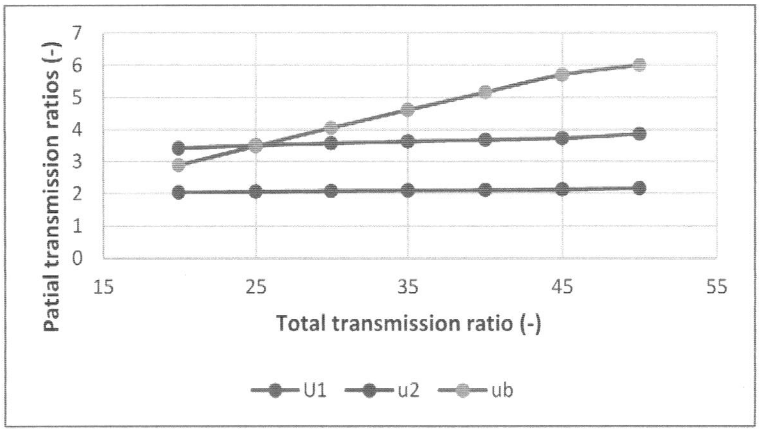

Figure. 3: *Partial ratios versus total ratio (With T_r=10^6 Nmm)*

Figure 3 shows the relation between the optimum partial ratios including the gear ratio of the first stage u_1, the gear ratio of the second stage u_2 and transmission ratio of the V-belt drive u_b and the total ratios of the system.

From the results of the optimization program, a regression equation (equation (27) was found for determination of the transmission ratio of the V-belt drive (with the coefficients of determination was $R^2 = 0.971$). Also, for calculation of the optimum gear ratios of the first and the second stages of the reducer, equations (28) and (29) were proposed (with the coefficients of determination of both equations are $R^2 = 1$):

$$u_b = 45.9344 \cdot [T_{out}]^{-0.3795} \cdot u_t^{0.8407} \tag{27}$$

$$u_1 = 0.9365 \cdot u_g^{0.667} \tag{28}$$

$$u_2 = 1.0678 \cdot u_g^{0.333} \tag{29}$$

Where, u_g is the total ratio of the reducer; u_g is calculated by $u_g = u_t / u_b$.

4. Conclusions

The minimum cross-sectional area of a mechanical drive system using a V-belt drive and a two-stage helical reducer can be found by using the optimum partial transmission ratios when designing.

Formulas for calculation of the partial transmission ratios of the V-belt drive and the gear-stages were proposed for getting the minimum cross-sectional area of the system.

IOP Conf. Series: Materials Science and Engineering **635** (2019) 012011 doi:10.1088/1757-899X/635/1/012011

The partial transmission ratios of the V-belt drive and the helical gear-stages can be calculated simply by using explicit formulas. The proposed formulas can be used for calculation of optimum reducer design in practice or in universities for student project.

Acknowledgements
The work described in this paper was supported by Thai Nguyen University of Technology for a scientific project.

References

[1] V.N. Kudreavtev; I.A. Gierzaves; E.G. Glukharev, *Design and calculus of gearboxes* (in Russian), Mashinostroenie Publishing, Sankt Petersburg, 1971.

[2] G. Milou; G. Dobre; F. Visa; H. Vitila, *Optimal Design of Two Step Gear Units, regarding the Main Parameters*, VDI Berichte No 1230 (1996), p. 227.

[3] Vu Ngoc Pi, *A method for optimal calculation of total transmission ratio of two step helical gearboxes,* Proceedings of the National conference on Engineering Mechanics, Ha Noi (2001), p. 12.

[4] A.N. Petrovski, B.A. Sapiro, N.K. Saphonova, *About optimal problem for multi-step gearboxes* (in Russian),Vestnik Mashinostroenie, No. 10 (1987), p. 13.

[5] Vu Ngoc Pi, Nguyen Khac Tuan, *Optimum determination of partial transmission ratios of Three-step helical gearboxes for getting minimum cross section*, Journal of Environmental Science and Engineering A 5 (2016) 570-573.

[6] Vu Ngoc Pi, Nguyen Dang Binh, Vu Quy Dac, Phan Quang The, *"Optimal Calculation of Total Transmission Ratio of Three-Step Helical Gearboxes for Minimum Mass of Gears."* (In Vietnamese) Journal of Science and Technology of 6 Engineering Universities (2006), p. 91.

[7] Vu Ngoc Pi, *A new study on optimal calculation of partial transmission ratio of three-step helical reducers,* The 3rd IASME / WSEAS International Conference on Continuum Mechanics, Cambridge, UK (2008), p. 23.

[8] Romhild I. , Linke H., *Gezielte Auslegung Von Zahnradgetrieben mit minimaler Masse auf der Basis neuer Berechnungsverfahren,* Konstruktion 44, 1992, pp. 229- 236.

[9] Vu Ngoc Pi, *Optimal determination of partial transmission ratios for four-step helical gearboxes with firt and third step double gear-sets for minimal mass of gears*, Applied Computing Conference (ACC' 08), Istanbul, Turkey, May 27-30, 2008.

[10] Le Xuan Hung, Vu Ngoc Pi, Nguyen Van Du, *Optimal calculation of partial transmission ratios of four-step helical gearboxes with second and fourth-step double gear-sets for minimal mass of gears,* The international symposium on Mechanical Engineering (ISME Ho Chi Minh city, Vietnam, 21-23, September 2009.

[11] Vu Ngoc Pi, Tran Thi Phuong Thao, Le Thi Phuong Thao, *A new study on optimum determination of partial transmission ratios of mechanical driven systems using a V-belt and two-step helical gearbox*, Vietnam Mechanical Engineering Journal, No. 10 (2015), p. 123.

[12] Vu Ngoc Pi, Nguyen Thi Hong Cam, Nguyen Khac Tuan, *Optimum calculation of partial transmission ratios of mechanical driven systems using a V-belt and two-step bevel helical gearbox,* Journal of Environmental Science and Engineering A 5 (2016), p. 566.

[13] Vu Ngoc Pi, Tran Thi Phuong Thao, Dang Anh Tuan, *Optimum determination of partial transmission ratios of mechanical driven systems using a chain drive and two-step helical gearbox*, Journal of Environmental Science and Engineering B 6 (2017), p. 80.

[14] Trinh Chat, Le Van Uyen, *Design and calculus of Mechanical Transmissions* (in Vietnamese), Educational Republishing House, Hanoi, 1998.

IOP Conf. Series: Materials Science and Engineering **635** (2019) 012012 doi:10.1088/1757-899X/635/1/012012

Determination of optimum gear ratios of a two-stage helical gearbox with second stage double gear sets

Vu Ngoc Pi[1], Nguyen Thi Hong Cam[1], Tran Thi Hong[2], Le Xuan Hung[1], Luu Anh Tung[1], Nguyen Khac Tuan[1], Hoang Thi Tham[1]

[1]Thai Nguyen University of Technology, Thai Nguyen city, Vietnam
[2]Nguyen Tat Thanh University, Ho Chi Minh city, Vietnam

vungocpi@tnut.edu.vn

Abstract. This paper introduces a study on the optimum calculation of the gear ratios of a two-stage helical gearbox with second stage double gear sets. In this study, for finding the optimum gear ratios of the gearbox, an optimization problem was conducted. In the optimization problem, the objective function was the acreage of the cross section of the gearbox. Besides, the effect of the input parameters including the total gearbox ratio, the wheel face width coefficient, the allowable contact stress and the output torque were considered. For investigation of the influence of these factors on the optimum gear ratios, an "experiment" was designed and a computer program was built for conducting the "experiment". Also, models for determination of the optimum gear ratios of the gearbox were proposed. By using these models, the gear ratios can be calculated accurately and simply.

1. Introduction

In optimum gearbox design, the determination of the optimum gear ratios is very important. The reason of that is the dimension, the mass, and the cost of a gearbox depend strongly on the gear ratios. Hence, until now there have been many researches on the determination of the gear ratios.

For a two-stage helical gearbox, V.N. Kudreavtev et al. [1] presented a graph method (Figure 1) for finding the gear ratio of the first stage u_1. It is imagined that the determination of u_1 by using the graph in Figure 1 is very complicated. In addition, for determining u_1, we must select the coefficient λc^3 (with λc^3 = 0.6 to 4) so it is quite complicated and the gear ratio can not reach to optimum values. G. Milou et al. [2] proposed "a practical method" in which the gear ratios were found from practical data. Based on the data from gearbox factories, it was noted that the two-stage gearbox weight is minimum if the ratio of two center distances a_{w2} / a_{w1} ranged from 1.4 to 1.6 [2]. Therefore, the optimal gear ratios were proposed in the tabulated form. In practice, the most common method for determining the optimum gear ratios is the model method in which the optimum gear ratio model is usually found based on the results of the optimization problem. In the optimization problem, the objective can be the minimum volume of gears [3], the minimum cross section dimension of the gearbox [4], the minimum gearbox mass [5] or the minimum mass of gears [6].

For two-stage helical gearbox with second stage double gear sets, Vu Ngoc Pi used the model method for determination of the optimum gear in order to get the minimum cross section [7] and the minimum gearbox length [8]. However, in these studies, the influence of the input parameters on the

IOP Conf. Series: Materials Science and Engineering **635** (2019) 012012 doi:10.1088/1757-899X/635/1/012012

optimum gears was still not evaluated.

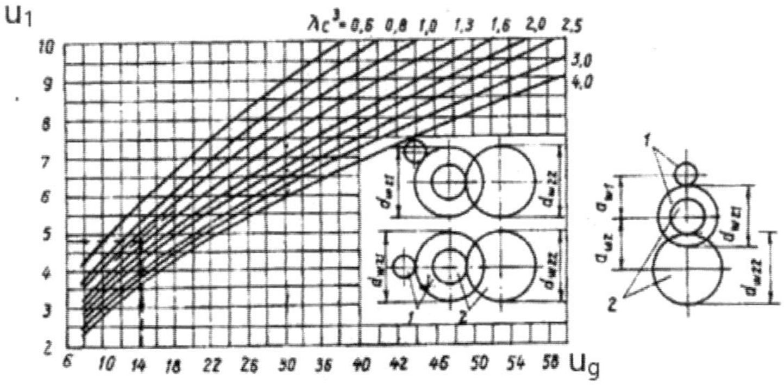

Figure 1. The gear ratio of the stage 1 versus the total gearbox ratio [1].

This paper introduces a study for the determination of the optimum gear ratios of a two-stage helical gearbox with second stage double gear sets. In this study, a optimization problem in which the objective is the minimum gearbox cross-sectional acreage was carried out. In addition, the effect of the input factors on the optimum gear ratios was investigated.

2. Optimization problem

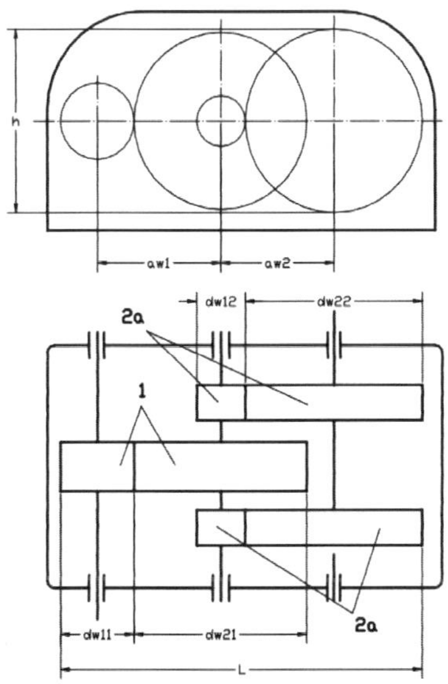

Figure 2. Calculation schema.

With a two stage helical gearbox with second stage double gear sets, the acreage of cross section is calculated by (Figure 2):

$$A = L \cdot h \tag{1}$$

Where, L and h are determined by the following equations:

$$L = d_{w11} / 2 + a_{w1} + a_{w2} + d_{w22} / 2 \tag{2}$$

$$h = \max\left(d_{w21}, d_{w22}\right) \tag{3}$$

In equations (2) and (3), a_{w1} and a_{w2} are the center distance of the first and the second stage; d_{w11}, d_{w21} and d_{w22} are pitch diameters (mm) of the first and the second stage, respectively. The diameters d_{w11}, d_{w21} and d_{w22} can be calculated as [9]:

$$d_{w11} = 2 \cdot a_{w1} / (u_1 + 1) \tag{4}$$

$$d_{w21} = 2 \cdot a_{w1} \cdot u_1 \cdot / (u_1 + 1) \tag{5}$$

$$d_{w22} = 2 \cdot a_{w2} \cdot u_2 / (u_2 + 1) \tag{6}$$

Also, for this gearbox we have:

$$u_2 = u_g / u_1 \tag{7}$$

In which, u_g is the total gearbox ratio; u_1, u_2 are the gear ratios of the first and the second stages. From (2), (3), (4), (5) and (6) the following equations are given:

$$L = f(a_{w1}, a_{w2}, u_1, u_g) \tag{8}$$

$$h = f(a_{w1}, a_{w2}, u_1, u_g) \tag{9}$$

Thus, the optimization problem is defined as

$$\text{minimize } A = L \cdot h \tag{10}$$

With the following constraints

$$5 \le u_g \le 30 \tag{11}$$

From equations (8), (9) and (10), it is known that in for solving the optimization problem it is necessary to determine the center distances of the first stage a_{w1} and the second stage a_{w2}.

2.1. Calculating the center distance of the first stage

The center distance of the first stage a_{w1} can be found by the following equation [9]:

$$a_{w1} = K_a \cdot (u_1 + 1) \cdot \left(\frac{T_{11} \cdot k_{H\beta}}{[\sigma_H]^2 \cdot u_1 \cdot \psi_{ba1}} \right)^{1/3} \tag{12}$$

Where,

- $K_{H\beta}$ is the contact load ratio for pitting resistance; for the first stage of the gearbox $k_{H\beta} = 1.02 \div 1.28$ [9]. Hence, we can chose $k_{H\beta} = 1.1$;
- $[\sigma_H]$ - the allowable contact stress (MPa); In practice, $[\sigma_H] = 350 \ldots 410$ (MPa);
- k_a is the material coefficient; Normally, the gear material is steel, k_a=43 [9];
- ψ_{ba} is the coefficient of wheel face width; for the for the first stage of the gearbox $\psi_{ba1} = 0.3 \ldots 0.35$;

From the moment equilibrium condition of the gearbox we have:

$$T_{out} = T_{11} \cdot \eta_{hg}^2 \cdot \eta_{be}^3 \cdot u_g \tag{13}$$

In which, η_{hg} is the helical gear transmission efficiency ($\eta_{hg} = 0.96 \ldots 0.98$ [9]); η_{be} is the transmission efficiency of a pair of rolling bearing ($\eta_{be} = 0.99 \ldots 0.995$ [9]). Choosing $\eta_{hg} = 0.97$ and $\eta_{be} = 0.992$ and substituting them into (13) gets:

$$T_{11} = 1.0887 \cdot T_{out} / u_g \tag{14}$$

Substituting (14) and $k_{H\beta} = 1.1$ into (12) we have:

$$a_{w1} = 45.6635 \cdot (u_1 + 1) \cdot \left(\frac{T_{out}}{[\sigma_{H1}]^2 \cdot u_1 \cdot u_g \cdot \psi_{ba1}} \right)^{1/3} \tag{15}$$

2.2. Calculating the center distance of the second stage

The center distance of the second stage a_{w2} can be determined by [9]:

$$a_{w2} = K_a \cdot (u_2 + 1) \cdot \left(\frac{T_{12} \cdot k_{H\beta}}{[\sigma_H]^2 \cdot u_2 \cdot \psi_{ba2}} \right)^{1/3} \quad (16)$$

In addition, for the second stage we have

$$T_{out} = 2 \cdot T_{12} \cdot \eta_{hg} \cdot \eta_{be}^2 \cdot u_2 \quad (17)$$

Choosing $\eta_{hg} = 0.97$ and $\eta_{be} = 0.992$ as in section 2.1 and substituting them into (17) gives

$$T_{12} = 0.5238 \cdot T_{out} / u_2 \quad (18)$$

Substituting (18), k_a =43 and $k_{H\beta} = 1.1$ (as in section 2.1) into (16) gets:

$$a_{w2} = 35.7812 \cdot (u_2 + 1) \cdot \left(\frac{T_{out}}{[\sigma_H]^2 \cdot u_2^2 \cdot \psi_{ba2}} \right)^{1/3} \quad (19)$$

2.3. Experimental work

To explore the influence of the input parameters on the optimum gear ratios, an "experiment" was designed and conducted. For the design of the experiment, a 2-level full factorial design was carefully chosen. Table 1 presents 5 input factors which were selected for the investigation. Consequently, there are $2^5 = 32$ number of tests for conducting. From equation (10) and (11), a computer program was built for performing the experiment. The various levels of input factors and the results of the output of the program (the optimum gear ratio of the first stage u_1) are shown in Table 2.

Table 1. Input parameters.

Factor	Code	Unit	Low	High
Total gearbox ratio	u_g	-	5	30
Coefficient of wheel face width of stage 1	x_{ba1}	-	0.3	0.35
Coefficient of wheel face width of stage 2	x_{ba2}	-	0.35	0.4
Allowable contact stress	AS	MPa	350	410
Output torque	T_{out}	Nmm	10^5	10^6

3. Optimization results and discussions

For evaluating the effect of input factors on the response and the relative strength of the effect, a graph of the main effect of each input factor is plotted in Figure 3. From the Figure, it is found that the optimum gear ratio u_1 depends strongly on the total gearbox ratio u_g. It increases significantly with the increase of the total gearbox ratio. Besides, it also effected by the coefficient of wheel face width of the first and the second stages (ψ_{ba1} and ψ_{ba2}). In addition, the optimum gear ratio is not effected by the allowable contact stress of the helical gear set AS, and the output torque T_{out}.

Table 2. Experimental plans and output response.

StdOrder	RunOrder	CenterPt	Blocks	u_g	Xba 1	Xba 2	AS (MPa)	Tout (Nm)	u_1
25	1	1	1	5	0.3	0.35	410	1000	2.17
6	2	1	1	30	0.3	0.4	350	100	6.86
21	3	1	1	5	0.3	0.4	350	1000	2.08
3	4	1	1	5	0.35	0.35	350	100	2.29

10	5	1	1	30	0.3	0.35	410	100	7.18
22	6	1	1	30	0.3	0.4	350	1000	6.86
14	7	1	1	30	0.3	0.4	410	100	6.86
9	8	1	1	5	0.3	0.35	410	100	2.17
...									
20	31	1	1	30	0.35	0.35	350	1000	7.56
28	32	1	1	30	0.35	0.35	410	1000	7.56

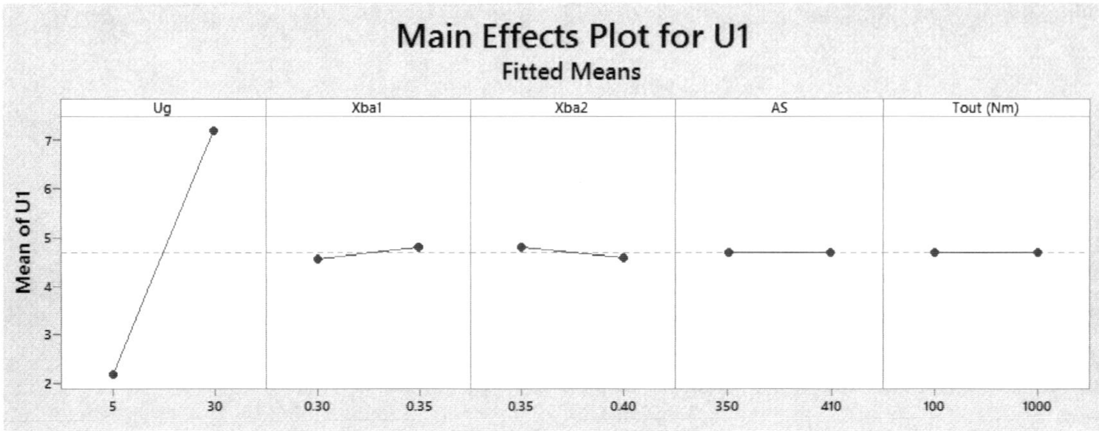

Figure 3. Main effects plot for the optimum gear ratio of the first stage.

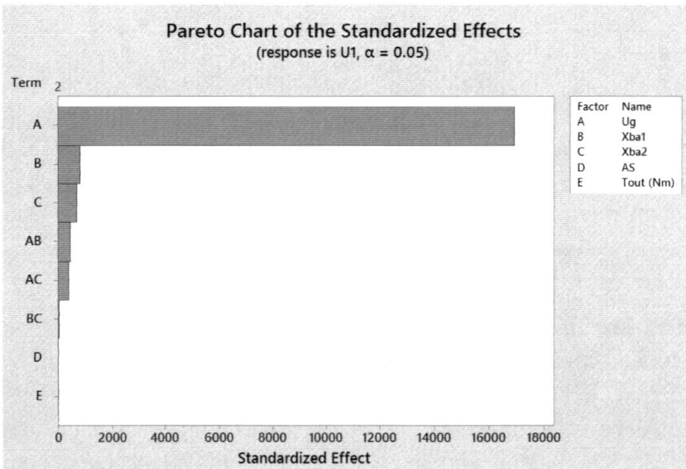

Figure 4. Pareto Chart of the Standardized Effects.

The Pareto chart of the standardized effects from the largest to the smallest effect is shown in Figure 4. It was found from this graph, the bars that represent factors including the total gearbox ratio (factor A), the coefficients of wheel face width of the first and the second stages (factors B and C) and the interactions between them (AB and AC) cross the reference line. Hence, these input factors are statistically significant at the 0.05 level with the response model (the gear ratio of the first stage u_1). Also, because the bars which represent the allowable contact stress of the helical gear set (factor D), and the output torque (factor E) do not cross the reference line, these factors (D and E) do not affect the response model.

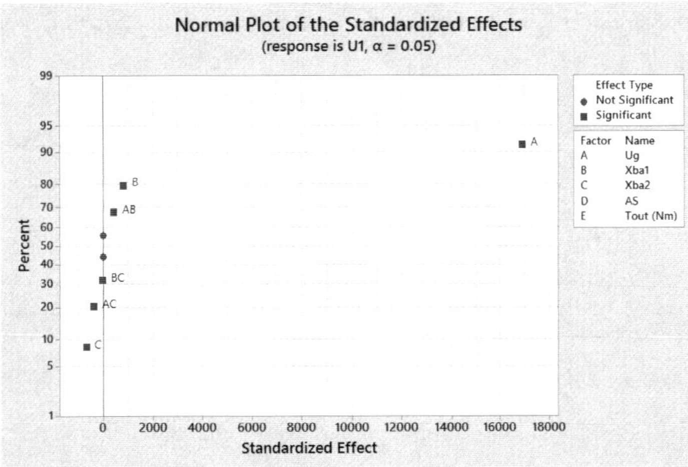

Figure 5. Normal Plot for the optimum gear ratio of the first stage.

Coded Coefficients

Term	Effect	Coef	SE Coef	T-Value	P-Value	VIF
Constant		2.30558	0.00007	32521.13	0.000	
Ug	2.05550	1.02775	0.00007	14496.78	0.000	1.00
Xba1	0.033430	0.016715	0.000071	235.77	0.000	1.00
Xba2	-0.028535	-0.014267	0.000071	-201.25	0.000	1.00
AS	0.000000	0.000000	0.000071	0.00	**1.000**	1.00
Tout (Nm)	-0.000000	-0.000000	0.000071	-0.00	**1.000**	1.00
Ug*Xba1	-0.033430	-0.016715	0.000071	-235.77	0.000	1.00
Ug*Xba2	0.028535	0.014268	0.000071	201.25	0.000	1.00
Xba1*Xba2	-0.000680	-0.000340	0.000071	-4.80	0.000	1.00

S	R-sq	R-sq(adj)	R-sq(pred)
0.0004010	100.00%	100.00%	100.00%

Figure 6. Estimated Effects and Coefficients for the optimum gear ratio of the first stage.

Figure 5 shows the Normal Plot of the standardized effects. This graph is used for describe which effects increase or decrease the response. From Figure 5, it was found that the total gearbox ratio (factor A) is the most significant factor for the optimum gear ratio. In addition, the total gearbox ratio and the coefficient of wheel face width of the first stage (factor B) have a positive standardized effect. If they change from the low level to the high level of the factors, the optimum gear ratio of the first stage increases. Also, the coefficient of wheel face width of the second stage (factor C) has a negative standardized effect. The optimum gear ratio decreases when it increases.

Figure 6 presents the estimated effects and coefficients for the optimum gear ratio u_1. From the figure, the total gearbox ratio u_g, the coefficient of wheel face width of the first and the second stage (ψ_{ba1} and ψ_{ba2}) and their interactions have P-values lower than 0.05. Therefore, these factors are significant to the optimum gear ratio. In addition, the relation between the optimum gear ratio and these significant effect factors can be described as follows:

$$u_1 = 0.5238 + 0.201277 \cdot u_g + 3.1952 \cdot \psi_{ba1} + 0.6585 \cdot \psi_{ba2} + \\ +0.205704 \cdot u_g \cdot \psi_{a1} - 0.179448 u_g \cdot \psi_{ba2} - 5.148 \cdot \psi_{ba1} \cdot \psi_{ba2} \quad (20)$$

The above estimated model fit the data very well because the adj-R2 and pred-R2 are in the high values (Figure 6).

IOP Conf. Series: Materials Science and Engineering **635** (2019) 012012 doi:10.1088/1757-899X/635/1/012012

Equation (20) is used to determine the optimum gear ratio of the first stage u_1. After having u_1, the gear ratio of the second stage can be found by $u_2 = u_g / u_1$.

4. Conclusions

The minimum cross-sectional acreage of a two-stage helical gearbox with second stage double gear sets can be obtained by determination of optimum gear ratios of the gearbox.

Models for determining the optimum gear ratios of a two-stage helical gearbox with second stage double gear sets were introduced for getting the minimum cross-sectional acreage of the gearbox.

The optimum gear ratios of the gearbox can be calculated accurately and simple as the estimated models are explicit.

Acknowledgements

The work described in this paper was supported by Thai Nguyen University of Technology for a scientific project.

References

[1] Kudryavtsev V N, Derzhavets Yu A and Glukharev E G 1971 *Design and calculation of gear reducers. Reference guide* (Saint-Petersburg: Mechanical Engineering) p 328

[2] G Milou, G Dobre, F Visa and H Vitila 1996 Optimal Design of Two Step Gear Units, regarding the Main Parameters *VDI Berichte* No 1230 p 227

[3] A N Petrovski, B A Sapiro and N K Saphonova 1987 About optimal problem for multi-step gearboxes (in Russian) *Vestnik Mashinostroenie* **No 10** p 13

[4] Vu Ngoc Pi 2001 A method for optimal calculation of total transmission ratio of two step helical gearboxes *Proc. of the National conf. on Engineering Mechanics* (Ha Noi) pp 12-15

[5] Romhild I Linke H Gezielte Auslegung 1992 Von Zahnradgetrieben mit minimaler Masse auf der Basis neuer Berechnungsverfahren *Konstruktion* **44** pp 229- 236

[6] V N Pi and V Q Dac 2004 A new and effective method for optimal calculation of total transmission ratio of two step helical gearboxes, *School of Computational Sciences and Engineering: Theory and Applications, March 3* (Ho Chi Minh city: Vietnam) pp 103-106

[7] Vu Ngoc Pi 2008 A Study on Optimal Determination of Partial Transmission Ratios of Helical Gearboxes with Second-Step Double Gear-Sets *Int. J. of Mathematical, Physical and Engineering Sciences Volume 2 Number 2* pp 99-102

[8] Vu Ngoc Pi, Optimal calculation of partial ratios of helical gearboxes with second-step double gear-sets, *Proc. of the 3rd IASME / WSEAS International Conference on CONTINUUM MECHANICS (CM'08)* pp 63-66

[9] T Chat and L V Uyen 1998 *Design and calculation of Mechanical Transmissions* (in Vietnamese), (Hanoi: Educational Republishing House) p 229

IOP Conf. Series: Materials Science and Engineering **635** (2019) 012013 doi:10.1088/1757-899X/635/1/012013

Dynamic analysis of a high speed motorized spindle for internal grinding of slender holes

Shengye Lin, Shaowen Zhang and Keke Geng [1]

School of Mechanical Engineering, Southeast University, Nanjing 211189, China

[1] Corresponding author: lsy100815@163.com

Abstract. The purpose of this paper is to investigate the dynamic characteristics of a high speed motorized spindle for internal grinding of slender holes. The modified transfer matrix method is used to establish the dynamic model of the motorized spindle rotor system including the additional stiffness of the non-contact rotary union. The linear perturbation method is adopted to derive the perturbation equations for the flow passing through the gap in the non-contact rotary union. The finite difference method is used to solve the Reynolds equation and the perturbation equations to obtain the four dimensional stiffness matrix of the non-contact rotary union. Finally, the effect of the non-contact rotary union on the dynamic behaviours of the motorized spindle are studied.

1. Introduction

High speed motorized spindles have been widely used in machine tools due to their advantages of eliminating the need for conventional power transmission devices [1, 2]. The machining productivity and finishing quality of workpieces are mainly determined by the dynamic performance of the motorized spindle. Therefore, it is essential to study the dynamic characteristics of the motorized spindle for reference in design and operation.

For internal grinding of normal holes, the external cooling method [3] is commonly used to supply the coolant into the grinding zone. However, for internal grinding of slender holes, such as the blind hole (ϕ3mm×45mm) in the nozzle of diesel injector, the inner cooling method is adopted to supply the grinding coolant [4, 5]. Consequently, a non-contact rotary union has to be specially designed for the motorized spindle [1]. When the motorized spindle operates at high speed, the hydrodynamic effect will produce in the gap between the water pipe and the rotating shaft. In other words, the non-contact rotary union will perform like a hydrodynamic bearing, providing additional stiffness to the rotor system. As the dynamic behaviours of the high speed motorized spindle are significantly affected by the supporting stiffness, the effect of the non-contact rotary union on the dynamic performance of the motorized spindle can not be ignored.

By far, considerable researches on the dynamic analysis and design for the motorized spindle have been conducted [1, 2, 6, 7]. Effects of various structural parameters on the dynamic characteristics of the motorized spindle have been studied systematically. However, the effect of the non-contact rotary union has not been investigated yet.

This paper presents a 4-DOF dynamic model for the motorized spindle by using the modified transfer matrix method. The stiffness coefficients of the non-contact rotary union are included. Based on the proposed model, the effect of the non-contact rotary union on the dynamic behaviours of the motorized spindle are studied.

IOP Conf. Series: Materials Science and Engineering **635** (2019) 012013 doi:10.1088/1757-899X/635/1/012013

2. Theoretical model

2.1. Dynamic model of the motorized spindle rotor system

Figure 1 illustrates the rotor schematic of the high speed motorized spindle for internal grinding of slender holes. The rotating shaft is supported by a pair of angular contact ball bearings with face-to-face arrangement. The grinding coolant is supplied by a water bump outside. A portion of the coolant passes through the non-contact rotary union to enter the channel in the shaft, and then flows out from the nozzles of the tool bar, and finally sprays into the grinding zone. The other portion of the coolant flows out from the gap between the water pipe and the rotating shaft. The real scene image of the motorized spindle is shown in Fig.2.

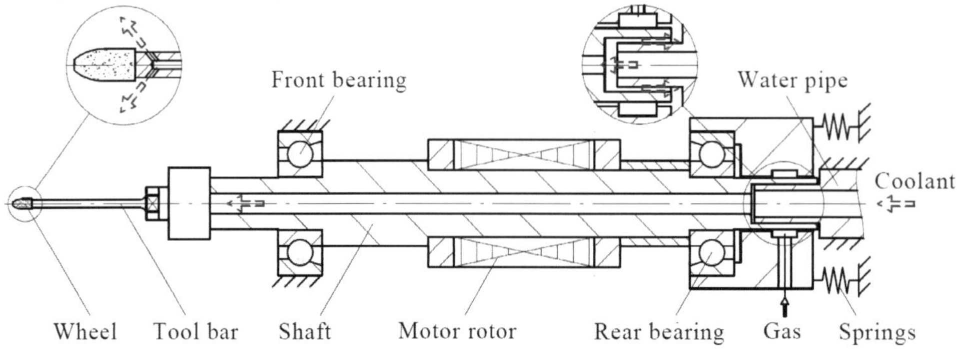

Figure 1. The rotor schematic of the motorized spindle

Figure 2. Real scene image of the motorized spindle

Figure 3 shows the lumped mass model of the motorized spindle rotor system. The shaft is divided into a series of sections, which are composed of lumped mass, rigid disk, and massless elastic shaft. Both the angular contact ball bearings and the non-contact rotary union are simplified as massless springs possessing 16 stiffness coefficients. By using the transfer matrix method [8, 9], the overall transfer matrix can be obtained and the critical speed of the rotor system can be calculated.

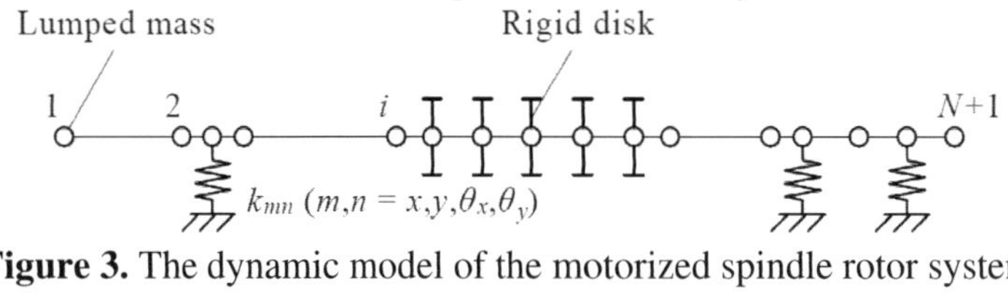

Figure 3. The dynamic model of the motorized spindle rotor system

2.2. Stiffness coefficients of the non-contact rotary union

Figure 4 illustrates the schematic of the non-contact rotary union. The flow is governed by using the Reynolds equation [10]:

$$\frac{\partial}{R\partial\psi}\left(\frac{h^3}{6\mu}\frac{\partial P}{R\partial\psi}\right)+\frac{\partial}{\partial z}\left(\frac{h^3}{6\mu}\frac{\partial P}{\partial z}\right)=\Omega\frac{\partial h}{\partial\psi} \tag{1}$$

where, P denotes the pressure; h denotes the film thickness; R denotes the radius of the shaft; μ denotes the viscosity; Ω denotes the angular velocity; ψ denotes the angular coordinate; and z denotes the axial coordinate.

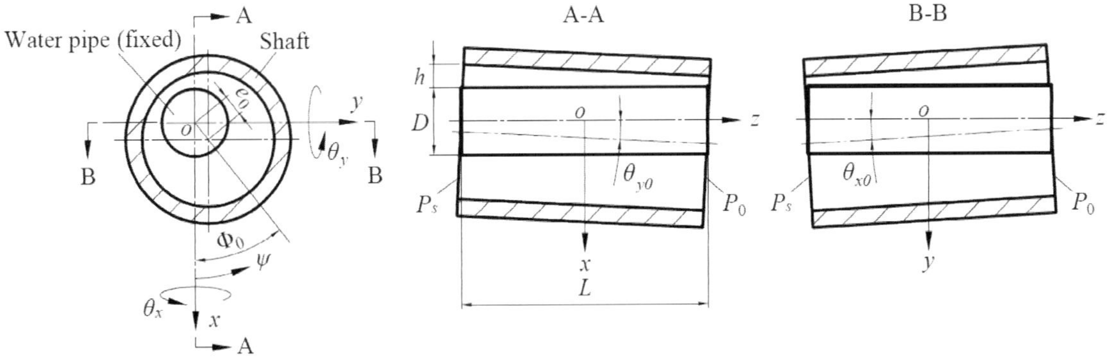

Figure 4. Schematic of the non-contact rotary union

From Fig.4, film thickness can be expressed as follows:

$$h = c + e_0 \cos(\psi - \Phi_0) + z(\theta_{y0} \cos\psi - \theta_{x0} \sin\psi) \tag{2}$$

where, c denotes the radial clearance; e_0 denotes the initial eccentricity ; Φ_0 denotes the initial attitude angle; θ_{x0} and θ_{y0} respectively denote the tilting angles of the rotating shaft around x-axis and around y-axis.

The linear perturbation method [11] is adopted to derive the perturbation equations, and then the finite difference method [12] is used to solve the steady and perturbed equations to obtain the perturbed pressure. The additional stiffness coefficients provided by the non-contact rotary union are expressed as follows:

$$\mathbf{K}^r = \begin{bmatrix} k_{xx} & k_{xy} & k_{x\theta_x} & k_{x\theta_y} \\ k_{yx} & k_{yy} & k_{y\theta_x} & k_{y\theta_y} \\ k_{\theta_x x} & k_{\theta_x y} & k_{\theta_x\theta_x} & k_{\theta_x\theta_y} \\ k_{\theta_y x} & k_{\theta_y y} & k_{\theta_y\theta_x} & k_{\theta_y\theta_y} \end{bmatrix} = \int_{-\frac{L}{2}}^{\frac{L}{2}} \int_0^{2\pi} \begin{bmatrix} -\cos\psi \\ -\sin\psi \\ z\sin\psi \\ -z\cos\psi \end{bmatrix} \begin{Bmatrix} P_x & P_y & P_{\theta_x} & P_{\theta_y} \end{Bmatrix} d\psi dz \tag{3}$$

where, P_ξ ($\xi = x$, y, θ_x, θ_y) denotes the perturbed pressure.

3. Results and discussion

3.1. Stiffness coefficients of the non-contact rotary union

The parameters of the non-contact rotary union are summarized in Table 1. Since the grinding force for manufacturing the slender hole is very small, the eccentricity is considered to be zero. The non-zero additional stiffness coefficients of the non-contact rotary union include: the force stiffness coefficients k_{xy}, k_{yx}, $k_{x\theta y}$ and $k_{y\theta x}$, and the moment stiffness coefficients $k_{\theta x\theta y}$, and $k_{\theta y\theta x}$. Fig.5 shows the non-zero stiffness coefficients of the non-contact rotary union with respect to the rotary speed. The negative sign represents the direction. It can be seen from Fig.5 that, k_{xy}, k_{yx}, $k_{\theta x\theta y}$, and $k_{\theta y\theta x}$ increase with the rotary speed; but $k_{x\theta y}$ and $k_{y\theta x}$ keep constant at all speed. This is due to the fact that, the convergent gap along axial direction is generated when the rotating shaft runs with an angular misalignment; and the axial velocity of the flow slightly changes with the rotary speed. It can be concluded that, the force stiffness coefficients $k_{x\theta y}$ and $k_{y\theta x}$ will significantly affected by the tilting angles rather than the rotary speed. Besides, the above coefficients are skew symmetric (k_{xy}=-k_{yx}, $k_{x\theta y}$=-$k_{y\theta x}$, $k_{\theta x\theta y}$=-$k_{\theta y\theta x}$) due to the concentric structure of the water pipe and the rotating shaft when the eccentricity is zero.

Table 1. Parameters of the non-contact rotary union

Length L (mm)	10
Diameter D (mm)	6
Radial clearance c (μm)	10
Viscosity μ (Pa·s)	0.001
Supply pressure P_s (MPa)	0.3
Ambient pressure P_0 (MPa)	0.1
Rotary speed (rpm)	0~200000

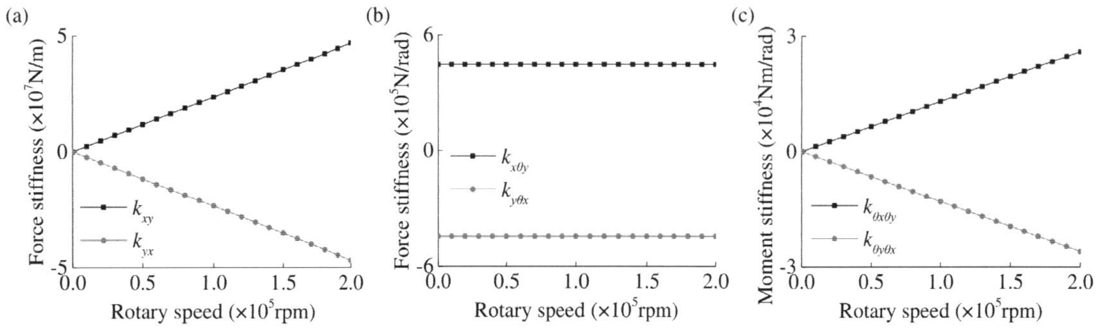

Figure 5. Non-zero stiffness coefficients of the non-contact rotary union at e_0=0

3.2. Effect of the non-contact rotary union on the spindle critical speed

In this section, a comparative study of the rotor dynamic behaviours of the motorized spindle with and without considering the non-contact rotary union is conducted. The parameters of the rotor system are listed in Table 2. The nonlinear stiffness of the angular contact ball bearing can be calculated by using the 5-DOF bearing model firstly established by Jones [13, 14]. In the absence of any external load, the bearing's non-zero stiffness coefficients include the radial stiffness k_{xx} (=k_{yy}), the angular stiffness $k_{\theta x\theta x}$ (=$k_{\theta y\theta y}$), and the cross stiffness $k_{x\theta y}$ (=-$k_{y\theta x}$=-$k_{\theta xy}$=$k_{y\theta x}$). Fig.6 gives the nonzero stiffness coefficients of the angular contact ball bearing with respect to the rotary speed. It can be seen that, the bearing stiffness soften phenomenon occurs when the rotary speed exceeds 50000rpm. This is caused by the increasing centrifugal force and the gyroscopic moment of the ball. Note that, the cross stiffness of the front bearing must be mapped to the rotor coordinate system [15].

Table 2. Parameters of the rotor system

Length of the rotating shaft (mm)	98
Length of the motor rotor (mm)	30
Inner diameter of the motor rotor (mm)	10
Outer diameter of the motor rotor (mm)	19.4
Bearing type	HC708-E-T-P4S
Bearing span (mm)	65
Axial preload (N)	40

IOP Conf. Series: Materials Science and Engineering **635** (2019) 012013 doi:10.1088/1757-899X/635/1/012013

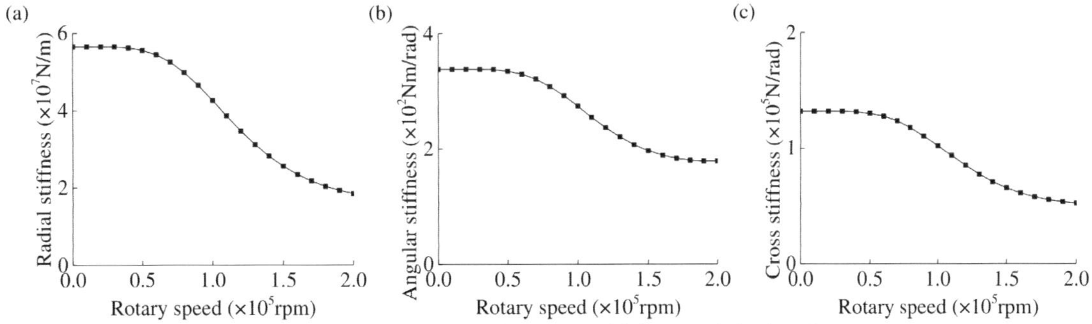

Figure 6. Non-zero stiffness coefficients of the bearing

Figure 7 shows the Campbell diagram of the motorized spindle rotor system with and without considering the non-contact rotary union. The simulated results show that, the first natural frequency of the rotor system considering the non-contact rotary union is larger than that without considering the non-contact rotary union, indicating that the non-contact rotary union is helpful to enhance the rigidity of the rotor system. The first critical speeds of the rotor system with and without considering the non-contact rotary union are 167474rpm and 158372rpm, respectively. The corresponding first mode shapes are plotted in Fig.8 and Fig.9. It can be seen that, the maximum relative amplitude locates at the middle of the bearing span. The centreline of the rotor is a space curve considering the non-contact rotary union, but it is a plane curve without considering the non-contact rotary union.

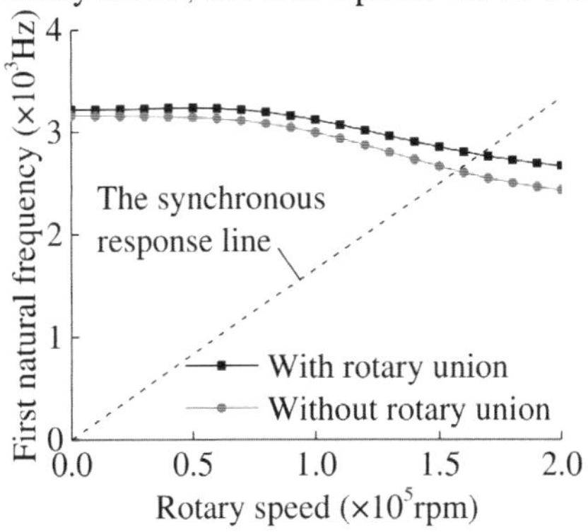

Figure 7. Non-zero stiffness coefficients of the bearing

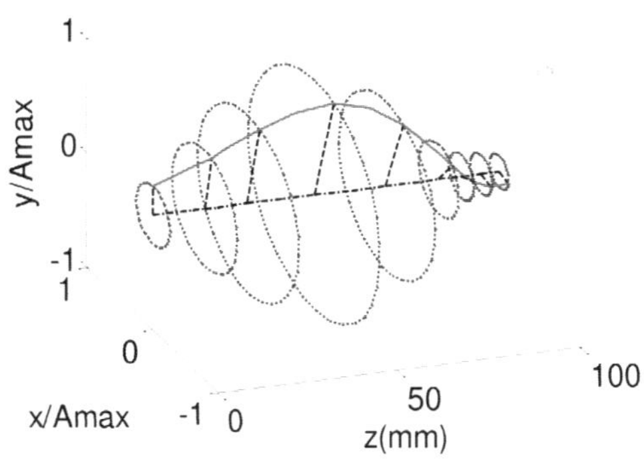

Figure 8. First mode shape of the rotor considering the rotary union

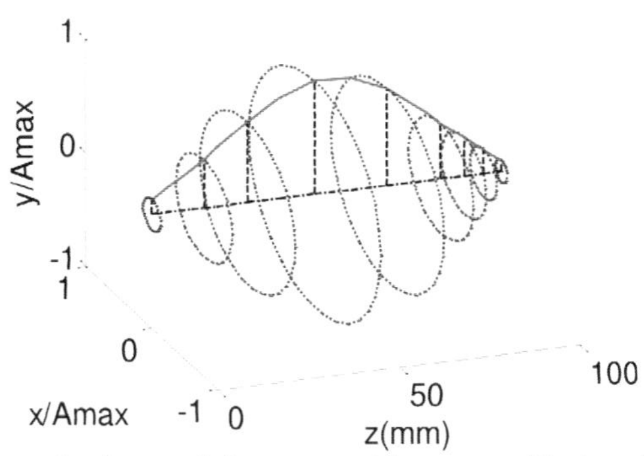

Figure 9. First mode shape of the rotor without considering the rotary union

4. Conclusions

(1) A dynamic model of the motorized spindle for the internal grinding of slender holes is developed including the additional stiffness of the non-contact rotary union, which laid a foundation for the analysis of the effect of the non-contact rotary union on the dynamic behaviours of the motorized spindle rotor system.

(2) The force stiffness coefficients $k_{x\theta y}$ and $k_{y\theta x}$ of the non-contact rotary union are almost constant at all speed, but they are closely related to the tilting angles θ_{x0} and θ_{y0};
(3) The additional stiffness coefficients k_{xy}, k_{yx}, $k_{\theta x\theta y}$, and $k_{\theta y\theta x}$ increase with the rotary speed, while the bearing stiffness keeps constant at a lower speed, but decreases when the rotary speed exceeds a certain value;
(4) The non-contact rotary union is helpful for improving the rigidity and the first critical speed of the motorized spindle rotor system.

5. References

[1] E. Abele, Y. Altinas and C. Brecher 2010 Machine tool spindle units *CIRP Ann. - Manuf. Techn.* **59** 781-802
[2] C. Lin, Y. Lin and C. Chu 2013 Dynamic models and design of spindle-bearing systems of machine tools: a review *Int. J. Precis. Eng. Man.* **14** 513-21
[3] J. Cao, Y. Wu, J. Li and Q. Zhang 2015 A grinding force model for ultrasonic assisted internal grinding (UAIG) of SiC ceramics *Int. J. Adv. Manuf. Technol.* **81** 8751-85
[4] K. Nadolny and A. S. M. Al-Obaidi 2016 A multi-criteria methodology for effectiveness assessment of internal cylindrical grinding process with modified grinding wheels *Int. J. Adv. Manuf. Technol.* **87** 625-37
[5] S. Jiang and S. Lin 2018 A technical note: an ultra-high-speed motorized spindle for internal grinding of small-deep hole *Int. J. Adv. Manuf. Technol.* **97** 1457-63
[6] Y. Li, X. Chen, P. Zhang and J. Zhou 2017 Dynamic modelling and modal experimental study of high speed motorized spindle *J. Mech. Sci. Technol.* **31** 1049-56
[7] J. Liu and P. Zhang 2018 Thermo-mechanical behaviour analysis of motorized spindle based on a coupled model *Adv. Mech. Eng.* **10** 1-12
[8] Kikuchi K. 1970 Analysis of unbalance vibration of rotating shaft system with many bearings and disks *Bull. JSME.* **13** 864-72
[9] H. Feng and S. Jiang 2017 Dynamic of a motorized spindle supported on water-lubricated bearings *Proc. IMechE. Part C: J. Mechanical Engineering Science.* **231** 459-72
[10] Dowson D. 1962 A generalized Reynolds equation for fluid-film lubrication *Int. J. Mech. Sci.* **4** 159-70
[11] G. H. Jang and Y. J. Kim 1999 Calculation of dynamic coefficients in a hydrodynamic bearing considering five degrees of freedom for a general rotor-bearing system *ASME J. Tribol.* **121** 499-505
[12] C. A. Heshmat, D. S. Xu and H. Heshmat 2000 Anaylsis of gas lubricated foil thrust bearings using coupled finite element and finite difference methods AS*ME J. Tribol.* **122** 199-204
[13] Jones A. B. 1960 A general theory for elastically constrained ball and radial roller bearings under arbitrary load and speed conditions *ASME J. Basic Eng.* **82** 309-20
[14] D. Noel, M. Ritou, B. Furet, and S. L. Loch 2013 Complete analytical expression of the stiffness matrix of angular contact ball bearings AS*ME J. Tribol.* **135** 1-8
[15] H. Li and Y. C. Shin 2004 Analysis of bearing configuration effects on high speed spindles using an integrated dynamic thermos-mechanical spindle model *Int. J. Mach. Tool. Manu.* **44** 347-64

Lubrication performance of conical spiral-groove bearings with rough surfaces

Shaowen Zhang[a], Shengye Lin

School of Mechanical Engineering, Southeast University, Nanjing 211189, China

[a]shaowen2004@163.com

Abstract. A mixed lubrication model of conical spiral-groove bearings is established by integrating partial fluid lubrication theory and G-W statistical theory. Considering the effect of bearing surface roughness, the equivalent average Reynolds equation is derived based on average flow model, which is solved by boundary fitted coordinate system and finite difference method. The G-W contact model is used to calculate the axial force and friction force produced by the contact of multi asperities. The coefficients of friction are calculated and measured under different rotational speed and the theoretical results are in accordance with the test results. The results show that comparing with bearings with deep groove, ones with shallower groove have lower transition speed and lower friction coefficient at low speeds.

1 Introduction

Conical spiral-groove bearings have been adopted for high-speed rotating machines, such as centrifuges, flywheels, machine spindles et al. because of their remarkable hydrodynamic effect, the capability of carrying both axial and radial loads, self-sealing performance, and good stability in the case of small eccentricities. Many studies on static and dynamic characteristics of this type of bearings are on the base of the assumption of full-film lubrication [1-4]. In fact, hydrodynamic bearings are usually in mixed lubrication, boundary lubrication or even dry friction during startup and shutdown transient operation. Analyzing the single lubricated state cannot fully reveal the evolution of different lubrication regime. There are, however, only a few reports that evaluate hydrodynamic performance of conical spiral-groove bearings throughout the entire range of operating conditions.

It needs to consider the combined effects of surface roughness and asperities contact when the spiral- groove bearings are in mixed lubrication at low speed. As for the influence of surface roughness on the hydrodynamic performance of spiral-groove bearings, a typical approach is to establish the average Reynold equation by means of average flow model, which is popular in dealing with rough plain bearings [5-7]. In addition, several literature have considered other effects on lubrication performance of spiral-groove bearings, such as fluid centrifugal inertia force, sleeve elastic deformation, cavitation occurrence et al [8-10].

In general, there are three methods to calculate the load carried by asperities. G-W contact model is a classical statistical model for analyzing elastic or elastic-plastic contact between rough and smooth surfaces, the disadvantage of which is that the shape and height distribution of asperities are too simple. On the base of G-W contact model, K-E contact model further improves the elastic-plastic contact. While M-B contact model uses fractal theory to establish the elastic-plastic contact model, which is on the base of fractal parameters and limited to the surface with fractal characteristics [11-16].

In this paper, combining with partial lubrication and asperities contact is proposed to analyze the lubrication performance of conical spiral-groove bearings. The average Reynold equation is used to calculate hydrodynamic pressure, while G-W contact model is adopted to compute the force of asperities contact. The calculated results of the friction coefficients are compared with the measured results to verify the validity of the proposed model. Then the effect of depth of groove on hydrodynamic performance is discussed.

2 Theoretical approach

2.1 Partial lubrication model

Figure 1 shows the geometry configuration of a conical spiral-groove bearings treated here. The physical coordinate system (r,θ,z) is fixed to the shaft with twelve spiral grooves, s is the meridian coordinate. In order to analyze the performance of spiral-groove bearings more accurately, the physical coordinate system (r,θ,z) could be transformed into the computational coordinate system (r,η,z) by means of boundary fitting coordinate, ξ is the meridian coordinate in computational coordinate system, and the transformation relation between these two coordinate systems is:

$$\begin{cases} \xi = s \\ \eta = \theta - \dfrac{1}{\sin\alpha \tan\beta} In(\dfrac{s}{s_g}) \end{cases} \tag{1}$$

Where, α is the half angle of the cone, β is the angle of spiral groove, s_g is the base radius of the spiral line.

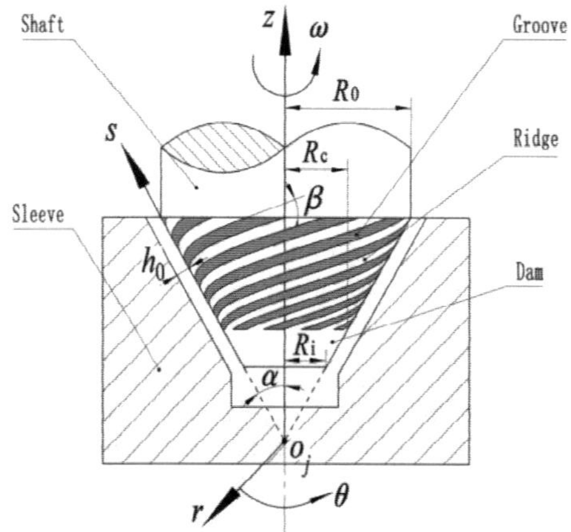

Figure 1. The geometry configuration of a conical spiral-groove bearing.

Considering the effect of surface roughness of bearings, the pressure flow factor and shear flow factor are introduced based on the Patir-Cheng average flow model [5,6]. It is assumed that the fluid is incompressible and in the condition of isothermal laminar flow. According to the relationship between these two coordinate systems, the average Reynold equation deduced in the computational coordinate system is given by

$$s_\xi \frac{\partial}{\partial \eta}\left[\phi_\theta \frac{h^3}{12\mu J}\left(s_\xi \frac{\partial p}{\partial \eta} - s_\eta \frac{\partial p}{\partial \xi}\right)\right] - s_\eta \frac{\partial}{\partial \xi}\left[\phi_\theta \frac{h^3}{12\mu J}\left(s_\xi \frac{\partial p}{\partial \eta} - s_\eta \frac{\partial p}{\partial \xi}\right)\right]$$
$$+ r\theta_\eta \frac{\partial}{r\partial \xi}\left[\phi_s \frac{rh^3}{12\mu J}\left(r\theta_\eta \frac{\partial p}{\partial \xi} - r\theta_\xi \frac{\partial p}{\partial \eta}\right)\right] - r\theta_\xi \frac{\partial}{r\partial \eta}\left[\phi_s \frac{rh^3}{12\mu J}\left(r\theta_\eta \frac{\partial p}{\partial \xi} - r\theta_\xi \frac{\partial p}{\partial \eta}\right)\right] \quad (2)$$
$$= \frac{\omega r}{2}\left(s_\xi \frac{\partial \bar{h}_T}{\partial \eta} - s_\eta \frac{\partial \bar{h}_T}{\partial \xi}\right) + \frac{\omega r}{2}\sigma\left(s_\xi \frac{\partial \phi_{sh}}{\partial \eta} - s_\eta \frac{\partial \phi_{sh}}{\partial \xi}\right) + J\frac{\partial h}{\partial t}$$

Where,

$$J = s_\xi (r\theta_\eta) - s_\eta (r\theta_\xi) \quad (3)$$

Here, h is the nominal film thickness, ρ is the density of the fluid, μ is the dynamic viscosity of the fluid, σ is the composite roughness of the surface profile of shaft and sleeve, $\bar{h}_T$ is the expected value of local film thickness. ϕ_θ and ϕ_s denote the pressure flow factors in $\theta-$ and $s-$direction, respectively, ϕ_{sh} denotes the shear flow factor in $\theta-$ direction, which can be calculated according to reference [5,6].

After the local hydrodynamic pressure p is determined from average Reynold equation which can be solved by means of finite volume and finite difference method, the axial load carried by hydrodynamic pressure is given by

$$F_{z,h} = \iint_A J(p - p_a)\sin\alpha d\eta d\xi + (p_{poc} - p_a)\cdot \pi R_i^2 \quad (4)$$

Where, p_{poc} is the static pressure in the closed cavity near the small end of bearings.

The hydrodynamic frictional torque is obtained from

$$T_h = \iint_S J\left[\frac{\mu r\omega}{h}(\phi_f + \phi_{fsh}) - \phi_{fp}\frac{h}{2}\frac{\partial p}{r\partial \theta}\right] r d\eta d\xi \quad (5)$$

Where, ϕ_f denotes the shear stress factor, ϕ_{fsh} and ϕ_{fp} denote shear stress correction factors defined on the basis of the average flow model [5,6].

2.2 Asperities contact model

It is assumed that only elastic deformation occurs on contact surfaces, and the contact is simplified as contact between elastic rough surfaces and flat surfaces. Here we adopt G-W contact model to build asperities contact model between rough surfaces of bearings [11]. The axial load carried by asperities contact can be given by

$$F_{z,a} = \sin\alpha\left[\frac{4}{3}D_{sum}A_n E' R_s^{1/2}\sigma_s^{3/2}F_{3/2}(h')\right] \quad (6)$$

Where, D_{sum}, R_s and σ_s denote the density, the average radius and the standard deviation of asperities, E' denotes composite elastic modulus, A_n is the nominal contact area, $F_{3/2}(h')$ is the probability cylindrical function [11,17].

The friction torque produced by asperities contact is obtained from

$$T_a = F_{z,a} f_a \hat{r} \quad (7)$$

Where f_a is the friction coefficient under boundary lubrication, which is measured by the test of flat journal bearings.

The total axial capacity of load is the sum of the hydrodynamic force $F_{z,h}$ and the contact force $F_{z,a}$, and the total friction torque is the sum of the contact friction T_h and the viscous resistance T_h.

3 Experiment

3.1 Apparatus and specimens

The experiments are carried out on the MM-W1A friction and wear test machine. A schematic diagram and a photograph of specimen fixture we designed are shown in figure 2. The upper specimen is attached to the spindle of friction test machine by the coupling, so the upper specimen could rotates about the axis. The lower specimen is put in the inside holder laid in the outside holder which is placed on the platform of the test machine and could be moved up and down along the vertical direction, the gap between inside and outside holders is filled with rubber rings.

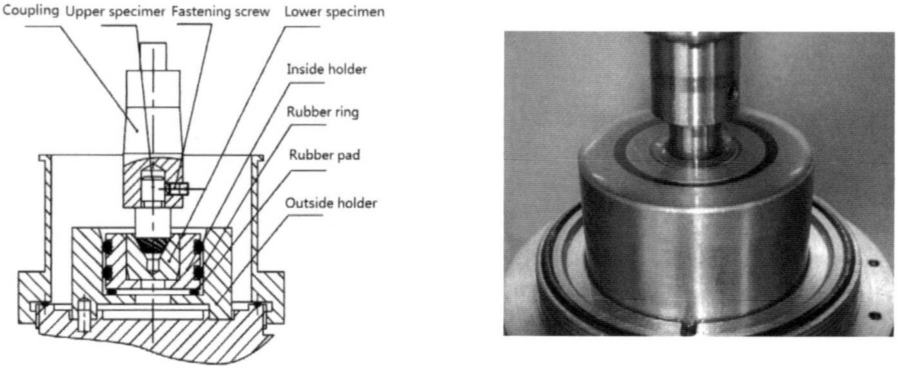

(a) (b)

Figure 2. The specimen fixture, (a) schematic diagram, (b) photograph.

As shown in Figure 3, the upper specimens are made of GCr15, the cone surfaces with hardness 750HV are lapped to a fine finish. The spiral grooves are ablated by laser marking machine. There are three types of upper specimens, including conical flat specimen, conical specimen with shallower spiral-groove(10um) and conical specimen with deep spiral-groove(50um). The lower specimen is made of aluminum bronze with surface hardness 250HV.

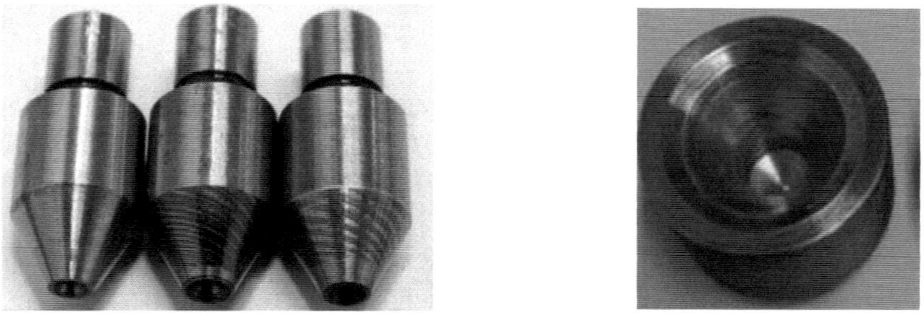

(a) (b)

Figure 3. The photograph of specimens, (a) the upper specimens (b) the lower specimen.

3.2 Procedure

The lubricant is SAE 100 oil without additives, and oil-bath lubrication is used in the test. In order to avoid excessive rise of oil temperature, after applying load, set upper specimen rotates at 200rpm for 2 minutes and then stop the machine until the oil is fully cooled. The next sequence increases 200rpm.To minimize the friction, each sequence is repeated several times, and the average value is taken as the final test results.

4 Results and discussion

Table 1 presents the input parameters for the numerical calculation.

Table 1. The parameters of the conical spiral-groove bearing for numerical calculation.

Items	Values(unit)	Items	Values(unit)
α	30(°)	$\sigma=\sigma_2$	2.68(μm)
β	20(°)	C^*	0.870(γ=9), 1.48(γ=1/9)
R_0	6(mm)	r^*	1.5 (γ=9), 0.42(γ=1/9)
R_c	3(mm)	D_{sum}	$2.21\times10^9(m^{-2})$
R_i	2(mm)	R_s	7.5(μm)
h_g	10,50(μm)	σ_s	2.41(μm)
μ	0.385(Pa·s)	$E(z_s^*)$	2.77(μm)

Figure 4(a) compares the results of numerical calculation and experiment for a specimen with shallower groove (10um). The curves demonstrate the agreement between the numerical calculation and experiment. The friction coefficient drops rapidly with the rotational speed at first and then slowly increases when the speed exceeds a specific value, which indicates the transition from mixed lubrication regime to hydrodynamic regime. The rotational speed responding to the minimum friction coefficient value is the transition speed. The transition speed predicted theoretically (about 600 rpm) is slightly smaller than experiment result (about 800 rpm). The difference between calculation and experiment below the transition speed is a little smaller than that above the transition speed.

Figure 4(b) is for a specimen with deep groove (50um). Comparing the results of figure 4(a) and 4(b), it implies that at low speeds, the shallower groove bearing has a lower friction coefficient than one with deep groove, but at high speed, it is opposite. Moreover, the bearing with shallower groove has lower transition speed.

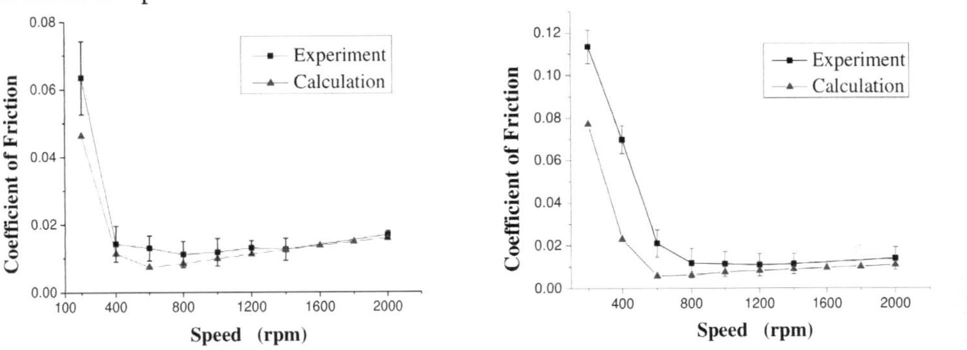

(a) (b)

Figure 4. Comparison between experiment and calculation results of COF and rotational speed relationship, (a) bearing with shallower groove, (b) bearing with deep groove.

Figure 5 (a) and (b) present variations in axial load and friction torque with speed, respectively. With the increase of speed, load carried by hydrodynamic pressure increases gradually until all of the external load, while load carried by contact force decreases gradually until zero, indicating that with increasing speed the hydrodynamic effect becomes stronger and the number of direct contact of asperities gets rarer. Meanwhile, with increasing speed the friction torque due to fluid viscous resistance increases gradually, while one due to asperities contact decreases. In general, friction coefficient of asperities contact is larger than that of fluid viscous resistance, so as shown in figure 5 (b), the friction torque due asperities contact under low speed is much larger than that due to fluid viscous resistance.

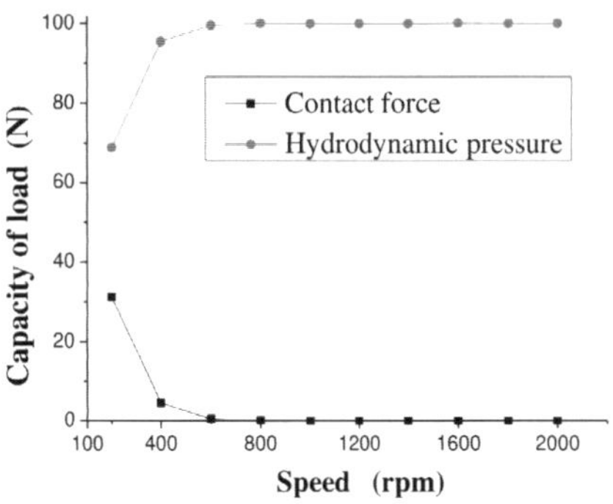

(a)

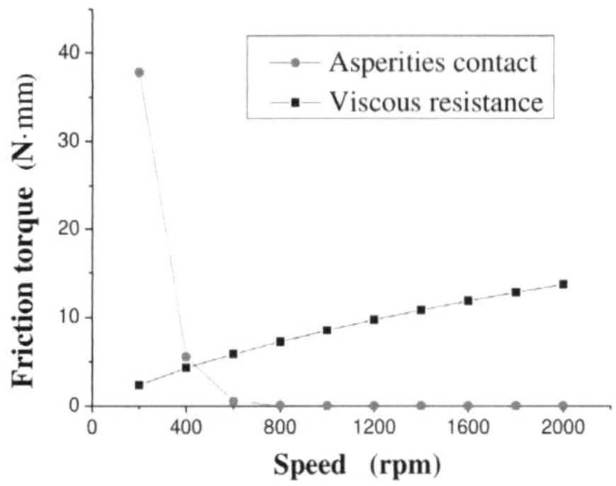

(b)

Figure 5. Calculation result of capacity of load, friction torque and rotational speed relationship for bearing with shallower groove, (a) capacity of load, (b) friction torque.

5 Conclusions

By means of experiments, the hydrodynamic effect produced by spiral-groove is obviously observed, and the transition speed from mixed lubrication regime to hydrodynamic regime can be detected.

The friction coefficient predicted by numerical calculation on the base of combining partial lubrication and asperities contact agrees well with the results of experiment.

At low speeds bearings with shallower groove has greater hydrodynamic effect, lower transition speed and friction. However, at high speeds bearings with deep groove has lower friction due to larger average film thickness.

References

[1] Muijderman E A 1967 *Tribol.* **1** 291
[2] Huang Y, Chen D G 1996 *J. Tribol. Int.* **29** 281-290
[3] Yoshimoto S, Oshima S, Danbara S, and Shitara T 2002 *J. Tribol.* **124** 398-405
[4] Pan C H T, Kim D 2015 *J. Tribol.* **129** 375-83
[5] Patir N, Cheng H S 1978 *J. Lubr. Technol.* **100** 12-17
[6] Patir N, Cheng H S 1979 *J. Tribol.* **101** 220
[7] Hirayama T, Shiotani H, Yamada K, Yamashita N, Matsuoka T, and Sawada H 2015 *J. Tribol.* **137** 011704
[8] Lin X, Jiang S, Zhang C and Liu X 2018 *J. Tribol. Int.* **119** 645-658
[9] Bai S, Ma C, Peng X and Wen S 2015 *J. Tribol.* **137** 0215021-10
[10] Hirayama T 2004 *J. Tribol.* **126** 140-44
[11] Greenwood J A, Williamson J B P 1966 *Proc. R. Soc. London* **295** 300-19
[12] Qiu Y, Khonsari M M 2011 *J. Tribol. Int.* **44** 1610-19
[13] Gong R, Zhou C, Che H Zhu M and Xu x 2015 *Proc. I. Mech. Eng. D J. Aut. Eng.* **229** 34-42
[14] Chang W R, Etsion I, Bogy D B 1988 *J. Tribol.* **110** 57-63
[15] Feng M, Kenjo T A 2007 *IEEE T. Magn.* **43** 3727-33
[16] Hu J B, Wei C, Li X Y 2013 *J. Tribol. Int.* **62** 8-17
[17] Mccool J I 1986 *J. Tribol.* **108** 380-86

IOP Conf. Series: Materials Science and Engineering **635** (2019) 012015 doi:10.1088/1757-899X/635/1/012015

Design optimization of worm gear drive with reduced power loss

Paridhi Rai[a], Asim Gopal Barman[b]

Department of Mechanical Engineering, National Institute of Technology Patna, Patna

E mail: [a] raiparidhi25@gmail.com, [b] asim@nitp.ac.in

Abstract. In worm gear mechanism, power loss is one of the critical issues, which requires attention from gear designers and researchers. This is due to power loss; the overall efficiency of worm gear drive gets reduced. In this article, Simulated Annealing (SA) is used to minimize the power loss of worm gear drives within the prescribed set of design constraints. Various design constraints included in the design optimization problem are linear pressure, bending strength and deflection of worm. Gear tooth number, coefficient of friction and helix angle are the design variables considered. Results obtained are compared and validated with the results obtained in previous research work found in the literature. It shows that SA is more efficient in attaining optimal design values for worm gear drive for reducing power loss.

1. Introduction

Worm gear drives are used in transmission system because of their ability to achieve high reduction ratio with compact size. The relative sliding motion between the gear teeth is the reason for poor efficiency of worm gear mechanism. Thus friction is high in these gears leading to high power loss. Very less attention has been paid by the gear researchers and designers on reducing the power loss of worm gears using evolutionary techniques.

There are many literatures available related to design optimization of various gear drives using conventional and non- conventional optimization techniques. Gologlu and Zeyveli [1] demonstrated the use of genetic algorithm for minimization of volume of a two-stage helical gear drive. Genetic algorithm was again used by Mendi and Baskal [2] for reduction of volume of gearbox which includes shaft bearing and gears. Weight of cylindrical parallel gear train has been minimized using GA by Caballero [3]. An approach developed by combining Finite Element Analysis, Artificial Neural Network and Genetic Algorithm was proposed by Su and Peng [4] for achieving optimal design values of a cylindrical worm with helical gear drive. Chong and Lee [5] proposed a technique for utilizing genetic algorithm for optimization of volume of two-stage gear drive and simple planetary gear. To reduce the weight of worm gear pair, modified artificial immune algorithm had been used by Padmabhanan et al. [6]. Stojanovic et al. [7] studied the effect of lubricant viscosity on the worm gear efficiency. Rai and Barman [8] utilized genetic algorithm and simulated annealing for performing design optimization of spur gear set. Yaman et al. [9] used genetic algorithm to minimize the power loss in worm gears and further compared it with analytical results to prove its efficiency. In continuation of this work, differential evolution and particle swarm optimization technique was used for optimizing the worm gear power loss by Sabrinath et al [10].

In the presented article, the problem of power loss of worm gear mechanism has been reported. Simulated annealing, which has not been utilized in the previous studies, has been used to minimize the power loss. The efficiency and performance of the algorithm is analysed by comparing it with the results obtained using other optimization algorithms like genetic algorithm (GA), differential evolution (DE), particle swarm optimization (PSO) etc.

2. Problem definition

This work aims at minimizing loss of power in worm gear drive. High friction between the gear teeth while in meshing is the reason for power loss. Therefore, power loss in worm gear drive can be expressed as [10]

$$\Delta P = P_1 - P_2$$

where P_1 is the input power and P_2 is the output power. In this case, the expression for output power is

$$P_2 = F_n(cos\alpha_n cos\emptyset_n - \mu sin\emptyset_n)\frac{m\omega_1 z_2}{2u}$$

where F_n is the normal force, α_n is the normal pressure angle, $\emptyset_n$ is the helix angle , m is the axial module, z_2 is the number of teeth on worm wheel and ω_1 is the angular velocity of worm. The schematic diagram for worm gear drive is shown in Fig 1.

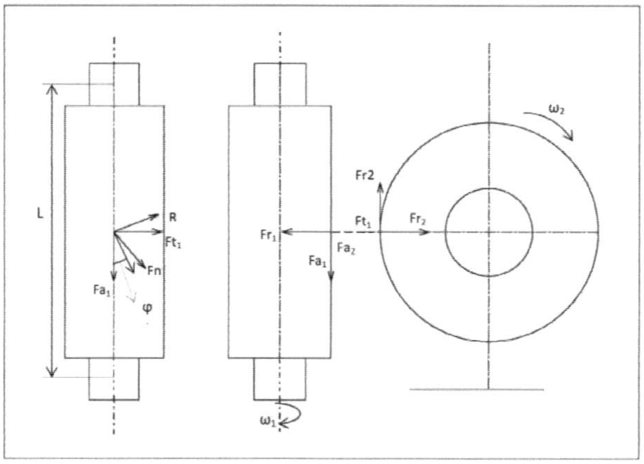

Fig 1. Schematic Diagram of worm gears

Here the material of worm and worm wheel is hardened steel and bronze respectively. The angle between the shaft is 90° and worm has three threads. Various input parameters used for performing design optimization of worm gears are shown in Table 1.

Table 1. Input parameters for worm gear design optimization [9]

Input Parameter	Unit	Value
Input Power	kW	11
Transmission ratio		15
Number of revolution for worm	rpm	720
Centre distance of worm gear pair	mm	200
Distance between of worm shaft bearing	mm	330
Module	mm	7
Number of worm teeth		3
Worm diameter	mm	71
Bottom of teeth diameter of worm	mm	55
Pressure angle	degree	22.5

Elasticity Modulus	N/mm^2	$21*10^4$
Inertia	mm^4	449000

2.1. Objective function
Thus the final expression for the objective function for power loss minimization in worm gear mechanism, is

$$\Delta P = P_1 - F_n(cos\alpha_n cos\emptyset_n - \mu sin\emptyset_n)\frac{m\omega_1 z_2}{2i}$$

2.2. Design variables
Three design variables are considered for performing design optimization of worm gear mechanism. They are gear tooth number, helix angle and coefficient of friction. The ranges for these design variables are given below

$$21 \leq z_2 \leq 80$$

$$0.03 \leq \mu \leq 0.05$$

$$15 \leq \emptyset_n \leq 25$$

2.3. Design constraints
Bending strength, contact pressure and linear deflection of worm are the design constraints considered while performing optimization for worm gear design problem. They can be expressed as

$$g_1 = \frac{F_{a2}}{b_2 m\pi} - 30 \leq 0$$

$$g_2 = \frac{F_{t2}}{b_2 m z_2} 2.5 - 3.6 \leq 0$$

$$g_3 = \frac{d_{f1}}{1000} - \frac{F_{rt}L^3}{48EI} \leq 0$$

Here, g_1 is the bending strength of the worm gear tooth, g_2 is the linear pressure of worm gear and g_3 is the linear deflection of worm shaft. Here F_{a2} and F_{rt} are the axial force on worm gear and total radial force respectively. b_2 and d_{f1} are the face width of worm gear and root diameter for worm gears respectively.

3. Optimization technique
Simulated annealing is a stochastic optimization technique which can solve both constrained and unconstrained optimization problems [11]. This technique is based on the principle of heating of a material and then letting it cool down, thereby minimizing the defects and thus further reducing the system energy. There is random generation of new point at every iteration in simulated annealing. The distance between new and current point can be measured with the help of probability distribution with a scale. This probability distribution is proportional to the temperature. When the value of objective function is lowered, the algorithm accepts candidate point for solution along with those points which raise the value with a certain probability. Thus, the algorithm will not be trapped in local minima and produces a global optimum solution.

4. Results and Discussions
Minimization of power loss of worm gear set problem is solved using simulated annealing algorithm. This algorithm has been run using MATLAB 2016 on PC compatible with Intel(R) Core(TM) i7-3770,

a 3.4 GHz processor and 4 GB of RAM. Various parameters of SA employed while performing the optimization of worm gear mechanism are shown in Table 2.

Table 2. Design parameters for SA

Design parameters	Value
Maximum run	100
Final stopping temperature	1e-2
Initial temperature	1
Cooling rate	0.8

The algorithm has been executed 10 times for verifying the convergence of SA. The results obtained using SA are compared with the results obtained in previous works (DE, PSO, GA and analytical method). The value of power loss for worm gear drive obtained is 0.2988 kW which is lower in comparison to power loss obtained using other algorithms as shown in Table 3. The percentage reduction of power loss is also shown in Table 3.

Table 3. Comparison of results obtained using SA and other evolutionary optimization techniques [10].

Optimization Algorithm	No of teeth on worm gear	Friction coefficient	Helix angle	Power loss (kW)	% reduction in power loss by SA
SA (present article)	45	0.0301	15.045	0.2988	-
DE algorithm [10]	45	0.0302	15.008	0.388	22.98%
PSO [10]	45	0.0302	15.002	0.369	19.02%
Binary GA[9]	44	0.0305	15.24	0.881	65.5%
Analytical method[9]	46	0.0390	16.280	1.362	78.02%

The optimum solution obtained using SA converges after 16th iteration. The convergence curve of power loss versus iteration is shown in Fig. 2

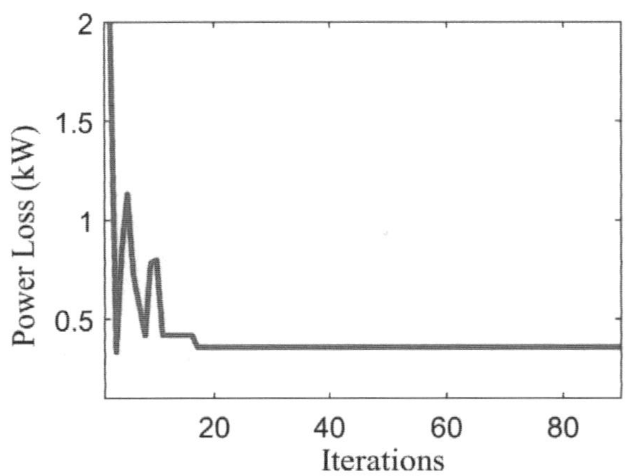

Fig 2. Convergence curve of Power loss and iterations for SA

The frictional coefficient value is less which results in better efficiency of worm gear pair. This is because of less heat generation while in mesh. There is no change in gear tooth number obtained using SA, but the value of helix angle is greater compared to other optimization techniques.

IOP Conf. Series: Materials Science and Engineering **635** (2019) 012015 doi:10.1088/1757-899X/635/1/012015

5. Conclusions

This study presents the power loss minimization of worm gear mechanism using simulated annealing. The optimal worm gear model obtained is found to be better in comparison to earlier used optimization techniques. The algorithm used in the design optimization of worm gear proves to be effective in avoiding the premature convergence in local minima. Thus, this shows good performance of the algorithm in implementing the design problem. It can be seen that the value of friction coefficient is lower in comparison to other optimization technique used earlier (DE, PSO and GA). Helix angle obtained using SA is greater in comparison with DE, PSO and GA. The percentage reduction of power loss using SA ranges from 78.02% to 22.98%. The optimized gear pair will have better service life. Hence, the attained results show that the performance of SA is better in comparison with other approaches used earlier for power loss minimization of worm gears. Thus, optimum design of worm gear drive has been achieved with reduced power loss. Further, the optimization problem can be formulated as multi-objective optimization problem.

6. References

[1] C. Gologlu and M. Zeyveli, 2009, A genetic approach to automate preliminary design of gear drives, *Computers & Industrial Engineering*, **57,** pp. 1043-1051

[2] F. Mendi, T. Başkal, K. Boran and F.E. Boran 2010, Optimization of module, shaft diameter and rolling bearing for spur gear through genetic algorithm. *Expert Systems with Applications*, **37**, pp. 8058-8064

[3] V.S. Caballero, R.C. Romero, A.P. Ferrando and S.M. Sanz, 2009, Cylindrical parallel gear trains design by means of genetic optimization. In *Selected Proceedings from the 12th International Congress on Project Engineering. Zaragoza, July 2008* Asociación española de ingeniería de proyectos. AEIPRO, pp. 150-160

[4] D. Su and W. Peng, 2008, Optimum design of worm gears with multiple computer aided techniques. ICCES, **6**, pp. 221–227

[5] T.H. Chong and J.S. Lee 2000, A design method of gear trains using a genetic algorithm. *International Journal of the Korean society of precision engineering*, **1**, pp.62-70

[6] S. Padmanabhan, M. Chandrasekaran, V. Srinivasa Raman, 2013 Worm gear drive optimization using modified artificial immune algorithm. Journal of Current Engineering Research , **3**, pp. 48–51

[7] B. Stojanovic, S. Radosavljevic, S. Velickovic, S. Miladinovic & M. Bukvic, 2017, The influence of lubricant viscosity on the efficiency of worm gear reducer. In *8th International Scientific Conference Research and development of mechanical elements and systems (IRMES 2017)*, pp. 7-9.

[8] P. Rai and A.G. Barman, 2018, Design optimization of spur gear using SA and RCGA. *Journal of the Brazilian Society of Mechanical Sciences and Engineering*, **40**, pp. 1-8.

[9] M. Yaman, H. Saruhan and F. Mendi, 2006, Power loss optimization of a worm gear mechanism by using genetic algorithm. In *11th international FIGES user conference for computer aided engineering and system modeling, Bolu, Turkey* pp. 13-15.

[10] P. Sabarinath M.R. Thansekhar and R. Saravanan, 2015, Performance Evaluation of Differential Evolution and Particle Swarm Optimization Algorithms for Optimizing Power Loss in a Worm Gear Mechanism. In *Power Electronics and Renewable Energy Systems*, pp. 433-441

[11] S. Kirkpatrick, C.D. Gelatt, and M.P. Vecchi, 1983, Optimization by simulated annealing, Science, 220, pp. 671-680.

IOP Conf. Series: Materials Science and Engineering **635** (2019) 012016 doi:10.1088/1757-899X/635/1/012016

Prospects for using the Bayes algorithm for assessing the technical condition of internal combustion engines

Farit Kh. Khaliullin[1], Alexander V. Matyashin[2], Rishat R. Akhmetzyanov[2], Vladimir M. Medvedev[2] and Maxim A. Lushnov[3]

[1]Department of Automobile Engines and Services of the Institute of Aviation, Land Transport and Energy Production, Kazan National Research Technical University named after A.N. Tupolev – KAI, 10 K. Marksa St., 420111, Kazan, Russia
[2]Department of Machines Operation and Repair of the Mechanization and Technical Service Institute, Kazan State Agrarian University, 65 K. Marksa St., 420015, Kazan, Russia
[3]Department of Machinery and Equipment in Agribusiness of the Mechanization and Technical Service Institute, Kazan State Agrarian University, 65 K. Marksa St., 420015, Kazan, Russia

khaliullin_kai_adis@mail.ru

Abstract. Appropriate assessment of the technical condition of internal combustion engines allows determining optimal values for standards of their technical operation, adjusting the parameters of maintenance and repair process operations, and, as a result, reducing the cost of their effective operation. The article is devoted to the use of the Bayes algorithm for assessing the technical condition of internal combustion engines. This algorithm allows using statistical data on repair and maintenance of internal combustion engines, wherein the missing information is determined by the dynamic systems modeling method. The choice of transitional functions of constituent elements of internal combustion engines as diagnostic parameters increases the reliability of obtained results.

1. Introduction

The choice of efficient methods and technologies for diagnosing internal combustion engines reduces the likelihood of making erroneous decisions on maintenance repair and overhaul.
Currently, there are quite a lot of methods and technologies for in-place diagnostics of internal combustion engines [1-3], which in varying degrees meet the objectives of assessing their technical condition.

First of all, it is necessary to note the vibroacoustic diagnostic methods, which allows for the registration, processing, and interpretation of vibration or noise indicators of the operation of internal combustion engines. The current state of development of vibroacoustic diagnostic technologies makes it possible to single out the following advantages [4-6]:

- flaw detection and the possibility of diagnosing failures at an early stage;
- forecasting the possibility of further operation of machine components or the machine as a whole, as well as adjustment and planning of the maintenance and repair volume and technology;

- the ability to quickly collect information on the technical condition of equipment and the mobility of diagnostic equipment.

At the same time, it is necessary to note the complexity of extracting the vibration signal caused by the presence of a malfunction [7-9], difficulties of applying Fourier transforms for non-stationary processes, as well as the dependence of vibration on the transient time, and increased requirements for processing the vibroacoustic signal.

Another common method is to analyze the composition and amount of engine exhaust gases. It allows determining the nature and conditions of the combustion process and assessing the maximum allowable wear of parts included in the cylinder-piston group (CPG). The main disadvantages of this method include the impossibility of in-process monitoring of the engine condition and the low capabilities for forecasting the residual engine life. The measurement of the flow rate of crankcase gases is a special case which, in fact, can be carried out in process. It has been established that in this method of analyzing the technical condition, the most significant factors influencing the flow rate of crankcase gases, in addition to the condition of the CPG itself, are the engine load and temperature, which allows for rather accurate assessment of the engine CPG condition in most steady modes of operation. At the same time, it is difficult to estimate the engine operation under unsteady load by this method due to large fluctuations in the flow rate of crankcase gases in such modes. In addition, an obvious disadvantage of this method is the ability to assess only the condition of the CPG, but not the general engine condition.

Also, there is an interesting method for assessing the technical condition of an internal combustion engine (ICE) based on monitoring the current values of wear product concentrations. During the ICE operation, wear products that come from various parts are accumulated in the engine oil, which makes it a carrier of information about the technical condition of the engine. However, the method also cannot be performed in the process and does not allow assessing the overall technical condition of the engine.

To assess the technical condition of the CPG, such an indicator as burning oil consumption is also used, which makes it possible to establish the type of necessary repairs in the case of proper accounting. The oil feed mode, level, and temperature significantly influence the life of the engine and the oil itself. However, during the period from the start of engine operation to overhaul, oil consumption increases by 3-5 times and depends on the engine operation conditions; therefore, the assessment of the CPG technical condition according to burning oil consumption requires operating the equipment under the same conditions.

The CPG technical condition is also determined according to the compression pressure. The disadvantage is that the extreme wear may cause a slight change in compression due to significant oil consumption for sealing the CPG connection [10]. It should be noted that the amount of compression is also influenced by not only the wear of the CPG but also by the faulty sealing of the gas distribution mechanism (GRM) valves. When assessing the technical condition, the compression value is not reliable, since the compression pressure depends not only on the CPG condition but also on the crankshaft speed, the constancy of which cannot be achieved when using a starter.

2. Procedures

The diagnostics of internal combustion engines uses dynamic methods, which involve analyzing the response of operation process parameters to standard signals. Due to the convenience of implementation, the Fourier method for analyzing the harmonic series is used, the essence of which is to identify the main harmonics and subsequently identify the amplitude and frequency spectra, which allow obtaining the spectrum of the signal under study and then identifying the malfunction [11]. However, the well-known analysis by the Fourier method has several disadvantages, among which the main ones are as follows:

- Lack of information and the almost complete lack of the ability to analyze based on the time of occurrence and the shape of amplitudes, low denoising capabilities.
- Fourier harmonic series cannot reliably reflect the shape of signals with a large steepness of harmonics, since this requires cumbersome construction of digital series.

Due to the fact that the above method has a sufficient number of limitations, the wavelet analysis of ICE characteristics, which is an alternative method, eliminates the disadvantages of the Fourier series analysis method and provides a more flexible signal processing technique, is used more and more frequently. This transform method is a detailed and time-weighted space argument through the representation of signals. A distinctive feature of the Fourier transform is that the wavelet transform of one-dimensional signals provides their two-dimensional representation, and at the same time, the frequency and coordinate are considered as independent variables, which allows simultaneous analysis of signals in two dimensions [12].

However, the wavelet analysis method is not widely used to date due to software and hardware limitations.

Due to simplicity and efficiency, the method based on the generalized Bayes formula is the most commonly used. This method has its disadvantages: a large amount of preliminary information and rare diagnoses. But if the volume of statistical information allows application of this method, it is reasonable to use it as one of the most reliable and efficient. This method is used when diagnosing the technical condition of tractors, automobiles, railway transport, or aviation equipment [13, 14].

It is also worth noting the perspectivity and usability of the Bayes algorithm for forecasting malfunctions of ICE due to the fact that this formula allows more accurate recalculation of the probability using either previously known statistical information or new test data.

3. Results

The authors have developed an in-place engine diagnostic method based on the Bayes algorithm. Considering the internal combustion engine as a dynamic system with stationary parameters, the authors used methods for identifying dynamic systems through dynamic characteristics. As diagnostic parameters, the intensity of change in parameters was chosen, namely, the transition functions of the fuel and air supply systems (Figure 1 and Figure 2), as well as the response of the system to a single pedal perturbation in terms of engine crankshaft speed (Figure 3) [15].

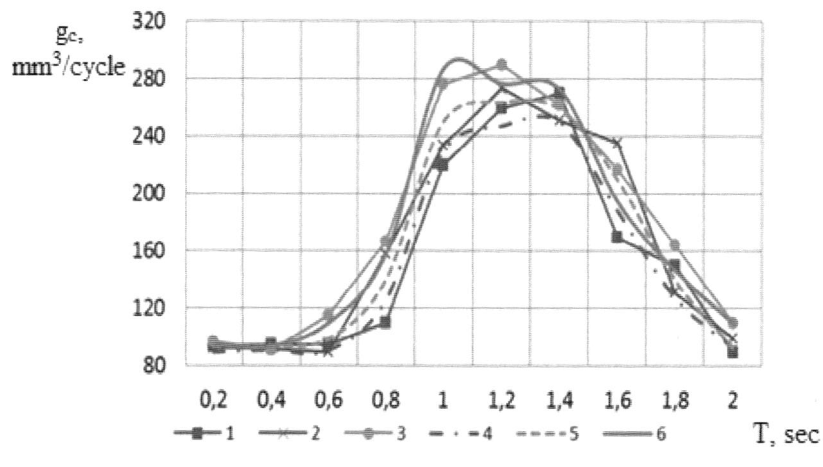

Figure 1. Cyclic feed transition functions g_c depending on the D-243 engine nozzle condition.

Note: 1, 4 – at a fuel injection pressure of 18.5 MPa, 2, 5 – at a fuel injection pressure of 16.5 MPa, 3, 6 – at a fuel injection pressure of 15.5 MPa.

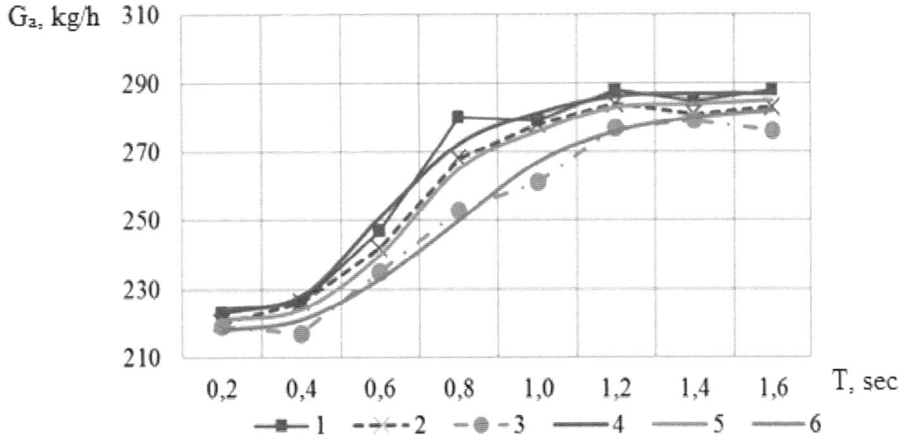

Figure 2. Transient functions of air flow G_a depending on compression k of the D-243 engine.

Note: 1, 4 – at compression of 2.7 MPa, 2, 5 – at compression of 2.4 MPa, 3, 6 – at compression of 2.1 MPa.

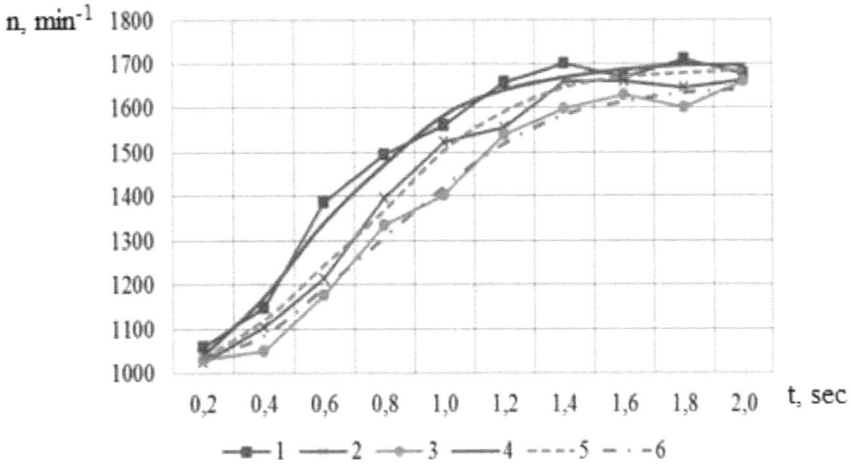

Figure 3. Transient functions of rotation speed n depending on compression k of the D-243 engine.

Experimental and approximating data: 1, 4 – at compression of 2.7 MPa,
2, 5 – at compression of 2.4 MPa, 3, 6 – at compression of 2.1 MPa.

The obtained values of diagnostic parameters depend on the technical condition of the engine and therefore, it is necessary to conduct preliminary calibration studies of the engine under test in its various conditions [16, 17] in order to deliver a relevant diagnosis. As a result, one can obtain three ranges for each diagnostic parameter: good (G), satisfactory (S) and bad (B) (Table 1).

Table 1. Gradation of diagnostic parameters

T_{gc}, sec (cyclic fuel feed)		
good (G)	satisfactory (S)	bad (B)
0.5-1	1-1.5	>1.5

T_n, sec (rotation speed)								
G	S	B	G	S	B	G	S	B
0.8-1.3	1.3-1.7	>1.7	0.8-1.3	1.3-1.7	>1.7	1.3-1.7	1.7-2.2	>2.2

T_{Ga}, sec (air flow)																							
G	S	B	G	S	B	G	S	B	G	S	B	G	S	B	G	S	B	G	S	B	G	S	B
1-1.5	1.5-2	>2	1-1.5	1.5-2	>2	1-1.5	1.5-2	>2	1-1.5	1.5-2	>2	1-1.5	1.5-2	>2	1-1.5	1.5-2	>2	1-2	2-2.5	>2.5	1-2	2-2.5	>2.5

4. Conclusion

The economic effect of implementing the developed method of engine diagnostics is achieved by reducing the process complexity. Compared with existing methods, the proposed method improves productivity by 30-40% and reduces complexity by 25-35%. Based on the developed method, the authors plan to continue research on determining and refining the diagnostic parameters for modern engines with various types of malfunctions, as well as to expand the field of application of the developed method for assessing the residual life of modern engines.

5. References

[1] Pestrikov V M and Eukarpiev V 2014 Features of modern vehicle diagnostics *Technical and Technological Service Problems* **4(30)** 15–16

[2] Gessenhardt C 2009 Hybrid endoscopes for laser-based imaging diagnostics in IC engines *SAE Tech. Pap.* **1** 11

[3] Snyder J A 2011 *Development and Application of Tracer-based Planar Laser-induced Fluorescence Imaging Diagnostics for HCCI Engines* (Palo Alto: Stanford University)

[4] Machalik S 2013 *Využití Obrazové Analýzy v Tribotechnické Diagnostice (Image Analysis in Tribodiagnostics)* (Pardubice: University of Pardubice) p 112

[5] Medvedev V M, Ahmetzyanov I R, Shiriyazdanov R R and Khaliullin F K 2014 Operating conditions of the D-240 Engine based dual fuel gas-diesel engine of the agricultural tractjr *European Conference on Innovations in Technical and Natural Sciences Proceedings of the 2nd International Scientific Conference (May 12, 2014)* (Vienna: East West Association for Advanced Studies and Higher Education GmbH)

[6] Wani M F 1999 Development of design methodology for mainatinability of mechanical systems. *Inidain Institute of Technology* 205

[7] Churilin A S 2011 Efficient vibroacoustic diagnostics of cars *Technical and Technological Service Problems* **17** 20–22

[8] Guzzomi F G, O'Neill P L and Tavner A C R 2007 Investigation of damper valve dynamics using parametric numerical methods *16th Australasian Fluid Mechanics Conference Crown Plaza*, Gold Coast, Australia 1123–30

[9] Benedetto J and Ferreira P J S G 2001 *Modern Sampling Theory* (Boston: Birkhauser) p 428

[10] Rizvi M A, Raza Q, Bhatti A, Zaidi S and Khan M 2012 Modeling and simulation of si engines for fault detection, internal combustion engines *InTech*, DOI: 10.5772/50487, http://www.intechopen.com/books/internal-combustion-engines/modeling-and-simulation-of-si-engines-for-fault-detection

IOP Conf. Series: Materials Science and Engineering **635** (2019) 012016 doi:10.1088/1757-899X/635/1/012016

[11] Pavlenko V B and Fomin A A 2014 Information Technology for Model Diagnostics of Nonlinear Dynamic Objects *East European Journal of Advanced Technology* **4/11(70)** 38–43

[12] Shcheglov V A 2015 On the possibility of diagnosing the cylinder-piston group of ICEs by the frequency-amplitude analysis of a vibration signal using wavelet analysis *Bulletin of Science and Education of the North-West of Russia* **1(3)** 1–8

[13] Lyapaev V G and Manfanovsky S B 2013 On developing the matrix of the technical condition of cars according to statistical data *Bulletin of the Pskov State University. Series: Economy. The law. Management* **3** 85–87

[14] Lukasov V V and Nikushkin N V 2007 Application of the trouble shooting method based on the theorem of hypotheses (Bayes method) in the diagnostics of aviation equipment *Bulletin of the Siberian State Aerospace University named after Academician M.F. Reshetnev* **4** 71–73.

[15] Khaliullin, F Kh and Akhmetzyanov I R 2014 Justification of the choice of diagnostic parameters of power plants of mobile machines *Bulletin of Kazan State Agrarian University* **2** 72.

[16] Khaliullin F Kh, Akhmetzyanov I R, Shiriyazdanov R R and Khaliullin A F 2014 Bench for studying operation processes of internal combustion engine in dynamic modes with the ability to simulate some engine faults *Utility model patent of Russia No. 151482.2014*. Rep. No. 10

[17] Khaliullin F Kh, Akhmetzyanov I R, Shiriyazdanov R R and Khaliullin A F 2014 Device for diagnosing internal combustion engine according to transient characteristics *Utility model patent of Russia No. 160474.2016*. Rep. No. 8

IOP Conf. Series: Materials Science and Engineering **635** (2019) 012017 doi:10.1088/1757-899X/635/1/012017

Determination of statistical data of conditional probabilities of the technical condition of internal combustion engines when compiling the Bayes diagnostic table

Farit Kh Khaliullin[1], Joseph K Aladashvili[1], Azat A Nurmiev[1], Gennady V Pikmullin[2] and Stanislav A Sinitsky[1]

[1]Department of Tractors, Automobiles and Power Plants of the Mechanization and Technical Service Institute, Kazan State Agrarian University, 65 K. Marksa St., 420015, Kazan, Russia
[2]Department of General Engineering Disciplines of the Mechanization and Technical Service Institute, Kazan State Agrarian University, 65 K. Marksa St., 420015, Kazan, Russia

khaliullin_ kgau_taeu@mail.ru

Abstract. One of attractive methods for in-place diagnostics of internal combustion engines is the method based on using the generalized Bayes formula of conditional probabilities. The article is devoted to the applicability improvement of the Bayes algorithm for assessing the technical condition of internal combustion engines. The use of this algorithm involves the compilation of a diagnostic table, the elements of which are conditional probabilities of selected diagnoses in case of a range of faults. Their relevant determination requires a large amount of statistical data, which is often lacking. To supplement deficient data and increase the reliability of obtained results, the authors propose to use a method for simulating patterns of changes in design parameters of the technical condition of internal combustion engines depending on operating time.

1. Introduction

Based on the accepted algorithm of the Bayes method, one can compile a diagnostic table, the elements of which are conditional probabilities of accepted diagnoses in case of a range of faults [1, 2]. These probabilities are generated on the basis of statistical data on maintenance and repair of internal combustion engines. Often, however, information about the technical condition of the engine, depending on its design parameters and operating time, is not enough or is contradictory. This is also caused by the currently common aggregate method for extending the operational life of an internal combustion engine. It results in a problem of a lack of relevant statistical information, which requires additional research. One of the solutions to the problem of determining the missing information for the Bayes diagnostic matrix is a numerical simulation of the influence of the operating time of an internal combustion engine on the intensity of changes in parameters of its technical conditions [3].

2. Methods

When determining the diagnosis of the technical condition of an internal combustion engine using the Bayes method, it is necessary to compile a diagnostic table (matrix), the elements of which are generated on the basis of previously collected statistical data. The diagnostic table contains conditional probabilities of digit orders of a set of signs for various technical conditions of the engine (diagnoses). Wherein, dimensions of the table are determined by the number of selected probable manifestations of signs of failures and fault conditions [4].

The diagnostic table includes only a priori probabilities of selected diagnoses. At the same time, it is important to provide for the possibility of refining and changing the table in the process of diagnostics, for which not only the probability values but also the following values are fixed:

- N – the total number of diagnostic objects used to compile the diagnostic matrix;
- N_t – the number of objects with the diagnosis D;
- N_{ij} – the number of objects with the diagnosis D examined in regards to k_j.

To obtain diagnostic results with a high probability of reliability, it is necessary to continuously record the technical condition of the engine during each case of routine maintenance with the registration of the condition of its systems and mechanisms, when compiling this matrix. According to the results of practical diagnostic studies, satisfactory convergence of results appears when the sample size is greater than 200. One of the issues that cause difficulties when compiling the Bayes diagnostic table is the fact that, according to the regulatory and technical documentation, not all the necessary information on the technical condition of engine components is recorded during these works.

Let us consider an example of compiling a diagnostic table according to the Bayes algorithm for the D-243 engine [5].

Experience in operating internal combustion engines and the content of papers in accordance with the regulatory and technical documents on their maintenance show that the current condition of the engine is primarily determined by the condition of the air supply system (ASS), fuel supply system (FSS), and cylinder-piston group (CPG) [6].

As a diagnosis, four main engine conditions are chosen:

- faults in the air supply system (ASS);
- faults in the fuel supply system (FSS);
- faults in the cylinder-piston group (CPG);
- engine in good condition.

3. Results

The correct choice of diagnostic parameters (signs) is quite a crucial point, as it affects not only the reliability of the diagnosis and the number of experiments to get it but also reduces the probability of errors of the first and second kind. Considering the internal combustion engine as a dynamic object with stationary characteristics, one can use identification methods for such objects. They reflect the dynamic characteristics of the system and depend on its technical condition. As a diagnostic sign, let's one can choose the intensity of change in engine performance after a single change in the regulator adjustment lever position – transient functions:

- rotation speed change intensity T_n $(k1i)$;
- fuel supply change intensity T_{gc} $(k2i)$;
- air flow change intensity T_{Ga} $(k3i)$.

Each sign has three condition levels: good, satisfactory and bad, corresponding to system faults:

- good change intensity k_{j1};
- satisfactory change intensity k_{j2};
- bad change intensity k_{j3}.

Determination of the range of change of indicators from good to bad values requires information on the correspondence of these data for engines with zero and maximum operating time [7]. These characteristics are determined by regulatory and technical documents or they can be obtained during bench testing of engines with different technical conditions, and serve to determine boundary conditions when conducting the diagnostics.

Baseline data for the engine being diagnosed are also:

- overhaul life;
- maintenance rate;
- current operating time.

A Bayes diagnostic table on engine failures and faults at selected diagnostic parameters is given in Table 1 below [8].

Table 1. Bayes diagnostic table.

Engine condition D_i	Signs k_j									$P(D_i)$
	Rotation speed k_1			Fuel consumption k_2			Air flow k_3			
	k_1			k_2			k_3			
	k_{11}	k_{12}	k_{13}	k_{21}	k_{22}	k_{23}	k_{31}	k_{32}	k_{33}	
Faults of the ASS D_1	$P(k_{1i}/D_1)$			$P(k_{2i}/D_1)$			$P(k_{3i}/D_1)$			$P(D_1)$
Faults of the FSS D_2	$P(k_{1i}/D_2)$			$P(k_{2i}/D_2)$			$P(k_{3i}/D_2)$			$P(D_2)$
Faults of the CPG D_3	$P(k_{1i}/D_3)$			$P(k_{2i}/D_3)$			$P(k_{3i}/D_3)$			$P(D_3)$
Good condition D_4	$P(k_{1i}/D_4)$			$P(k_{2i}/D_4)$			$P(k_{3i}/D_4)$			$P(D_4)$

where $P(D_i)$ – a priori probability of the hypotheses D_i;
$P(k_j/D_i)$ – probability of the hypotheses k_j upon the occurrence of the event D_i (a posteriori probability). These values are related by the Bayes formula.
It is necessary to determine the probability of the engine condition diagnosis D_i, if there is one of the possible joint manifestations of signs k_{1i}, k_{2i} and k_{3i}.
According to the requirements of regulatory and technical documents, the service life of an internal combustion engine is determined by the service life of its basic parts, which are the cylinder block and the crankshaft. It is usually considered that the reason for the engine overhaul is the extremely worn condition of the cylinder block or crankshaft.
According to studies [9], the operational performance of the engine in normal operating conditions is reduced according to the law of smooth continuous function with different intensities (Figure 1).

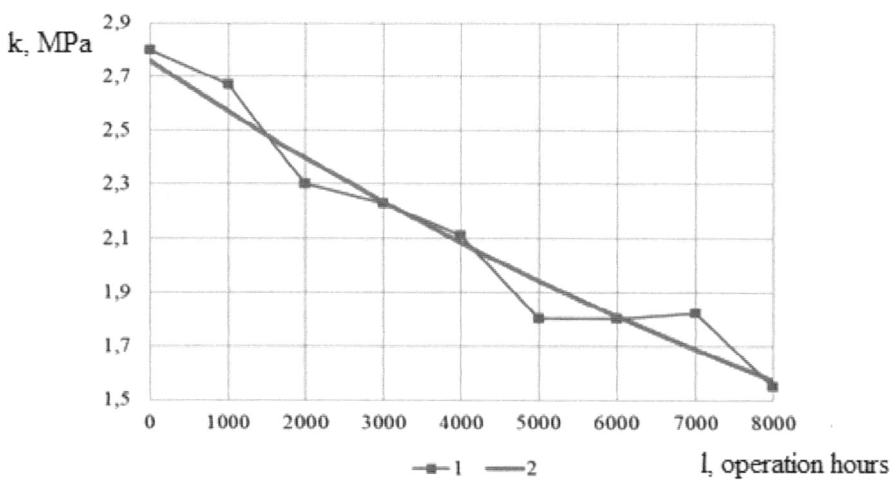

Figure 1. Dependence of compression *k* on the D-243 engine operating time *l*

Note: 1 – experimental data, 2 – approximated data.

The parameters of this function are determined by operating modes of the engine, the quality of maintenance works carried out and the consumables used. This pattern of change in the effective indicators of engine operation is the result of a violation of adjustments, pollution of fuel and air supply elements, as well as the inevitable occurrence of changes in design parameters of the technical condition (Figure 2) [10].

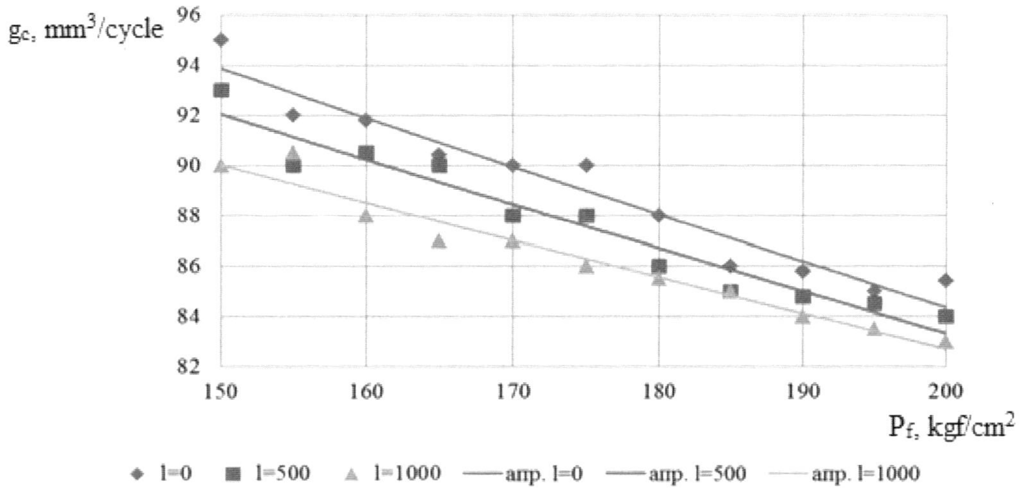

Figure 2. Dependence of cyclic fuel supply g_c, mm^3/cycle, on fuel injection pressure P_f, kgf/cm^2, at different operating time l, operation hours

Taking into account the probabilistic nature of patterns of change in design parameters of the technical condition and the normal distribution law of random variables, one can assume that the probability of good condition of the CPG when the compression changes from K_{min} to K_{max} changes exponentially [11, 12], as shown in Figure 3.

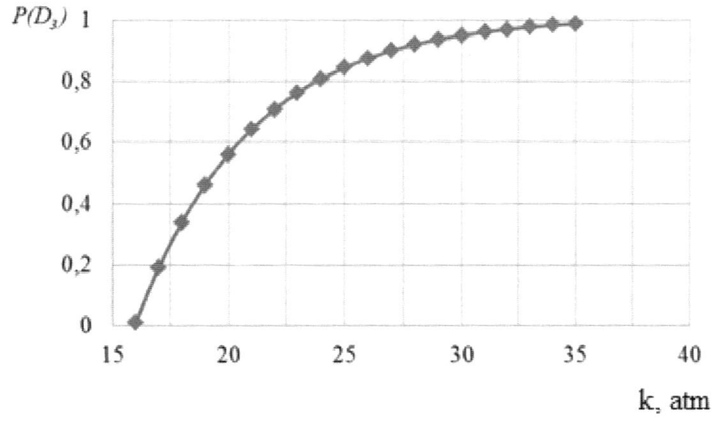

Figure 3. Dependence of cyclic fuel supply g_c, mm^3/cycle, on fuel injection pressure P_f, kgf/cm^2, at different operating time l, operation hours

Then the law of change in the probability of good condition of the engine $P(D_4)$ has the form shown in Figure 4.

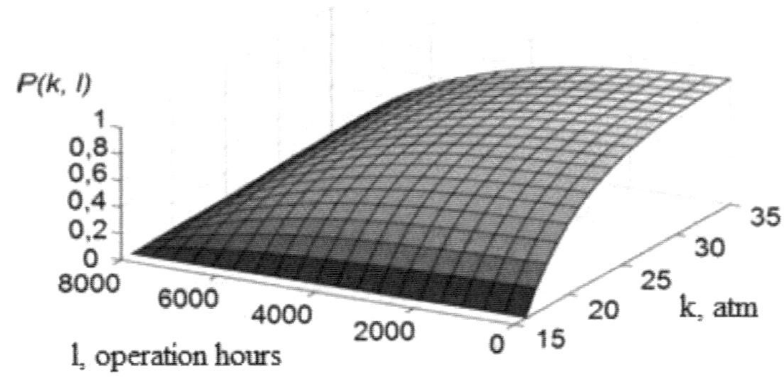

Figure 4. The probability of the CPG good condition, depending on the compression and operating time

The proposed methods for modeling probabilities allows simplifying the procedure of using the Bayes algorithm for in-place diagnostics of internal combustion engines and can be applied in the practical implementation of this algorithm.

The significance of the regression model is verified using the Fisher F-criterion, the calculated value of which is determined as the ratio of the variance of initial series of observation of the studied indicator and the unbiased estimate of the variance of residual sequence for this model (Table 2).

Table 2. Fisher criteria for regression equations.

The parameter being determined	Parameter level	Regression equation	Regression coefficient		F	F_{table}
			a	*b*		
n from t at	P_a=2 kPa	$n = a\,e^{bPa}$	7.0776	0.2361	17.81	5.32
	P_a=4 kPa		7.0231	0.2519	27.83	5.32
	P_a=7 kPa		6.9599	0.2764	50.95	5.32
n from t at	P_f=18.5 MPa	$n = a\,e^{bPf}$	7.079	0.2346	16.14	5.32
	P_f=16.5 MPa		7.0132	0.2589	27.61	5.32
	P_f=15.5 MPa		6.962	0.273	47.9	5.32
n from t at	k=2.7 MPa	$n = a\,e^{bk}$	7.0273	0.2523	25.28	5.32
	k=2.4 MPa		6.9536	0.2798	37.2	5.32
	k=2.1 MPa		6.9129	0.2888	59.72	5.32

According to the Fisher distribution tables, the table value of the Fisher criterion for a given level of significance is determined taking into account that the number of degrees of freedom for the total sum of squares (large variance) is 1, and the number of degrees of freedom for the residual sum of squares (smaller dispersion) with linear regression is n-2. The table value of the Fisher criterion is the maximum possible value of the criterion for these degrees of freedom and the level of significance α under the influence of random factors. The significance level α is the probability of rejecting the correct hypothesis, if it is true. Accept α equal to 0.05.

Since, in this case, the actual value of $F > F_{table}$, the determination coefficient is statistically significant, which means that the estimated value of the regression equation is statistically reliable.

IOP Conf. Series: Materials Science and Engineering **635** (2019) 012017 doi:10.1088/1757-899X/635/1/012017

4. Conclusion

The analysis shows that the proposed method for determining the statistical data of conditional probabilities of the technical condition provides relevant results.

5. References

[1] Krasheninnikov S V 2013 Modern approaches to the diagnosis of diesel internal combustion engines *Bulletin of the Novosibirsk State Pedagogical University* **12** 59–68

[2] Pestrikov V M and Eukarpiev V E 2014 Features of modern vehicle diagnostics *Technical and Technological Service Problems* **4(30)** 15–16

[3] Khaliullin F Kh, Khaliullin A F, Akhmetzyanov I R and Gilmutdinov I I 2017 Features of using the Bayes algorithm for in-place diagnostics of internal combustion engines *Modern High Tech Technologies* **8** 75–80

[4] Medvedev V M, Ahmetzyanov I R, Shiriyazdanov R R and Khaliullin F K 2014 Operating conditions of the D-240 Engine based dual fuel gas-diesel engine of the agricultural tractjr *European Conference on Innovations in Technical and Natural Sciences Proceedings of the 2nd International Scientific Conference (May 12, 2014)* (Vienna: East West Association for Advanced Studies and Higher Education GmbH)

[5] Gessenhardt C 2009 Hybrid endoscopes for laser-based imaging diagnostics in IC engines *SAE Technical Papers* **1** 11

[6] Guzzomi F G, O'Neill P L and Tavner A C R 2007 Investigation of damper valve dynamics using parametric numerical methods *16th Australasian Fluid Mechanics Conference Crown Plaza*, Gold Coast, Australia 1123–30

[7] Machalik S 2013 *Využití Obrazové Analýzy v Tribotechnické Diagnostice (Image Analysis in Tribodiagnostics)* (Pardubice: University of Pardubice) p 112

[8] Khaliullin F Kh and Akhmetzyanov I R 2016 Features of compiling the Bayes diagnostic matrix for in-place diagnostics of internal combustion engines *International Science and Research Journal* **5** 205

[9] Molodan A A 2011 Improving the accuracy of assessing the technical condition of the cylinder-piston group by separating crankcase gas flows *Papers of Tavrichesky State Agrotechnological University* **11** 292–303

[10] Suranov G I 2011 *Improving the Durability of Transport Engines* (Ukhta: USTU) p. 335

[11] Snyder J A 2011 *Development and Application of Tracer-based Planar Laser-induced Fluorescence Imaging Diagnostics for HCCI Engines* (Stanford University)

[12] Tausenev E M and Svistula A E 2012 Reliability, major faults and causes of failures of high-pressure pumps of diesel accumulator fuel systems *Science and Education* **9** 81–96

Research on computation and simulation technology of battlefield complex electromagnetic environment

Fan Yang, Zhenghong Dong, Xiaobo Duan

Space Engineering University, Beijing, China, 101416

Email: sarah0824@hotmail.com

Abstract. In view of the current fierce competition for electromagnetic power among countries, it is necessary to accelerate the simulation of frequency equipment in complex electromagnetic environments, and to construct complex electromagnetic environment simulation scenarios under joint combat conditions. How to construct the complex electromagnetic environment that meets up with the requirement of battlefields, and to improve combat command and training capabilities are the current research problems. Thus, this paper studies the simulation and calculation of the complex electromagnetic environment in the battlefield, and proposes and designs a system for the calculation and analysis of the complex electromagnetic environment of battlefield. The system can be used to estimate radio wave propagation of each frequency equipment, to analyse and compute complex electromagnetic situation of the battlefield, and visualize them. The above estimation, analysis, and computation are obtained according to the displacement of frequency equipment, and battlefield environment, such as geography information and meteorological information. The proposed system can obtain the impact of electromagnetic environment of the battlefield on various types of frequency equipment, and provide battlefield electromagnetic information protection for equipment application.

1. Introduction

In previous wars, the US military used various means of electronic interference, and radar equipment, etc. for investigations and counter-reconnaissance, interference and anti-interference, and anti-radiation destruction [1,2], to quickly won absolute victory of the war. Currently, U.S. military concerns a lot about the performance evaluation of frequency equipment in complex electromagnetic environments [2]. The main methods include the construction of complex electromagnetic environments, recording, analysis and monitoring of test parameters, attention to the leakage of electromagnetic parameters and participation in actual equipment, and the use of real electronic devices and a large number of simulators to generate complex electromagnetic environments, and using interactive simulation techniques to construct complex electromagnetic environments [3]. For the future war, information is the key factor that restricts the victory of the war. The frequency equipment of future battlefield is highly intensive, and the electromagnetic environment is extremely complex. Constructing a complex electromagnetic simulation environment that meets the requirements of battlefield joint operations and equipment performance standards, and making troops and informational equipment to training and practicing during such a simulation environment can improve combat capabilities, which also a practical action of actual battle-warning. This paper studies the calculation and simulation of the complex electromagnetic environment in the battlefield, and proposes and designs a system for the calculation and analysis of the complex electromagnetic environment on the battlefield. The system can be used to estimate radio wave propagation of each frequency equipment, to analyse and compute complex electromagnetic situation of the battlefield, and visualize them. The above estimation, analysis, and computation are according to the displacement of frequency equipment, and according to battlefield environment, such as geography information and meteorological information. The proposed system can obtain the impact of electromagnetic environment of the battlefield on various types of frequency equipment, and provide battlefield electromagnetic information protection for equipment application.

2. Related Work

2.1. Electromagnetic Situation Computation

Currently, for the calculation of electromagnetic situation [4], some studies are limited to the calculation of the electromagnetic situation of a single frequency equipment [3]. The calculation of the electromagnetic situation of the entire battlefield can be very complicated. The early calculation method is slow. When a large number of stations involved in the calculation, the requirement on calculation speed is very important. As shown in Fig. 1, when calculating the electromagnetic situation in a certain area, it is necessary to calculate the electromagnetic field intensity generated by multiple stations at each point in 3D space [5]. As the 3D space is relatively large, the amount of data after grid division is very large. Every relevant station participating in the calculation must undergo such a large number of calculations, which will greatly reduce the operating speed of the system and seriously affect the availability of the system. It is necessary to study new methods for calculating the electromagnetic situation of space, which is accurate and having faster computational efficiency.

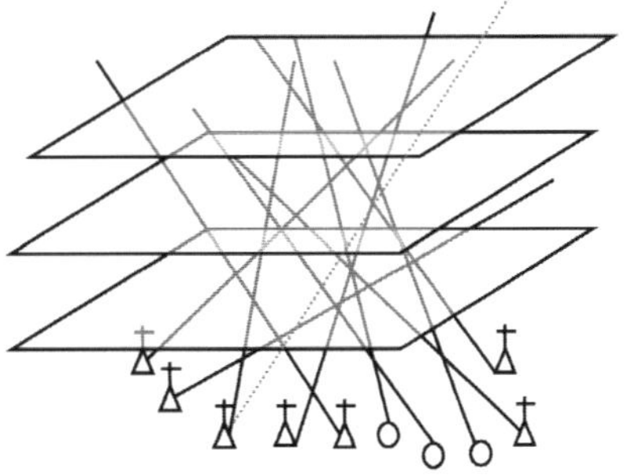

Fig.1 Three-Dimensional Electromagnetic Computation Complexity Schematic

According to the above model, there is still a problem that the calculation time is too long. Especially in practical applications, hundreds of stations are often analysed at the same time, and the calculation speed is difficult to meet the requirements. If we only calculate electromagnetic environment data within 30 km of four provinces (assuming the area is 500 × 500 km^2 for each province), this algorithm will calculate the synthetic field strength of nearly 30 million points on the ground and in space, and average calculation time at each point is about 6 seconds (assuming there are 2,000 stations in the area), even if just doing a rough calculation (assuming a point of 1,000 meters), the time required for preliminary estimation is: ((500+500) × (500+500) × 30 × 6S)/(3600S/H×24H/D×30D/M×12M/Y) ≈5.8Y, where S is short for second, H is short for hour, D is short for day, M is short for month, and Y is short for year. It takes almost 6 years, and the entire calculation time is simply unbearable. Therefore, this algorithm is only suitable for the analysis and calculation of small-area and small number of stations. It is necessary to study new algorithms and methods, to improve the speed of analysis and calculation.

2.2. Wave propagation prediction

The propagation modes [6] involved in the whole frequency band mainly include waveguide mode propagation, ground wave propagation, sky wave propagation (ionospheric reflection propagation), free space propagation, spatial wave propagation, barrier diffraction propagation, smooth ground diffraction propagation, and Tropospheric scatter propagation, etc. The establishment of corresponding forecasting methods for the above propagation modes is a key point to carry out simulation research.

The wave propagation analysis [7,8] is the core technology of the battlefield electromagnetic situation analysis and display. The basic content is to pre-analyse and calculate the electromagnetic field intensity distribution in a 2D and 3D space of a combat area based on station equipment data and geographic information data, to form electromagnetic situation. Electromagnetic situation information can be further used to display and frequency equipment impact analysis. The basic idea of the radio wave propagation analysis and calculation is to calculate space electromagnetic field coverage generated by stations according to station's location, power, antenna directionality, frequency, etc., and combine the geographical information on the propagation path, with the radio wave propagation model that is suitable for the local area and suitable for the corresponding frequency band. By analysing and calculating all stations in this region, the electromagnetic field distribution data at each spatial point and at each frequency can be obtained. These data can be used to analyse and display the electromagnetic situation of the battlefield.

3. System design and key techniques

3.1. System design

The proposed system has constructed a set of equipment test simulation software platform integrating experimental planning, whole-process electromagnetic numerical simulation analysis, 2D and 3D visualized deduction and comprehensive evaluation. It can carry out equipment test training mission planning, confrontation quantitative analysis of equipment testing and training, equipment test training confrontation deduction, equipment testing training comprehensive assessment, regional background situation analysis, point spectrum analysis, frequency equipment range of action analysis, equipment model management, shown as Fig.2.

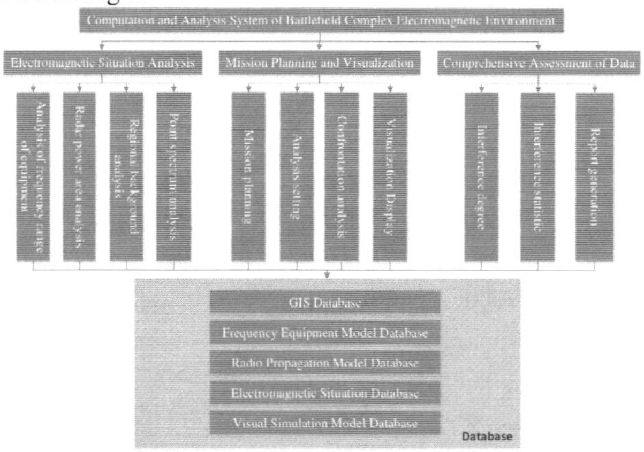

Fig.2 Battlefield Complex Electromagnetic Environment Calculation and Analysis System Diagram

(1) *Electromagnetic situation analysis:* Electromagnetic environment analysis mainly includes analysis of frequency range of equipment, radar power area analysis, regional background analysis, point spectrum analysis.

(2) *Mission planning and visualization:* The system equipment models are divided into mobile platform equipment and frequency equipment. User deploys different equipment models in the desired combat space, sets the operational parameters of the equipment and meteorological regions, and constructs a specific battlefield electromagnetic environment. The functions of mission planning include the deployment of frequency equipment, the deployment of mobile platform platforms, the scheduling of platform operations, etc.

(3) *Comprehensive* assessment of data: Evaluate the effective interference counted by the both confrontation parties, and compare them according to the degree of interference (mild, moderate, and severe).

And the database includes various models and data supporting the system's operation, including Geography information database, Frequency Equipment Model database, Radio propagation model database, Electromagnetic Situation Database, and Visual Simulation Model database.

3.2. Contributions
The main contributions of the system are to address the problems of lacking of quantitative calculations and information supporting methods in experimental training. The proposed system highlights the application of simulation calculations, and constructs a set of equipment simulation systems for electromagnetic environmental information. The system integrates the whole process of numerical simulation calculation, visualized deduction, and comprehensive evaluation. It pays attention to calculation of battlefield environment model, interactive revision of role parameters, and results evaluation of human-in-the-loop, and intensifies the training of command skills based on information technology. The following 2 contributions are proposed:

3.2.1. Electromagnetic situation calculation algorithm
The computation of electromagnetic situation space data requires comprehensive consideration of equipment technical data, antennas, geographic information data, and a variety of possible propagation mechanisms. The process is extremely complex, and the amount of calculation is large, and with the number of frequency equipment increases, the amount of calculation rapidly increases. Thus, this paper studies and solves the rapid matching calculation technology that matches with antenna models and terrain features, the electromagnetic situation raster technology, dynamic refresh, rapid request and extraction, and situational synthesis technologies. This paper also implemented rapid analysis and calculation of the electromagnetic situation, which provides strong support for the visualization of the electromagnetic situation. This paper has studied the calculation method of electromagnetic situation from different aspects and achieved good results, shown as following:

(1) *Electromagnetic Situation Grading.* In order to increase the speed of calculation, this paper has adopted a functional hierarchy. The functions of different levels correspond to the needs of different levels. Correspondingly, calculations with different degrees of accuracy are used to increase the speed of calculation. The specific measures are as follows:

(a) *Functional classification.* The electromagnetic situation is divided into overall situation, regional situation, and single-station situation according to the ranges of calculation. The overall situation focuses on area larger than 500 km^2. Other situational analysis focuses on area less than 500 km^2. The overall electromagnetic state function can help user understand the basic electromagnetic situation of the entire combat area.

(b) *Optimization of overall electromagnetic situation.* The purpose of the overall electromagnetic situation is to provide users with a macroscopic electromagnetic situation. Its characteristic is that the area of calculation and the number of stations are large, and the requirements on local details are not very high. The calculation method can be greatly simplified to improve the calculation speed.

(c) *Optimization of regional electromagnetic situation.* What we concerned most is the distribution of local electromagnetic environment within the region, so we need a more detailed analysis and calculation. This paper has adopted the following methods to improve the accuracy and speed of the system.

- Select a relative small area for analysis and calculation to reduce the amount of computation.
- Meshing. According to the experiments on the algorithm, combined with the research results and experience of relevant units, this paper sets the grid of the regional situation to 100-1000 meters.
- For data management, situation data is saved as volume data for those stations that are participated in the regional situation analysis. If the volume data already exists in the next calculation and the station parameters do not change, then the existing data is used directly. In this way, as the number of calculations increases, the system will be faster and faster.

IOP Conf. Series: Materials Science and Engineering **635** (2019) 012018 doi:10.1088/1757-899X/635/1/012018

(d) *Optimization of the electromagnetic status of a single station.* The analysis of a single electromagnetic situation adopts the most elaborate analysis method. Because the analysis of a single station, even if the highest accuracy is selected, the calculation time is acceptable. The optimization of a single station calculation is mainly from the aspect of data management, storing the situation volume data of a single station, and combined with the operation of the database. As long as the parameters of the station remain unchanged, its volume data is saved without repeating calculations. In this way, many key stations can be calculated and stored in advance.

(2) *Algorithm of automatic fast matching between radio service and antenna pattern.* Antenna patterns used by different communication services have different characteristics. The antenna of the earth station generally has a certain elevation angle and the directional pattern is relatively sharp. The directional pattern of the alert radar antenna is generally relatively sharp in order to accurately determine the position of the target. And the fixed service, especially the fixed service of the microwave frequency band, is required to obtain a better result. High gain antenna patterns are usually sharp; broadcast services use non-directional or wide-directional antennas in order to achieve a large coverage area. Therefore, the antenna pattern should match its service. In actual calculation process, when the data of the actual antenna pattern is missing, the error is large. The system must establish a spatial three-dimensional antenna pattern based on the minimum information such as the antenna gain, so as to establish a 3D spatial distribution of field strength and display 3D electromagnetic situation. The establishment of matching antenna pattern models for various communication services is also a challenge for the system. The electromagnetic situation analysis and display system includes almost all radio communication services. Thus, we have established a technique for matching antenna patterns with radio services for the application of this system. This technique has two advantages:

(a) *Solved the problem of missing antenna pattern data in the equipment database.* In the database design of this system, a special antenna data table is designed, which includes the description of the antenna directional data. However, in the data collection process, this data is difficult to collect, resulting in a large number of device antennas have no direction data. The accuracy of the calculation cannot be guaranteed. Using the proposed method, it supports for the calculation of equipment without this data, and the accuracy of the calculation is guaranteed.

(b) *Increased calculation speed.* When we choose to automatically match the antenna pattern, the directional pattern is simplified when the accuracy of the system is satisfied. This simplification improves the speed of calculation and improves the system's operating efficiency. For example, for microwave fixed communication service, in the frequency range of 1 GHz to 70 GHz, the antenna directional pattern model of a fixed wireless system can usually be expressed as Eq. 1.

$$G(\phi) = \begin{cases} G_{max} - 2.5 \times 10^{-3}\left(\frac{D}{\lambda}\phi\right)^2, & 0^\circ \le \phi < \phi_m \\ G_1, \phi_m \le \phi \le \max(\phi_m, \phi_r) \\ 29 - 25\, log\, \phi, & \max(\phi_m, \phi_r) \le \phi < 48^\circ \\ -13, & 48^\circ \le \phi \le 180^\circ \end{cases} \tag{1}$$

$$G(\phi) = \begin{cases} G_{max} - 2.5 \times 10^{-3}\left(\frac{D}{\lambda}\phi\right)^2, & 0^\circ \le \phi < \phi_m \\ 39 - 5\, log\frac{D}{\lambda} - 25\, log\, \phi, & \phi_m \le \phi < 48^\circ \\ -3 - 5\, log\frac{D}{\lambda}, & 48^\circ \le \phi \le 180^\circ \end{cases} \tag{2}$$

When $D/\lambda > 100$, the antenna pattern of fixed wireless systems can be expressed as Eq. 2.where, D——Antenna diameter; λ——wavelength;φ——Deviation from the antenna's main axis, degrees; θ——Antenna beam width, degree; G_{max}—— Gain in the direction of the antenna's spindle, dBi; $G_1 = 2 + 15\, log\frac{D}{\lambda}$; First side lobe gain; $\varphi_m = \frac{20}{D/\lambda}\sqrt{G_{max} - G_1}$,degree; $\varphi_r = 15.85(D/\lambda)^{-0.6}$, degree.

If we build the antenna pattern model according to the above ideas, it will encounter many difficulties. Of course, the operation speed will be relatively slow, and it may be too slow to be tolerated when a large number of spatial data calculations are required. Thus, we conducted extensive research to match different antenna patterns for different communication services. These models can automatically adjust the shape of the lobes based on the actual antenna gain and the half-power angle width. For example, we simplified the directional pattern of a C-band radar as Eq. 3:

$$G(\theta,\phi) = G\,exp\left(-\frac{b\phi^2}{(\Delta\phi)^2}\right) \tag{3}$$

where the antenna gain parameter G , the half-power angle width parameter $\Delta\varphi$ and the form factor b can be adjusted on the input interface. The model of antenna directionality maps for fixed radio communications was modified as Eq. 4-6:

$$G(\phi) = \begin{cases} G - 0.000425 \times 10^{0.1G}\vartheta^2, & 0^\circ \le \vartheta < \vartheta_m \\ 40.925 - 0.25G - 25\,log\,\vartheta, & \vartheta_m \le \vartheta < 48^\circ \\ -1.075 - 0.25G, & 48^\circ \le \vartheta \le 180^\circ \end{cases} \tag{4}$$

$$\vartheta_m = 48.532\sqrt{G - G_1} \times 10^{-0.05G} \tag{5}$$

$$G_1 = 0.75G - 3.775 \tag{6}$$

These modifications and simplifications must comply with actual situation of relevant communication service antenna, and also take into account the requirements for quick calculation and high computational speed. To configure a compliant antenna pattern model for all involved radio communication services, the difficulty of its implementation is quite huge.

In the realization of the algorithm of radio wave propagation, given the characteristic that different frequency carries out propagation calculation in different applications, this paper optimizes the level of algorithm design. On the premise of satisfying the accuracy of the system calculation, this paper selects as simple and efficient algorithm as possible. For example, for the ultra-short-wave band, this paper chooses more accurate propagation for small area-wide propagation calculations. A quick table look-up algorithm was used to implement the situation calculation for a large area, and the system determines the most suitable algorithm. This way greatly improves the calculation speed of the electromagnetic situation.

(3) *Calculation according to the coordinates of the principal axis of the antenna and implementation of the conversion of 3 coordinates.* The common field strength is to calculate field strength or reception level at certain points on the ground or space satellite orbit. We have solved the problem of spatial field strength mutual conversion among 3D geocenter coordinates (Fig. 4), 3D ground coordinates (Fig. 3), and 3D antenna principal axis coordinates (Fig. 5). The system can give a 3D spatial distribution of the field strength in any coordinate system.

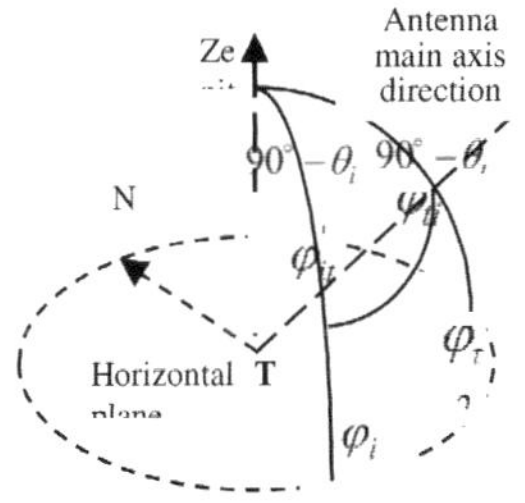

Fig.3 3D ground coordinates

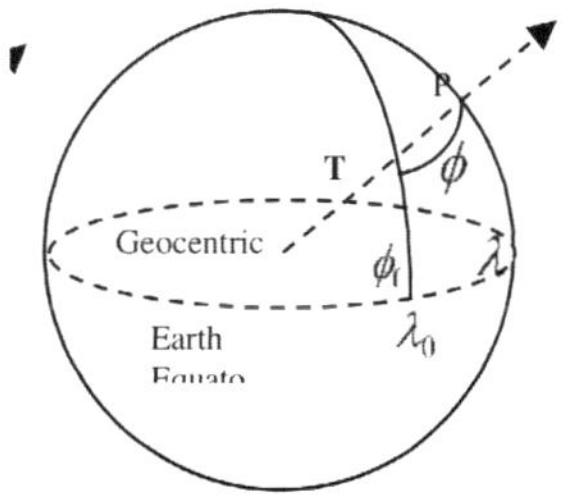

Fig.4 3D geocentric coordinates

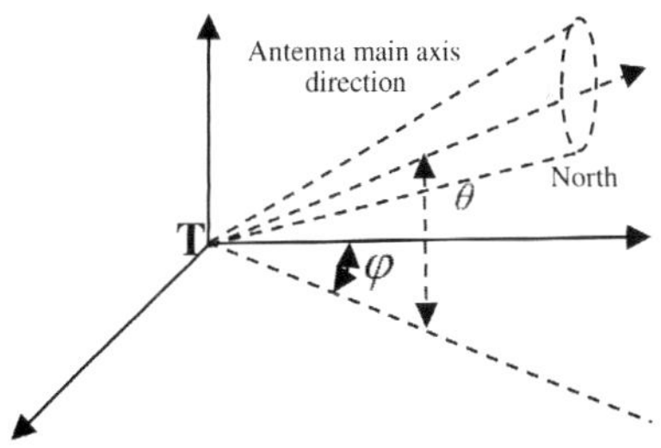

Fig.5 3D antenna polar coordinates

In the aspect of system development, this method improves the speed of calculation and simplifies programming. In the aspect of practical application, this method is relatively simple and universally achieves compatibility with different calculation requirements, making it possible for the system to analyse and calculate for a variety of battlefield frequency equipment. The propose method breaks through the limitations of current similar software focusing on the communications, and more adapt to the requirements of operational command applications. Both aeronautical navigation and aeronautical mobile communications require knowledge of the distribution of field strengths in specific spatial regions in order to understand, design and control the effective area of control of the aircraft. As long as the real parameters of the ground-based navigation station and the ground-based aeronautical communication station are given, using the method established by this system, it is easy to plan and design the effective control area of the aircraft. For example, ground equipment and the trajectories of air targets can be represented by geocentric coordinates as Eq. 7:

$$\lambda_{a1}, \varphi_{a1}, h_{a1}; \quad \lambda_{b1}, \varphi_{b1}, h_{b1}; \lambda_{a2}, \varphi_{a2}, h_{a2}; \quad \lambda_{b2}, \varphi_{b2}, h_{b2}; \cdots\cdots\cdots \quad \cdots\cdots\cdots; \lambda_{ai}, \varphi_{ai}, h_{ai}; \quad \lambda_{bi}, \varphi_{bi}, h_{bi}; \cdots\cdots\cdots \quad \cdots\cdots\cdots; \lambda_{an}, \varphi_{an}, h_{an}; \quad \lambda_{bn}, \varphi_{bn}, h_{bn}; \tag{7}$$

Or as Eq. 8-9,

$$i = 1,2,\cdots,n; \lambda_{ai} = \lambda_a + i \times \Delta\lambda_a; \varphi_{ai} = \varphi_a - i \times \Delta\varphi_a; \ h_{ai} = h_a + i \times \Delta h_a; \tag{8}$$

$$\lambda_{bi} = \lambda_a + i \times \Delta\lambda_b; \varphi_{bi} = \varphi_a - i \times \Delta\varphi_b; h_{bi} = h_a - i \times \Delta h_b \tag{9}$$

where $\lambda_a = 118.330°$, $\varphi_a = 28.220°$, $h_a = 9000$, $\Delta\lambda_a = 0.025$, $\Delta\varphi_a = 0.01$, $\Delta h_a = 0$, $\Delta\lambda_b = 0.050$, $\Delta\varphi_b = 0.020$, $\Delta h_b = 180$, $n = h_a/\Delta h_b$. The coordinate parameters of the lower footprint zone *a* are associated with the ground equipment, and the coordinate parameters of the lower footprint zone *b* are associated with the air targets. However, the field strength at the observation point should be a function of the polar angle, azimuth and distance of the polar coordinates of the antenna, as $E = E(\theta_i, \phi_i, D_i)$, where the azimuth of the spatial grid $\phi_{ijk} = \begin{cases} \phi'_{ijk}, & \lambda \le \lambda_{ijk} \\ 360° - \phi'_{ijk}, & \lambda > \lambda_{ijk} \end{cases}$, where $\phi'_{ijk} = arccos\left(\frac{sin\varphi_{ijk} - cos\alpha_{ijk} \sin\varphi}{sin\alpha_{ijk} \cos\varphi}\right)$, $\alpha_{ijk} = arccos\left[\sin\varphi \, sin\varphi_{ijk} + \cos\varphi \, cos\varphi_{ijk} \cos(\lambda - \lambda_{ijk})\right]$.
The distance of the spatial grid can be expressed as Eq.10.

$$D_{ijk} = \sqrt{(R+h)^2 + (R+h_{ijk})^2 - 2(R+h)(R+h_{ijk})cos\alpha_{ijk}} \tag{10}$$

It can be seen that this calculation method meets the requirements for electromagnetic situation analysis of a variety of different radio services, and can also be used for analysis of airborne mobile frequency equipment.

3.2.2. Full band radio wave propagation prediction method

In the field of radio wave propagation analysis and calculation algorithms, this paper has integrated a large number of theoretical models and statistical models of ITU. This paper has considered the geographical environment, the actual deployment of station sites, and the performance of equipment, and realized the scientific calculation of space wave propagation. The algorithm research is mainly based on the actual needs of battles to select the appropriate radio wave propagation model, and according to the actual needs of the system to modify and improve the propagation model. The radio wave propagation model is a description of the propagation characteristics of radio waves in specific environments and frequency bands. The radio wave propagation model is divided into two types, one is a theoretical model, derived from theoretical analysis, and can be expressed as a formula; the other is a statistical model, which is statistically obtained from a large number of actual test data, and is generally expressed as an approximate formula, a chart, etc. Since the propagation phenomenon in the near-Earth range is very complex, there are a large number of propagation models that involve different propagation modes, different propagation environments, and different frequency bands. Electromagnetic situation analysis shows that the system, as a system oriented to operational command applications, must find the location in terms of technology and practicality. It is necessary to ensure the accuracy of calculations and remove unnecessary details. It is necessary to demonstrate the role of operational decision-making aids. Moreover, according to the characteristics of local areas, this paper localized the radio wave propagation model. For this reason, this paper conducted tests and tests, obtained a large number of conclusions, and applied some research results of relevant units, and finally determined eighteen kinds of radio wave propagation models, and its applicable parameters. In the end, this paper applies radio wave propagation prediction methods proposed by ITU-R, URSI, ITS and other agencies, general experience and semi-empirical methods, combines radio wave propagation prediction methods that have been successfully tested and widely used by domestic measurement data, and use basic data of local areas. After a high degree of integration, various types of radio wave propagation prediction models applicable to this project were obtained.

4. System application

4.1. System Application Process

The main process of the simulation system software is divided into the following steps (Fig.6 upper left):

(1) Adding frequency equipment models and user data through the resource management subsystem;
(2) By scenario sub-systems, drawing up the intended deployment of the scenario, selecting the frequency equipment model and setting the deployment location and equipment operating parameters; selecting the deployment of the equipment platform and implementing the planning of the trajectory through 2D map icons. And set the running time and loading equipment operating parameters;
(3) Through electromagnetic situation calculation and electromagnetic environment analysis subsystem, analyze the electromagnetic characteristics of equipment, and adjust the parameter information of frequency equipment;
(4) Analyse the red-and-blue confrontation process through the counter simulation subsystem, and push the result to the visual display subsystem;
(5) Through the task scheduling and monitoring subsystem, access to the desired tasks generated by ordinary users, and can specify the red and blue attributes of technical operators, to merge any two operator's scenarios, and conduct a confrontational analysis on the post-merged scenario;
(6) Finally, the task can be analysed and evaluated.

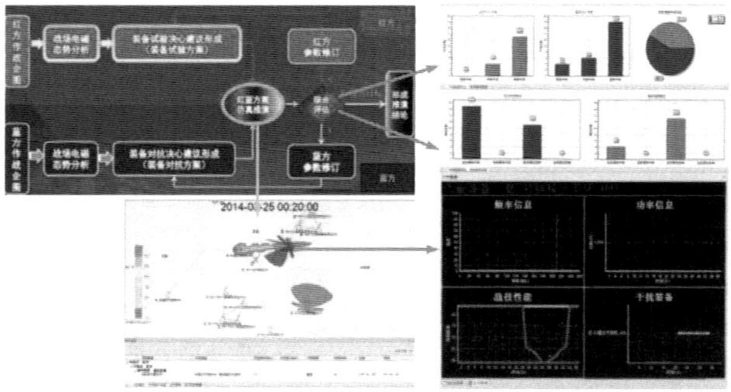

Fig.6 System application process.

4.2. Application mode

Both confrontation parties firstly conduct electromagnetic situation analysis based on their respective operations, and analyse frequency and power deployed by the battlefield, the scope of the interference equipment, and radar power areas (Figure 6 bottom right), and perform calculation of electromagnetic parameters for the combat process of mobile combat platforms. Secondly, the "back-to-back" countermeasure simulations are performed for equipment operational programs of both confrontation parties (Figure 6, bottom left). According to a comprehensive calculation of the complex electromagnetic environment, both confrontation parties battlefield frequency plan is comprehensive evaluated (Figure 6, upper right). If the expected goal is not achieved, both confrontation parties can revise the frequency plan of their respective programs and re-simulate until the expected goal is achieved, and forming an optimal battlefield frequency countermeasure scheme.

5. Conclusion

This study uses a complex electromagnetic environment confrontation and derivation platform, focusing on the complex electromagnetic environment, frequency equipment, electromagnetic situation, the organization and implementation of the mission planning, equipment application and effect assessment in the battlefield environment, comprehensively improves organization and command capabilities, equipment utilization capabilities, situational analysis capabilities, and planning and confrontational capabilities under the complex electromagnetic environment. This system is aimed at the lack of quantitative calculation and information support methods in the existing experimental training, highlights the use of simulation calculations, and constructs a set of electromagnetic environment information equipment test simulation system based on equipment model calculations, forming computable, deducible, interactive, and connectable experimental simulation capabilities.

References

[1] Junruo Liu, The main method for the US military to implement electronic warfare in a complex electromagnetic environment, Foreign Military Information War, 2:22-25,2010.

[2] Lianjie Zhu, The Basic Situation and Development Trend of U.S. Military Combined Training under Complex Electromagnetic Environment, Foreign Military Information War, 4:22-24,2010.
[3] Jianqing Cheng, Simulation Modeling Technology of Complex Electromagnetic Environment under Informationized Battlefield Conditions, Ship Electronic Engineering, 28(8):157-161, 2008.
[4] Fumin Yang, Research on 3D visualization of Complicated Battlefield Electromagnetic Environment, Engineering of Surveying and Mapping, 21 （2） :35-38,2012.
[5] Qingguo Liu, Battlefield Electromagnetic Situation Information and Its Measurement, Intelligence Command Control System and Simulation Technology, 4:27-34,2000.
[6] Li Li, Antenna and radio wave propagation （2nd Edition） , Science Press, 2016.6.
[7] Minghua Jia, A Simulating Method for Predicting Radio Wave Propagation, Computers Simulation, 27 （11） :91-94,2010.
[8] Ying Wang, Research and stimulation of radio propagation model over the ocean, Electronic Test, 9:15-20, 2010.

IOP Conf. Series: Materials Science and Engineering **635** (2019) 012019 doi:10.1088/1757-899X/635/1/012019

Improving the mechatronic system for automatic control of the reversing stands of mill 5000

A A Radionov[1], V R Gasiyarov[1], A S Karandaev[1], B M Loginov[2], V R Khramshin[3]

[1]South Ural State University, 76, Lenin Avenue, Chelyabinsk 454080, Russia
[2]PJSC Magnitogorsk Iron and Steel Works, 93, Kirov St., Magnitogorsk 455000, Russia
[3]Nosov Magnitogorsk State Technical University, 38, Lenin Avenue, Magnitogorsk 455000, Russia

gasiyarovvr@gmail.com

Abstract. It is noted that when extending the range of products made by plate rolling mills, it becomes relevant to improve the algorithms of the automatic parameters control systems. We hereby present a structure that explains the SMS Demag AG concept of automatic control over stand roll gauge and gaps. This concept is implemented in the automatic gauge control system (AGCS) of Mill 5000, Magnitogorsk Iron and Steel Works. We describe the drawbacks of the AGCS that manifest when rolling <10 mm plates. The most serious drawback is the tear of metal pieces from the trailing edge, which is caused by the inappropriate functioning of the gauge assignment and adjustment system. We present the structural diagram of the gauge calculation system. We also describe the functions of a system for dynamic compensation of disturbances. We provide rationale for excluding the roll counter-bending and deformation correction signals. The paper proposes a method to control hydraulic screwdowns, which is essentially about a quick increase in the inter-roll gap at the workpiece "trailing edge" during the last pass, which must be done when rolling thin plates. We analyze the oscillograms of signals describing the gauge interference when the proposed changes are implemented. We prove the technological efficiency of using the proposed control method on Mill 5000.

1. Introduction

Mill 5000 of PJSC Magnitogorsk Iron and Steel Works ("Mill 5000") is one of Europe's most productive plate rolling units. According to its design, it is capable of producing 1,500 thousand tons of (10–50)×(2000–4600)×(<18500) mm per annum. At the same time, the mill is currently efficiently used to roll an extended range of products, including 7- to 10-mm plates. Generally, changing the rolling profile requires, aside from solving some technological problems, an additional configuration and improvement of the automatic parameters control system. To that end, state-of-the-art mills use algorithms run by special automatic process control systems. This is why solving this problem boils down to developing and implementing new control methods and algorithms.

The most important mechatronic automation systems of Mill 5000 are the automatic gauge and gap control systems. Their combination uses the gauge adjustment concept developed by SMS Demag AG for plate rolling mills. This concept is discussed in papers [1-4] and is fully used by the mill.

The roll gauge and gap control principles is shown in the functional diagram, see Figure 1 [5]. It consists of the following basic systems (modules):

- hydraulic gap control (HGC);
- automatic gauge control (AGC);
- roll axial shifting control (RAC);
- dynamic disturbance compensation (DDC).

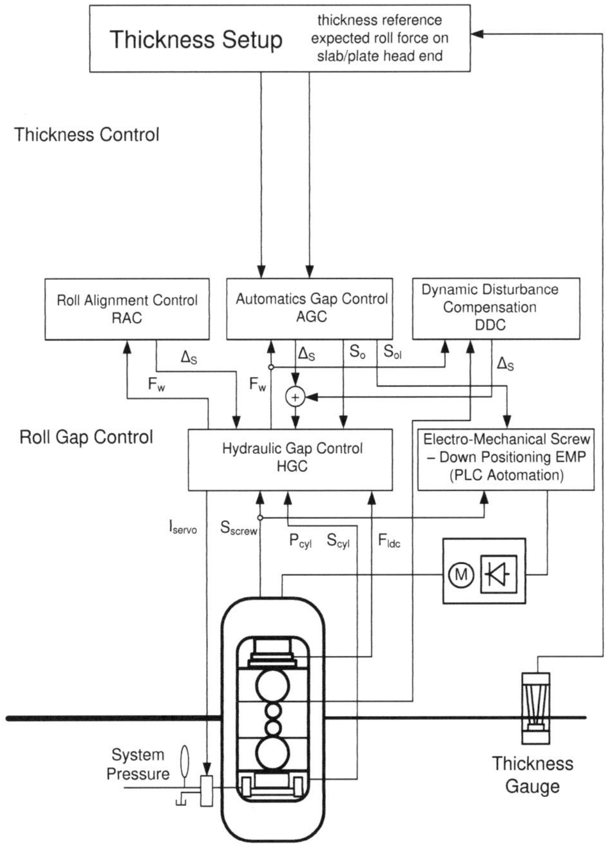

Figure 1. Explaining the concept of automatic gap and gauge control on a Mill-5000 stand

Fundamental to the structure are the two control circuits: the internal circuit to control the positioning of hydraulic screwdowns (HS), and the external gauge control system. The ultimate function is to compensate the stand deformations and various disturbances caused by deviations in the process parameters, which might occur during rolling. Aside from these modules, the structure includes systems for electromechanical pressure screw (EMP) positioning and conicity control; however, this paper does not dwell upon such systems.

The structure and configurations of the former two systems are described in a few papers, including [6-9]. Configuring the systems of the HGC, AGC, and RAC controller of Mill 5000 is described in [10]. Below is a summary of the AGC system and a description of the DDC functions.

2. Statement of Problem

Automatic Gauge Control (AGC). When using plate rolling mills, there frequently occur gauge defects caused by various factors. Some defects are due to the slab condition, including its lengthwise gauge interference and temperature difference caused by skidding. This indicates devitions in the material hardness.

Automatic gauge control is intended to maintain a constant roll gap when altering the rolling force by affecting the HS. The system can operate in the absolute mode or in the relative mode. In the absolute mode, the roll gap is adjusted by setting the operator-determined gauge or by computing the rolling schedule in a Level 2 automation system. In the relative mode, the slab/plate gauge attained at 0.5 s to 1 s after the rolling starts is kept constant over the entire length.

Since it is impossible to measure the actual roll gap during rolling, the system computes it. To that end, use the gap equation:

$$h = S + g(F_W)$$

Therefore, the roll gas h depends on the hydraulic cylinder piston position S and the stand stretching g, which is a function of the rolling force F_W. The AGCS is tasked to compensate the stand stretching fluctuations caused by the rolling process according to the equation

$$\Delta h = \Delta S + \Delta g(F_W) = 0 .$$

Gauge fluctuations resulting from roll eccentricity can be avoided by means of special adaptive filtering procedures. An adaptive notch filter is used to passively compensate gap eccentricities [10].

Successful use of such systems on Mill 5000 proves their reliability and high adjustment precision. At the same time, extending the Mill-5000 product range by rolling thin plates reveals the following drawbacks of the designed AGCS (plates are considered thin if < 10 mm):

1. There are deviations in the actual gauge measured by the thickness gauge after the last pass, whereas the indirect gauge signal used by the AGCS reveals no such deviations.

2. When rolling thin plates, the required gauge is attained by automatically setting the system to a minimum gap (in some cases, rolls are pre-converged with force). Since the plate gauge is small in last passages, the trailing edge has time to cool down. To compensate temperature losses, the gauge controller automatically reduces the gap by converging the rolls. As a result of strong pressure, some pieces are torn away from the trailing edge; such pieces have considerable kinetic energy. This damages the equipment while also endangering the personnel.

AGCS malfunctioning deteriorates the product quality, reduces the mill reliability, while also complicating the operator's life.

In order to eliminate such drawbacks, we had to improve the AGCS structure and algorithms. Below we discuss the proposed solutions and the outcomes of implementing them. We preliminarily describe the structural diagram of the gauge computation system used by the Mill-5000 AGCS.

3. Research Essentials

3.1. Structure of Gauge Assignment in Mill-5000 AGCS

Figure 2 shows the diagram of computing and generating the gauge assignment signal. The external AGCS circuit is the automatic gauge control circuit. It is formed by the gauge controller, the input of which receives the assignment signal sent by the Level 2 model (with corrections) as well as the actual gauge signal. The actual gauge is the total of mean gaps created by the hydraulic and electromechanical screwdowns (HS and and EMS) plus the stand stretching. Stretching-proportional signal is generated as a non-linear function of the total rolling force, defined by the stand rigidity [11]. The gauge controller output signal is summed with the disturbance compensation signals and the gap correction signals for the leading and trailing edge. It then goes to the intensity adjuster input, which limits the HS motion speed. The intensity adjuster output signal is the input signal for the slave HS positioning control circuit (not shown in Figure 2). Paper [10] describes the AGCS and gap control systems in detail.

Zero Point Corr is the gauge correction generated by comparing the actual gauge-metered gauge against the assigned gauge (it compensates the static error of indirect gauge computation); Agc Morgoil comp is the compensation for the oil-film thickness in the backup-roll oil-film bearings; Agc Roll Flat Comp is the compensation for roll flattening; Roll Wear comp is the compensation for roll wear (from the Level 2 model); ThermExpComp is the compensation for the thermal expansion of rolls (from the roll heating model); CvcShiftComp is the compensation for the axial roll shift; WrbBendComp is the compensation for the roll counterbending; EmpShpindleComp is the compensation for the spindle weight.

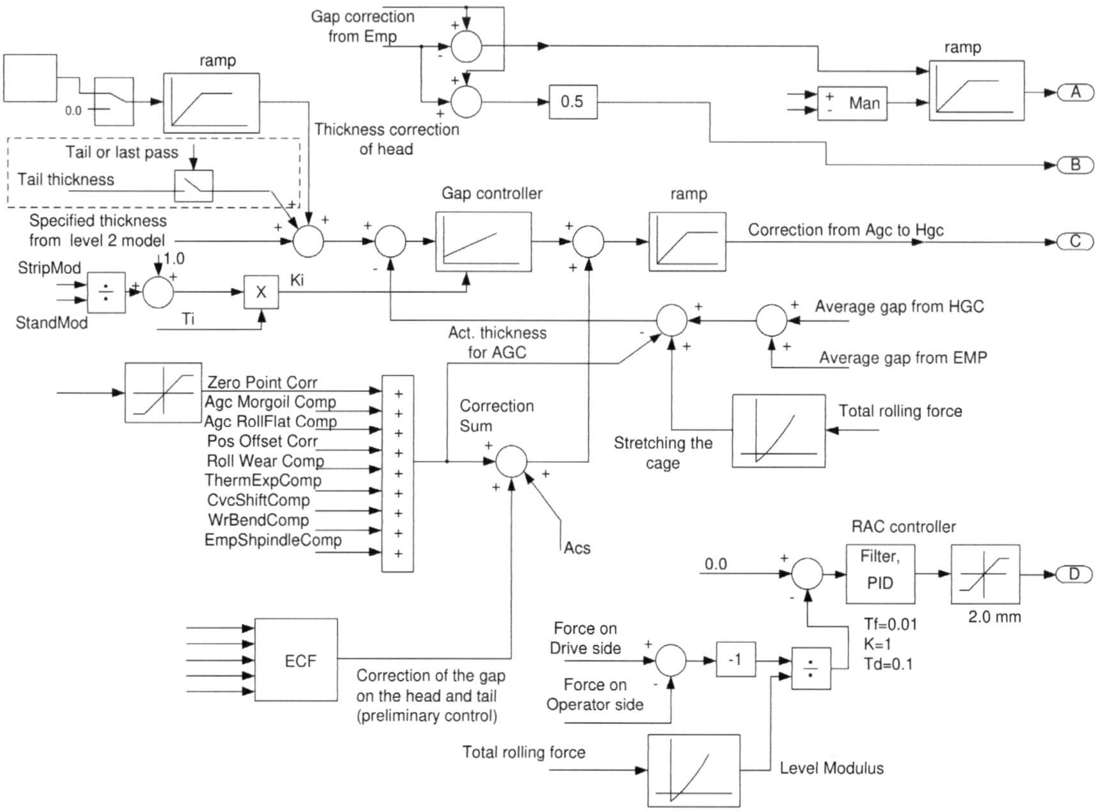

Figure 2. Structural diagram of gauge assignment in the Mill-5000 AGCS:

Before metal goes to the rolls, preliminary gap is set as defined on the basis of the expected rolling force taking into account the non-linearity of the stand rigidity curve.

The RAC controller, the diagram of which is shown in Figure 2, is to align the stand stretching for different forces exerted by the screwdowns on the drive side and on the operator side [10]. It aligns the resulting skew by affecting the HS cylinders.

3.2. Dynamic Disturbance Compensation (DDC)

During rolling, there occur various defects associated with gauge deviations. The DDC system compensates for such defects by correcting the roll gap. Each type of defects is compensated by an independent function.

Lift of roll necks in Morgoil-type oil-film bearings.

The input parameters are the measured force and rotation speed, which are taken into account when computing the actual oil-film thickness inside Morgoil bearings. It is known that at higher rolling speed and lower force, the lift of backup-roll necks in bearings is more pronounced. The actual oil-film thickness is used to compute the corrections for the gap ΔS, see Figure 1, 2.

Roll Heat Crown and Wear.

These parameters are computed cyclically by the Level 2 profile and flatness control system. The DDC module (see Figure 1) computes the appropriate adjustments for the gap ΔS.

Roll Bending.

The effect of roll bending is compensated taking into account the ΔS/ΔWRB derivative, which is computed by the profile and flatness control system and is transmitted to the controller together with the stand configuration. It and the actual bending force are used to compute and generated the adjustment of the gap ΔS.

For more details on the dynamic disturbance compensation, see [12,13].

Figure 3 shows the oscillograms of adjusting forces in the DDC system as measured on Mill 5000. These and other signals are used in the system for computing the gauge shown in Figure 2.

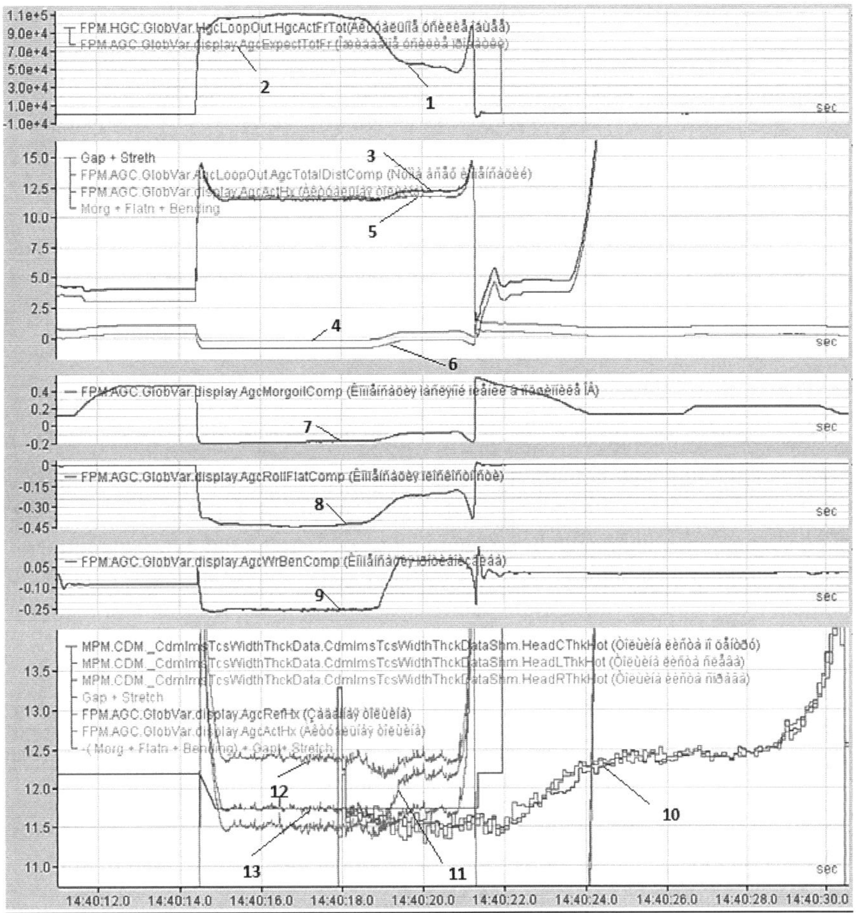

Figure 3. Oscillograms of signals used in gauge computing: 1, 2 are the assigned and the expected rolling force values; 3 is the total gap and stand stretching signal; 4 is the total of gauge compensations; 5 is the indirectly measured gauge; 6 is the total compensation signal for the oil film, roll deformation and bending; 7, 8 are the compensation signals for the oil film and the roll deformation; 9 is the roll bending compensation; 10 is the gauge-metered gauge; 11 is the total of gap and stand stretching; 12 is the total of all correction signals; 13 is the total signal of indirectly measured gauge.

Their decoding is given in the Figure caption. Independent correction signals are sent to the adder block input. The block output is summed up with the gap correction signal for the leading edge and the trailing edge, which signal is received from the EFC output; it is also summed up with the ACS signal controlling the motion of HS during profiled rolling [14,15]. The total AGC and HGC correction signal is sent to the inputs of closed circuits adjusting the positioning of HS on the drive side and the operator side (not shown in Figure 2).

Oscillograms are shown for the case where roll deformations and counterbending compensation signals are computed on the basis of the actual force. This is due to the fact that real signals are computed based on the actual rolling force (f(Frl.act), signals 8 and 9). When rolling thin plates, the 13-gauge signal in the initial circuit does not respond to the actual gauge deviations, as can be seen from the comparison of oscillograms 10 and 13. For the AGCS to compute these signals correctly, we propose using the Level 2 model-predicted values rather than actual values. Experiments have shows this improves the gauge adjustment quality.

3.3. Improving the AGCS for Thin-Plate Rolling

As noted above, rolling thin plates might result in tearing metal pieces away from the trailing edge during the last pass: To avoid such situations, we propose an HS control method that essentially consists in rapidly increasing the roll gap by 0.5 mm to 2.5 mm depending on the gauge. Roll retraction must be performed if the following conditions are met:

- plates being rolled are <10 mm thick;
- the mill is running the last pass;
- the unrolled portion of the plate remaining in the stand is less than 1.0 meters long.

To ensure fast roll retraction, we propose that the controller program generates signals:

- to increase the assigned gauge at the gauge adjustment circuit input;
- to move the HS to "bypass the gauge controller".

Additional blocks that are placed to meet these conditions are outlined in Figure 2.

Figure 4a shows the oscillograms obtained when testing this method. Window 1 shows the "trailing-edge" position oscillogram (the oblique line) relative to the stand (the horizontal line). One meter before the metal leaves the stand (Window 2, timepoint ~13.35.29.75) the gauge assignment signal is boosted. As a result, we can observe the repositioning of hydraulic cylinder pistons on the drive side and on the operator side (Window 3) in the time interval from 13.35.29.75 to ~13.35.30.00. The total rolling force (Window 4) decreases, which also decreases the chance of tearing away the "trailing edge".

4. Discussion of Results

Figure 4b presents the gauge alteration oscillograms based on the thickness-gauge signals. Gauge deviation resulting from additional roll retraction does not exceed 1 mm or 10% of the final plate gauge. This is proven by the assigned and actual gauge oscillograms shown in Window 2, Figure 4a. Besides, the oscillograms shown in Figure 4b allow us to conclude that the gauge on the plate edges and in the plate center alters in a near-uniform fashion. This proves that the RAC-controller is configured appropriately [10].

The trailing edge of a deviating gauge (~ 0.4 meters long) is cut by the finishing line. Experience shows that such an edge of the plate is cut in nearly all cases, as it often has defects. Therefore, the proposed method hardly increases the loss of metal that could be attributed to cutting.

The proposed method improves mechanism reliability as well as personnel safety.

5. Conclusion

The papers [1-9,12-14] mentioned herein present the structure and the results of mathematical modeling of system that use the known concepts of automatic gauge control. However, there are virtually no publications on the configuration and use of such systems. This paper seeks to fill that gap. We do not claim to have made any improvement to the concept of SMS Demag AG developed for plating rolling mills, the one we covered in the first part hereof. We herein discuss the experience of configuring and using the Mill-5000 AGCS in the context of constantly changing product range. In combination with improvements in CVC systems and roll counterbending systems [16–19], the obtained results become generalizing in nature. This research might be deemed comprehensive, as it addresses the problems of improving the profile and flatness of the rolled metal produced by modern plate rolling mills.

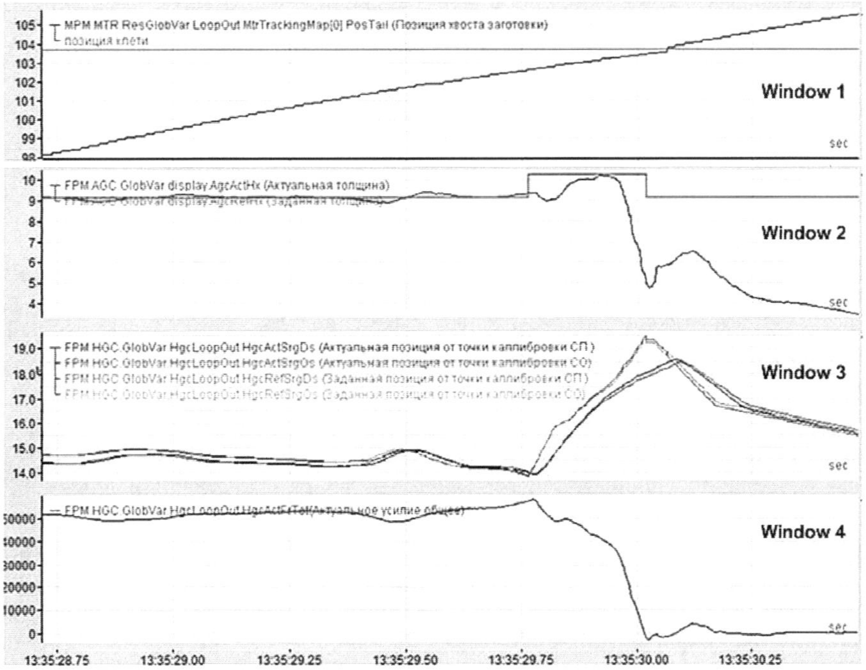

a)

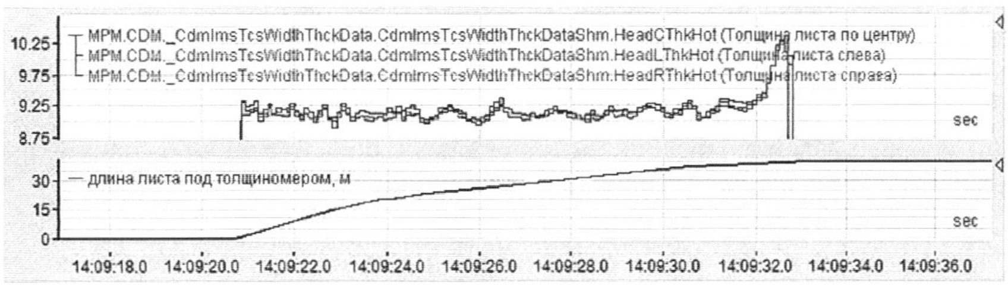

b)

Figure 4. HS parameters oscillograms, (a) and gauge over the plate length, (b) when implementing the proposed changes: Window 1 shows the trailing edge position (from the tracking system) relative to the stand position; Window 2 shows the actual gauge in the AGCS; Window 3 shows the actual position relative to the calibration point; Window 4 shows the rolling force

Acknowledgments
South Ural State University acknowledges financial support from Ministry of Education and Science of the Russian Federation (grant No 13.9656.2017/8.9)

Reference

[1] Peter Kucsera and Zsolt Beres 2015 Hot Rolling Mill Hydraulic Gap Control (HGC) gaugez control improvement *Acta Polytechnica Hungarica* **12 (6)** pp 93–106

[2] Fei Zhang, Yongjun Zhang, Jianxin Hou and Binbin Wang 2015 Thickness control strategies of plate rolling mill *International Journal of Innovative Computing, Information and Control* **11 (4)** pp 1227–1237

[3] Fei Zhang, Yongjun Zhang and Handan Chen 2016 Automatic Gauge Control of Plate Rolling Mill *International Journal of Control and Automation* **9 (2)** pp 143–156 http://dx.doi.org/10.14257/ijca.2016.9.2.14

[4] LI Xu, CHEN Shu-zong, Du De-shun, Chen Hua-xin, Zhang Dian-hua 2011 Simulation and Analyzing on Model Parameters Effect of BISRA-AGC *Physics Procedia* **22** pp 571–576

[5] *Fascinating World of Steel Plates (Charuyushchiy mir tolstolistovoy stali). SMS-group* http://sms-group-eaportal.com/brochures/R/A-319R.pdf

[6] Sun Jie, Chen Shu-zong, Han Huan-huan, Chen Xing-hua, Chen Qiu-jie, Zhang Dian-hua 2015 Identification and optimization for hydraulic roll gap control in strip rolling mill *J. Cent. South Univ.* **22** pp 2183–2191 DOI: 10.1007/s11771-015-2742-0

[7] John P, Nicholas S S, Marwan A S 2010 A new strategy for optimal control of continuous tandem cold metal rolling *IEEE Transactions on Industry Application* **46(2)** pp 703–711

[8] Roland Heeg, Andreas Kugi, Olivier Fichet, Laurent Irastorza 2005 Christophe PelletierModeling and control of plate thickness in hot rolling mills *IFAC Proceedings Volumes* **38 (1)** pp 13-18 https://doi.org/10.3182/20050703-6-CZ-1902.01681

[9] Prinz K, Steinboeck A, Müller M, Ettl A, Kugi A 2017 Automatic gauge control under laterally asymmetric rolling conditions combined with feedforward *IEEE Transactions on Industry Applications* **53 (3)** pp 2560–2568 DOI: 10.1109/TIA.2017.2660458

[10] Shubin A G, Loginov B M, Khramshin V R, Evdokimov S A, Karandaev A S 2015 System of Automatic Control of Hydraulic Screw-down Mechanisms of Plate Mill Stand *Proceedings of 2015 International Conference on Mechanical Engineering, Automation and Control Systems (MEACS)* p 6 DOI 10.1109/MEACS.2015.7414858

[11] Karandaev A S, Radionov A A, Khramshin V R, Andryushin I Yu, Shubin A G 2014 Automatic Gauge Control System with Combined Control of the Screw-Down Arrangement Position *12th International Conference on Actual Problems of Electronic Insrument Engineering (APEIE-2014)* **1** pp 88–94 DOI 10.1109/APEIE.2014.7040794

[12] Diego Alvarez, Alberto B Diez, Faustino Obeso 1988 Slab curvature compensation in hot rolling mill by means o Fuzzy Control *Paper at the III seminar on rolling Mill Rolls* (Maxico: Institute Latinoamericano del Fierro v Acero)

[13] Ji Yafeng, Zhang Dianhua, Chen Shuzong, Sun Jie, LI Xu, Di Hongshuang 2014 Algorithm Design and Application of Novel GMAGC based on Mill Stretch Characteristic Curve *Journal of Central South University* 21 (3) pp 942–947

[14] Khramshin V R, Khramshina E A, Karandaev A S, Gasiyarov V R, Voronin S S 2017 Control Methods and Systems Providing Reduced Consumption Index at Rolled Product Manufacture at Plate Mill *Proceedings of the IEEE NW Russia Young Researchers in Electrical and Electronic Engineering Conference (EIConRus)* pp 1540–1544 DOI: 10.1109/EIConRus.2017.7910865

[15] Karandaev A S, Khramshin V R, Andryushin I Yu, Petryakov S A, Khramshin R R 2013 Method for correction of gauge interference of the head-strip section in a system for automatic controlling of the thickness of a broad-strip hot-rolling mill *Russian Electrical Engineering* **84 (8)** pp 441–445

[16] Karandaev A S, Loginov B M, Gasiyarov V R, Khramshin V R 2017 Force limiting at roll axial shifting of plate mill *Procedia Engineering* **206** pp 1780–1786 DOI: 10.1016/j.proeng.2017.10.713

[17] Karandaev A S, Loginov B M, Radionov A A, Gasiyarov V R 2017 Setting Automatic Roll Axial Shifting Control System of Plate Mill *Procedia Engineering* 206 pp 1753–1750 DOI: 10.1016/j.proeng.2017.10.709

[18] Voronin S S, Usaty D Yu 2013 Roll Bending as a Factor of Transverse Gauge Interference of Plates and a System to Control the Counterbending of Plate Hot-Rolling Mills *Russian Internet Journal of Electrical Engineering* 13 (1) pp 51–55

[19] Voronin S S, Usaty D Yu, Gasiyarov V R, Radionov A A 2015 Using Axial Roll Sliding to Adjust the Roll Gap of Plate Hot-Rolling Mills *Russian Internet Journal of Industrial Engineering* **3 (1)** pp 45–48

Comparative analysis of 3L-NPC converter output voltage with selective harmonic elimination PWM

V R Gasiyarov, A A Radionov and A S Maklakov

South Ural State University, 76, Lenin Avenue, Chelyabinsk 454080, Russia

E-mail: alexandr.maklakov.ru@ieee.org

Abstract. The main purpose of this article is to show the importance of a comparative analysis of a three-level neutral point clamped (3L-NPC) output voltage with selective harmonic elimination PWM (SHEPWM). As compared to other PWM technique, the SHEPWM method has a lower dynamic power loss and a better THD at the same switching frequency. The objective of this analysis is to provide a modelling results of THD and a voltage/current spectrum of the output voltage with a various SHEPWM implementation for the 3L-NPC converter. A mathematical model for this research was developed in Matlab/Simulink. A practical use of the comparative analysis results can be applied to power electronic systems of high power wind generators.

1. Introduction

An improvement and a development of new pulse width modulation (PWM) algorithms are the priority research for power conversion by power semiconductor converters. The common topologies of the power converters are applied into energy and industrial applications, such as high voltage dc transmission (HVDC) and flexible ac transmission system (FACTS), renewable energy sources (RES) and electric drives. The utilising of the identical topologies leads to a non-effective use of PWM methods, as this depends on load or grid connection performances, especially at different electromagnetic compatibility requirements [1]. There are several variants for implementation of one PWM method, and at the selection of an optimal PWM method, it is possible to achieve the current/voltage THD, which will correspond to the THD reference requirements [2].

Neutral-point clamped (NPC) converter is an essential part of high-power energy applications. The specific purpose of NPC is to solve the problems of the voltage limit capability of semiconductor devices, and they are attractive for high-power energy control due to their superior performance in power semiconductor technology [3]. High quality of the current and voltage waveforms both at the input and output terminals of the converter are needed to be improved as much as possible, and this requirement is affected by the topology we used, the control algorithm and the actual application [4]. NPC converters are the most established and commercialized topology. New means of boosting the three level NPC converter energy efficiency are required to develop and study applied power circuits, modulation algorithms and control systems [5-7]. Figure 1 shows a typical power circuit of high-power drive system based on the NPC converters.

Currently there are a lot of scientific research focuses on selective harmonic elimination (SHE) and space vector pulse width modulation (SVPWM) for 3L-NPC converters. The article [8] describes hybrid PWM based on sinusoidal (SPWM) and SHEPWM, which provides comparative analysis and algorithm of a smooth transition between them for power quality improving. The article [9] presents a performance of hybrid PWM with a transition between SV and SHEPWM, which allows to make the power losses less.

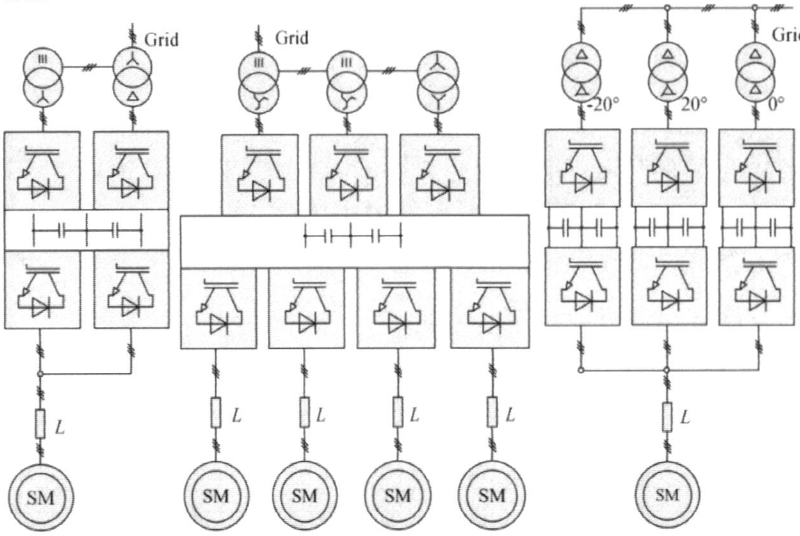

Figure 1. Typical industrial circuits are based on 3L- NPC converters.

2. Problem definition

The amount of dynamic losses, which depend on the sample frequency of switching, is the main indicator reducing the efficiency of the power converters. Reduction in the number of switching of the converter allows increasing power, reliability and efficiency, but it can provide a significant effect on the quality of the converted current and voltage. Using one method of modulation over the full frequency range and the output voltage and current magnitudes has not able to achieve the maximum efficiency of the converter and maintain the required power quality. The comparative analysis of converter output voltage with selective harmonic elimination PWM is needed to solve this task. It can assess voltage/current THD and spectrum for several SHE patterns and provide the information about an optimal pattern.

The objective of this analysis is to provide a modelling results of THD and a voltage/current spectrum of the output voltage with a various SHEPWM implementation for the 3L-NPC converter.

3. Mathematical description of 3L-NPC converter

The NPC has already applied in many conventional high-power ac motor drives, and a back-to-back topology for NPC makes it successfully used in regenerative applications, like the grid interfacing of renewable energy sources. In addition, the topology can be extended to any level by adding more diodes and capacitors, but the main problem of the topology is how the neutral point voltage balanced or maintained on the DC-link, that is, to keep DC-side voltage stable [10,11].

A mathematical model of three level NPC (Figure 2) was developed via discrete switching functions $S_{abc12rv}$ of power modules for grid side:

$$\gamma_{abc} = \begin{cases} 1, \left(S_{abc1}\ and\ S_{abc2}\right) = 1\ and\ \left(S_{abc3}\ and\ S_{abc4}\right) = 0 \\ 0, \left(S_{abc2}\ and\ S_{abc3}\right) = 1\ and\ \left(S_{abc1}\ or\ S_{abc4}\right) = 0 \\ -1, \left(S_{abc3}\ and\ S_{abc4}\right) = 1\ and\ \left(S_{abc1}\ and\ S_{abc2}\right) = 0 \end{cases}; \quad (1)$$

These switching states γ_{abcrv} create the discrete logic functions F.

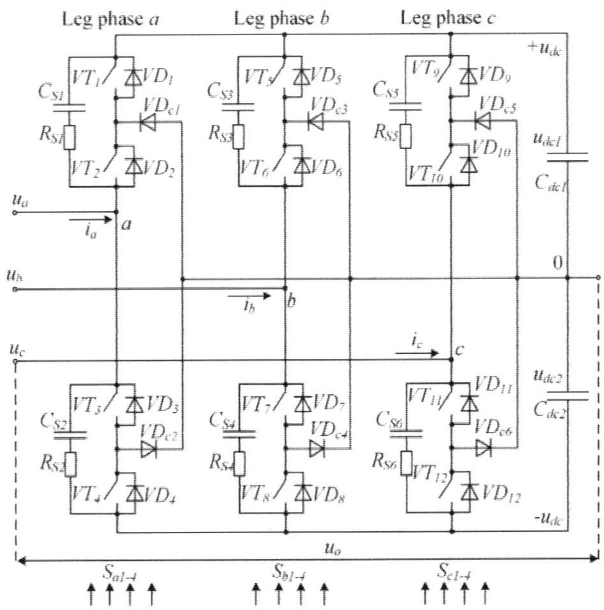

Figure 2. Power circuit of 3L-NPC converters.

$$\begin{cases} F_{abc1} = \dfrac{\gamma_{abc} \cdot (\gamma_{abc} + 1)}{2} \\ F_{abc2} = \dfrac{\gamma_{abc} \cdot (\gamma_{abc} - 1)}{2} \end{cases}. \quad (2)$$

Using the functions F and Kirchhoff`s laws, an equation system describing the electromagnetic processes of 3L-NPC was written as:

$$\begin{cases} u_{abc} = u_{dc1} \cdot \left(F_{abc1} - \dfrac{1}{3} \cdot \sum_{n=a}^{b.c} F_{n1} \right) + u_{dc2} \cdot \left(F_{abc2} - \dfrac{1}{3} \cdot \sum_{n=a}^{b.c} F_{n2} \right) \\ u_{dc1} = \dfrac{F_{a1} \cdot i_a + F_{b1} \cdot i_b + F_{c1} \cdot i_c}{p \cdot C_{dc1}} \\ u_{dc2} = \dfrac{F_{a2} \cdot i_a + F_{b2} \cdot i_b + F_{c2} \cdot i_c}{p \cdot C_{dc2}} \end{cases}, \quad (3)$$

where i_a, i_b and i_c are 3L-NPC converter phase currents.

4. Description of SHE pulse-with modulation

The selective harmonic elimination PWM has able to eliminate the defined harmonics from 3L-NPC converter output phase voltage. And the total amount of eliminated harmonics is limited the switching frequency of the power modules as

$$k = \frac{f_{sw\max}}{2 \cdot f_1} - 1 = N - 1 \quad (4)$$

where f_{swmax} is the max. switching frequency; N – is amount of the power module switching in per quarter of the voltage time period, k – is amount of the eliminated harmonics.
A typical phase voltage of SHE this a quarter-wave symmetry [12], as shown in Figure 3.

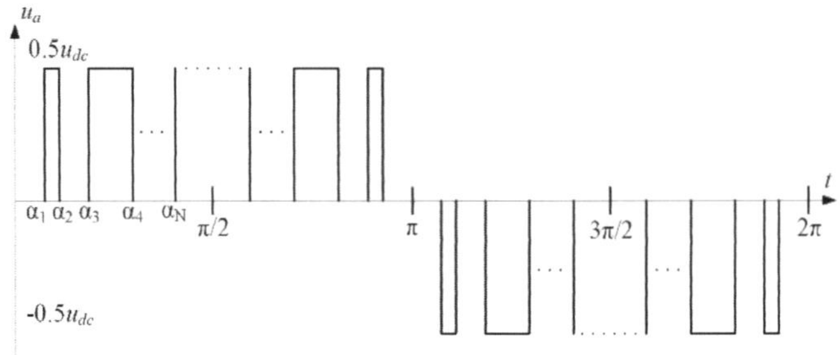

Figure 3. Typical 3L-NPC converter output voltage of under SHE with quarter-wave symmetry.

The switching angles and the spectrum of the 3L-NPC converter output voltage with the SHE by means of a system of nonlinear equations [13]

$$\begin{cases} \sum_{k=1}^{N}(-1)^k \cdot \cos(\alpha_k) = \frac{\pi}{4} \cdot \frac{2}{u_{dc}} \cdot U_m \\ \sum_{k=1}^{N}(-1)^k \cdot \cos(n \cdot \alpha_k) = 0, n = 5,7,11,... \end{cases} \tag{5}$$

where the pick voltage U_m is calculated [14, 15] by

$$U_m = \frac{4}{\pi}\int_0^{\pi/2} u(\omega t) \cdot \sin(n \cdot \omega t) d(\omega t) = \frac{4}{\pi \cdot n} \cdot \frac{u_{dc}}{2} \cdot \sum_{k=1}^{N}(-1)^k \cdot \cos(n \cdot \alpha_k) \tag{6}$$

where $u(\omega t) = \sum_{n=1,3,...,\infty}^{\infty} \frac{2 \cdot u_{dc}}{\pi \cdot n} \cdot \left[\sum_{k=1}^{N}(-1)^{k+1} \cdot \cos(n \cdot \alpha_k)\right] \cdot \sin(\omega t \cdot n)$, in which α_k – a switching angle at elimination of k harmonics [16].

5. Modelling results

Using the mathematical descriptions of the three-level converter and the SHE modulation algorithm, a research model was developed in the Matlab/Simulink. For the comparative analysis, presented in Table 1 four patterns of the SHE implementation were calculated. The results of calculation of the switching angles from the modulation factor are shown in Figure 4. The dependences of the THD voltage and current of the converter for the SHE patterns are shown in Figure 5.

The simulation results determined the non-linear dependence of voltage and current THD on the modulation factor for different patterns of SHEPWM. The analysis of the obtained results in Figure 5 shows that the voltage THD depends strongly on a converter modulation factor. At the maximum values of the modulation factor, the voltage THD becomes equal, therefore, it will be optimal using of the SHE pattern with a low switching. However, a comparative analysis of the current THD provides better results at the SHE pattern with a high switching.

Table 1. Patterns for selective harmonic elimination PWM

SHE	k	Patterns
SHE_1	6	5, 7, 11, 13, 17 and 19
SHE_2	8	5, 7, 11, 13, 17, 19, 23 and 25
SHE_3	10	5, 7, 11, 13, 17, 19, 23, 25, 29 and 31
SHE_4	12	5, 7, 11, 13, 17, 19, 23, 25, 29, 31, 35 and 37

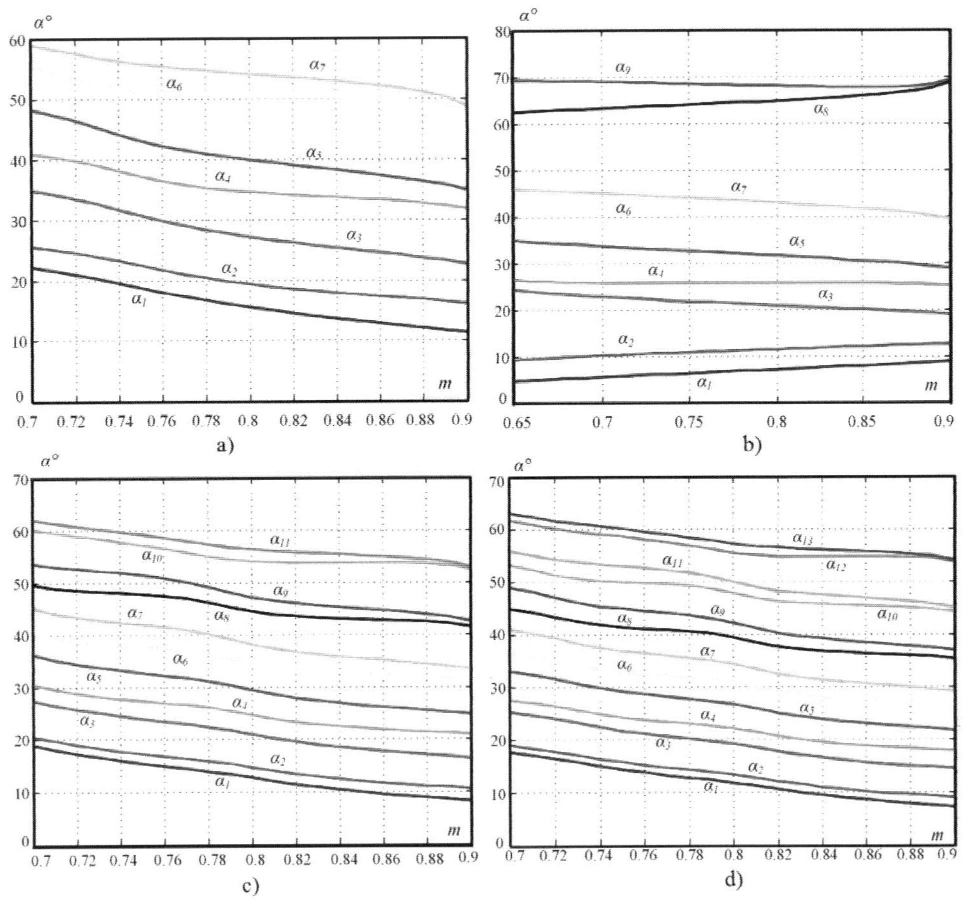

Figure 4. Switching angles for SHE_1(a), SHE_2(b), SHE_3(c), SHE_4(d).

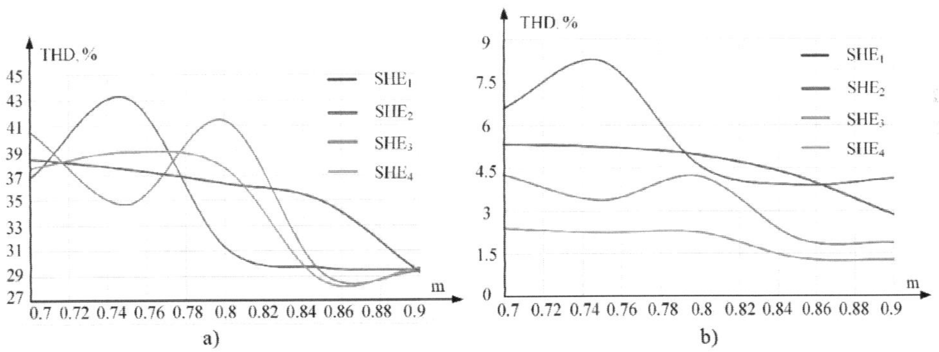

Figure 5. Comparative analysis of voltage(a) and current(b) THDs.

6. Conclusion

In this paper the comparative analysis of 3L-NPC converter output voltage with selective harmonic elimination PWM has been considered. Mathematical description and of 3L-NPC converter and selective harmonics elimination method were provided. The calculation results of voltage and current THDs at the four patterns of the SHE implementations were obtained. The results show a possibility of using a low switching SHE pattern at a high modulation factor for a voltage THD.

IOP Conf. Series: Materials Science and Engineering **635** (2019) 012020 doi:10.1088/1757-899X/635/1/012020

References

[1] Kouro S, Rodriguez J, Wu B, Bernet S and Perez M 2012 Powering the Future of Industry: High-Power Adjustable Speed Drive Topologies *IEEE Industry Applications Magazine* **18 (4)** pp 26–39

[2] Abu-Rub H, Bayhan S, Moinoddin S, Malinowski M, Guzinski J 2016 Medium-Voltage Drives: Challenges and existing technology *IEEE Power Electronics Magazine* 3 (2) pp 29–41

[3] Rodriguez J, Lai J-S and Peng F Z 2002 Multilevel inverters: A survey of topologies, controls, and applications *IEEE Trans. Ind. Electron* 49 (4) pp 724–738

[4] Nabae A, Takahashi I, Akagi H A 1981 Neutralpoint clamped PWM inverter *IEEE Transactions on Industry Applications* **1A-17 (5)** pp 518–523

[5] Remus Teodorescu, Marco Liserre, Pedro Rodrıguez 2011 *Grid converters for photovoltaic and wind power systems* (UK: John Wiley & Sons, Ltd.)

[6] Martin Gendrin, Jean-Yves Gauthier, Xuefang Lin-Shi 2016 A Predictive Hybrid Pulse-Width-Modulation Technique for Active-Front-End Rectifiers *IEEE Transactions on Power Electronics* **32 (7)** pp 5487–5496

[7] Jayanta Biswas, Meenu D Nair, Vivek Gopinath, Mukti Barai 2016 An Optimized Hybrid SVPWM Strategy Based on Multiple Division of Active Vector Time (MDAVT) *IEEE Transactions on Power Electronics* **32 (6)** pp 4607–4618

[8] Wang Y, Wen X, Guo X et al 2011 The smooth transition research of different PWM modulations for vector control of induction motor in medium voltage high power *in Proc. IEEE Elect. Mach. and Syst. conf.* pp 1–5

[9] Zhang Y, Zhao Z, and Zhu J A 2011 Hybrid PWM applied to high-power three-level inverter-fed induction-motor drives *IEEE Trans. Ind. Electron* **58 (8)** pp 3409–3420

[10] Konstantinou G, Agelidis V G 2014 On re-examining symmetry of two-level selective harmonic elimination PWM: Novel formulations, solutions and performance evaluation *Electric Power Systems Research* **108** pp 185–197

[11] Zaragoza J, Pou J, Ceballos S, Robles E, Ibaez P and Villate J L 2009 A comprehensive study of a hybrid modulation technique for the neutralpoint-clamped converter *IEEE Trans. Ind. Electron* **56 (2)** pp 294–304

[12] Hramshin T R, Krubtsov D S and Kornilov G P 2013 Evaluation of methods PWM voltage active rectifiers rolling mills *Russian Internet Journal of Industrial Engineering* **2** pp 48–52

[13] Grahame Holmes D, Lipo Thomas A 2003 *Pulse Width Modulation for Power Converters: Principles and Practice* p 744

[14] Jingling Cheng, Dongdong Chen, Yaowei Hu, Guozhu Chen 2017 An improved SHE algorithm and filter design method for high power grid-connected converter under unbalanced and harmonic distorted grid *2017 IEEE 26th International Symposium on Industrial Electronics (ISIE)* pp 594–599

[15] Cungang Hu, Grahame Holmes, Weixiang Shen, Xinghuo Yu, Qunjing Wang, Fanglin Luo 2017 Neutral-point potential balancing control strategy of three-level active NPC inverter based on SHEPWM *IET Power Electronics* **10 (14)** pp 1943–1950

[16] Andrey A. Radionov, Vadim R. Gasiyarov, Alexander S. Maklakov 2016 Hybrid PWM on the basis of SVPWM and SHEPWM for VSI as part of 3L-BtB-NPC converter *IECON 2017 - 43rd Annual Conference of the IEEE Industrial Electronics Society* pp 1232–1236

Acknowledgments

South Ural State University acknowledges financial support from Ministry of Education and Science of the Russian Federation (grant No 13.9656.2017/8.9).

IOP Conf. Series: Materials Science and Engineering **635** (2019) 012021 doi:10.1088/1757-899X/635/1/012021

Parallel Operation of Three-phase Self-excited Induction Generators Driven by Renewable Energy Sources

Y J Wang and B R Chen

Department of Electrical Engineering, National Yunlin University of Science and Technology, 123 sec. 3 University Road, Touliu, Yunlin 64002, Taiwan, ROC

wangyj@yuntech.edu.tw

Abstract. Squirrel-cage induction machines, because of their low prices and rare demand for maintenance, can work as self-excited induction generators (SEIGs) in remote areas to convert wind or pico-hydro power to electricity, enhancing the use of different forms of renewable energy. However, with the increase of load, a single SEIG may not provide enough capacity to meet the demand, so parallel operation of several SEIGs is often required, necessitating a technique to analyze the characteristics of this kind of operation. This paper presents steady-state analysis of two SEIGs based on the equivalent circuit of the induction generator. The two-machine model allows the performance and load sharing features to be evaluated when the two generators undergo different rotating speeds.

1. Introduction

Squirrel-cage induction machines working as self-excited generators (SEIGs) are robust, simple in structure, and relatively low-cost compared with synchronous machines, so they are suitable for generating electricity in remote areas to convert renewable energy sources (e.g., wind, bio-gas, pico-hydro, etc.) to electricity [1]. A single SEIG sometimes does not afford to meet increasing demand for electric power. Several SEIGs operating in parallel provide a solution to this problem.

The analysis of an SEIG is already complicated because it involves solving a circuit containing nonlinear magnetizing reactance, let alone solving several SEIGs in parallel. In the literature, analysis method of a single SEIG includes the loop impedance method [2], the nodal admittance method [3], and the two-step method [4] that allow the output frequency and the working point of the SEIG to be determined. When two or more SEIGs operate in parallel, the T equivalent circuit [5, 6] and the Γ equivalent circuit [7] have been used to solve for the system responding to different values of rotating speeds, loads, and excitation capacitances. In this paper, we modify the existing nodal admittance method so that it can consider the situation of two generators. Of course, the presented method can be further extended to include any number of SEIGs in parallel. The proposed two-generator model will show how the generators share their load when they undergo different rotating speeds.

2. Equivalent circuit analysis of a single generator

Figure 1 is the equivalent circuit of an induction generator where F refer to the per unit frequency, v the per unit rotating speed, R_r and X_r the rotor resistance and reactance, X_m the magnetizing reactance which is nonlinear, R_s and X_s the stator resistance and reactance, X_c the reactance of the excitation capacitor, R_L the load resistance, V_g the air-gap voltage, V_L the load voltage, and I_m the magnetizing current.

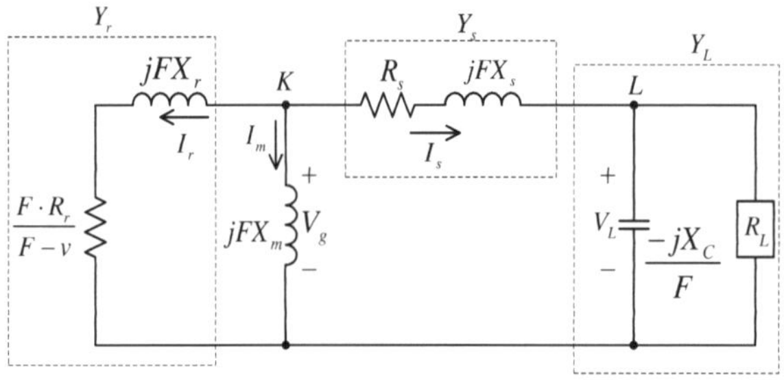

Figure 1. Equivalent circuit of a single induction generator.

2.1. Formulation

In figure 1, the rotor admittance Y_r, stator admittance Y_s, and load admittance Y_L can be written as

$$Y_r = [FR_r / (F - v) + jFX_r]^{-1} \tag{1}$$

$$Y_s = (R_s + jFX_s)^{-1} \tag{2}$$

$$Y_L = (-jX_c / F)^{-1} + (R_L)^{-1} \tag{3}$$

The nodal voltage equations for nodes K and L can be given by

$$V_g Y_r + I_m + (V_g - V_L) Y_s = 0 \tag{4}$$

$$V_L Y_L + (V_L - V_g) Y_s = 0 \tag{5}$$

The current-voltage relation of the magnetizing reactance is modeled using the logistic function

$$V_g(I_m) = \frac{a}{1 + (I_m / b)^{-c}} \cdot F \tag{6}$$

The magnetization curve has been obtained experimentally on a Δ-connected, 0.5 Hp, 220 V, 60 Hz, 4-pole cage induction motor. The parameters in equation (6) have been determined by the regression analysis. In this study, the air-gap voltage V_g is taken as the reference phasor, which implies that the magnetizing current I_m must be purely imaginary. The per unit frequency is by definition a real number. Therefore, two more equations need to be added to allow the voltages to be correctly solved.

$$\mathrm{Re}(I_m) = 0 \tag{7}$$

$$\mathrm{Im}(F) = 0 \tag{8}$$

2.2. Analysis results

Resolution of equations (1)-(8) allows the generator's output characteristics to be found. The Mathcad program has been used to solve this equation system. The generator terminal voltage V_L as a function of the output power P_L when the generator rotates at 1800 rpm is depicted in figure 2(a) in which the two curves correspond to two excitation capacitances of 12 and 14 μF. Figure 2(b) shows the variations of the per unit frequency F against P_L at the constant speed 1800 rpm and two excitation capacitances. It can be seen that a higher excitation capacitance brings about a higher terminal voltage for the same load. The influences of the rotational speed are shown in figures 2 (c) and (d) when the excitation capacitance is kept unchanged at C=12μF. Figure 2(c) clearly illustrates that a higher rotating speed renders a higher output voltage. The rotating speed has the same effects on the output frequency as shown by figure 2(d). The curves shown in figure 2 demonstrate that the nodal equations (1)-(8) are capable of analyzing an SEIG correctly. This method is a modification of the existing nodal admittance method, and can be extended to the analysis of two or more SEIGs working together.

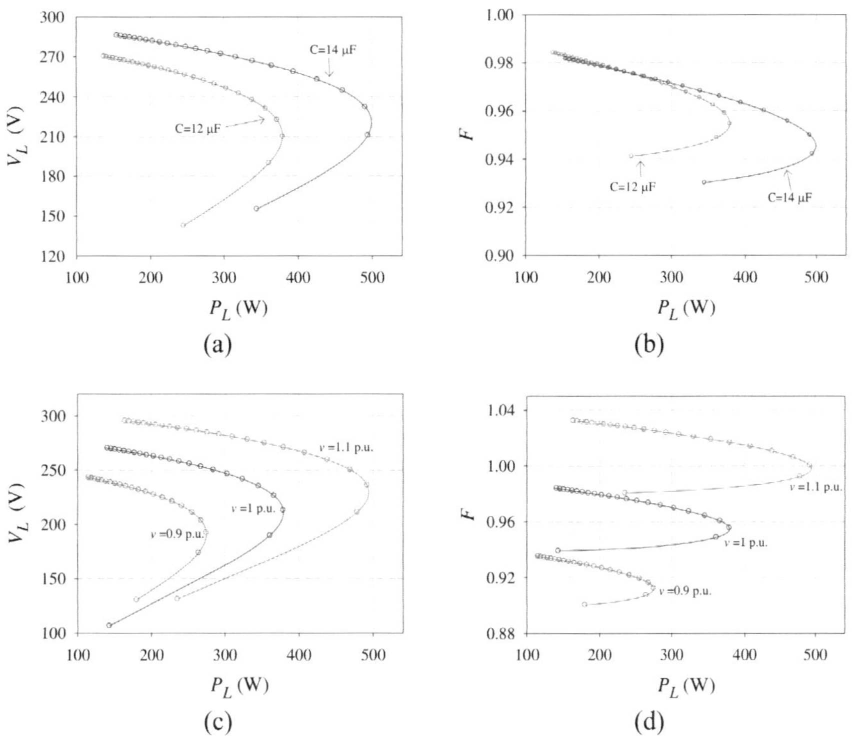

Figure 2. Analysis results of a single SEIG. (a) Load voltage versus output power, (b) Per unit frequency versus output power, for two values of excitation capacitance. (c) Load voltage versus output power, (d) Per unit frequency versus output power, for three rotating speeds, for C=12μF.

3. Formulation of two SEIGs operating in parallel

The equivalent circuit model shown in figure 1 can easily be extended to include two or more SEIGs in parallel. For clarity, this paper limits itself to the case of two SEIGs. The equivalent circuit model of two SEIGs is sketched in figure 3 where the means of the symbols are self-explanatory. The subscript 1 refers to the first generator and subscript 2 to the second generator. The rotor, stator and load admittances are given by

$$Y_{r1} = [FR_{r1} / (F - v_1) + jFX_{r1}]^{-1} \tag{9}$$

$$Y_{s1} = \left(R_{s1} + jFX_{s1}\right)^{-1} \tag{10}$$

$$Y_{r2} = [FR_{r2} / (F - v_2) + jFX_{r2}]^{-1} \tag{11}$$

$$Y_{s2} = \left(R_{s2} + jFX_{s2}\right)^{-1} \tag{12}$$

$$Y_L = (-jX_c / F)^{-1} + R_L^{-1} \tag{13}$$

The nodal voltage equations at nodes K, L, and M can be written as

$$V_{g1}Y_{r1} + I_{m1} + (V_{g1} - V_L)Y_{s1} = 0 \tag{14}$$

$$V_{g2}Y_{r2} + I_{m2} + (V_{g2} - V_L)Y_{s2} = 0 \tag{15}$$

$$V_L Y_L + (V_L - V_{g1})Y_{s1} + (V_L - V_{g2})Y_{s2} = 0 \tag{16}$$

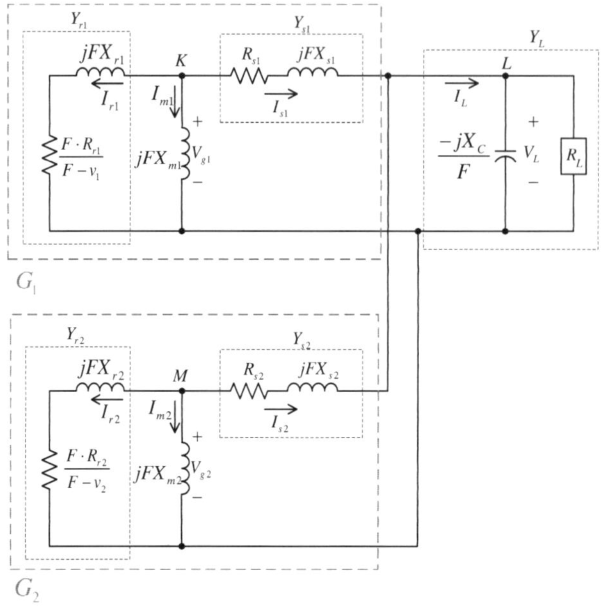

Figure 3. Equivalent circuit of two induction generators connected in parallel.

Again, we take voltage V_{g1} as the reference phasor, so the magnetizing current I_{m1} must be a pure imaginary number; the per unit frequency is naturally a real number, and the air-gap voltage V_{g2} must leads the magnetizing current I_{m2} by 90 degrees, which gives three more equations

$$\mathrm{Re}\left(I_{m1}\right)=0 \tag{17}$$

$$\mathrm{Im}\left(F\right)=0 \tag{18}$$

$$\arg\left(V_{g2}\right)=\arg\left(I_{m2}\right)+\pi/2 \tag{19}$$

where arg refers to the function that returns the (phase) angle of a complex number.

4. Analysis results of the two-generator model

Solving equations (9)-(19) using the Mathcad program based on the Newton-Raphson algorithm gives the voltage and frequency characteristics. In this paper, we assume that the two SEIGs have the same rating and the same parameters so that the influence of each generator is visible. We first check the curves of V_L and F when the two generators have the same speed. This is a basic test because these characteristics must be comparable to the case of a single generator. Figure 4(a) shows the curves of V_L against P_L. We can check when the excitation capacitance is 28 μF, the maximum load power is 1000 W, being double the peak load power of 500 W for C=14 μF shown in one of the curve in figure 2(a).

When the two SEIGs run at different speeds, the contour plots of F and P_L for R_L=500 Ω and C=30 μF are depicted in the v_1-v_2 plane in figures 5 (a) and (b). For the same values of R_L and C, the output voltage V_L is expressed by a 3D plot as a function of v_1 and v_2 in figure 5(c). Figure 5(d) is the contour plot of V_L over the v_1-v_2 plane. The contour plots of figures (a), (b) and (d) are symmetric about the 45-degree line, showing equal contribution of the two SEIGs since they have the same rating and same parameters.

An interesting characteristic is the active power shared by each generator when their speeds vary. The ratio of powers P_1/P_2 is a good indicator showing the distribution of active over the two generator. The contour of P_1/P_2 over the v_1-v_2 plane is depicted in figure 6(a) in which P_1/P_2=1.0 is the 45-degree line, conveying that the active power is uniformly shared by the two generator when they run at the same speed. To the left of the 45-degree line denotes the situation of $v_1<v_2$ where the power ratio $P_1/P_2<1$. On the other hand, to the right of the 45-degree line is the area for $v_1<v_2$ where the power ratio is greater than one.

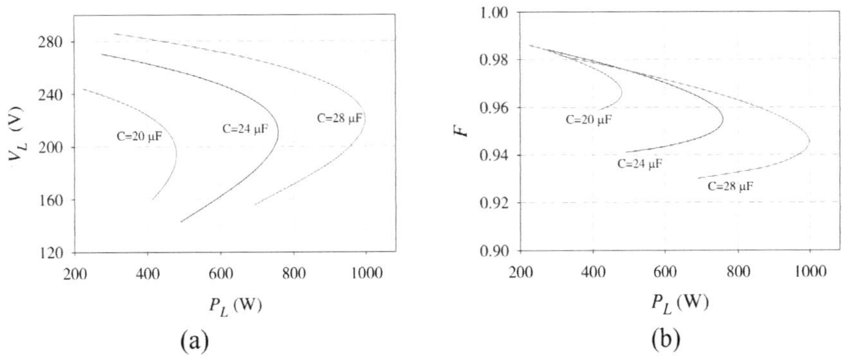

Figure 4. (a)Voltage and (b) frequency characteristics, when both SEIGs rotate at 1 pu.

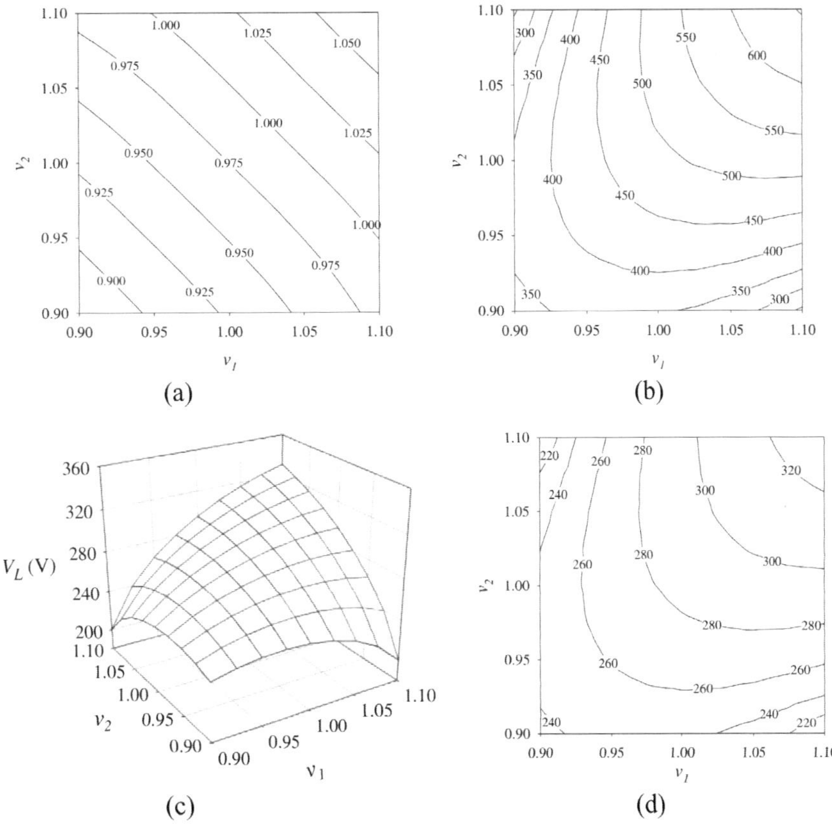

Figure 5. Output characteristics when the two SEIGs rotate at different speeds. (a) contour plot of the per unit frequency, (b) contour plot of the output power P_L, (c) 3D plot of V_L, (d) contour of V_L.

Figure 6(b) is another method to express how the active power is shared between the two generators. In figure 6(b) the power ratio P_1/P_2 is expressed as a function of v_2 while taking v_1 as the parameter. It can be seen that all the curves intersect the line $P_1/P_2=1$ when the two speeds v_1 and v_2 are equal. For example, the curve representing $v_1=1.04$ p.u. intersects the line $P_1/P_2=1$ at $v_2=1.04$ p.u. It is worth to note that when the curves in figure 6(b) drop below $P_1/P_2=0$, the generator G_1 becomes too slow to generate electricity. In this case, G_2 runs much faster than G_1. G_1 then works as a motor, absorbing electric power from G_2 and no longer able to generate electricity.

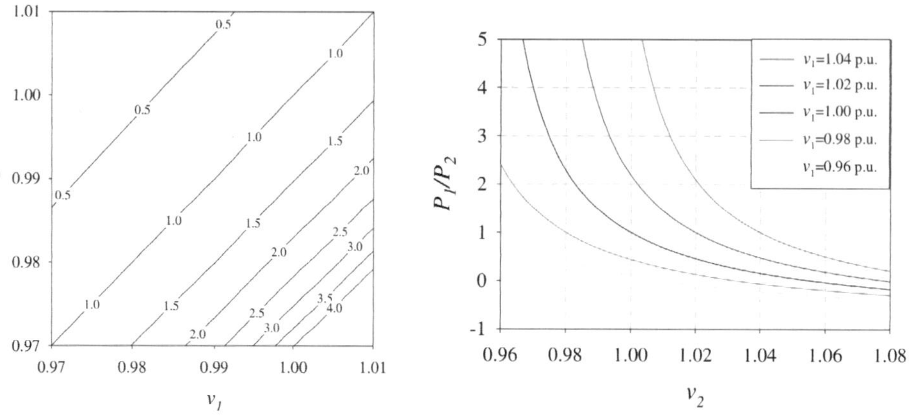

Figure 6. Output characteristics of the ratio P_1/P_2 when the two SEIGs have different speeds. (a) contour plot of P_1/P_2, (b) ratio P_1/P_2 as of function of v_2 taking v_1 as the parameter.

5. Conclusion

The equivalent circuit model of two induction generators operating in parallel has been proposed. Mathematical formulation of one single generator is given first. This is followed by extending the formulation to the case of two generators. Resolution of a set of linear and nonlinear equations allows the characteristics such as the output voltage, frequency, total active power to be observed through the analysis results. Sharing of the total active power between the two generator when they rotate at different speeds can also be effectively analyzed. A more extreme case of one of the generators turning from generating mode to motoring mode can also be simulated and observed. The formulation presented in this paper can easily be extended to a system containing more induction generators working in parallel.

6. References

[1] Bansal R C 2005 Three-phase self-excited induction generators : an overview *IEEE Transactions on Energy Conversion* **20** pp 292–9
[2] Haque M H 2012 Capacitance requirement in a three-phase SEIG under no-load and load conditions *IEEE Intl. Conf. on Power System Technology (New Zealand)*, p 1–6
[3] Selmi M and Rehaoulia H 2013 A simple method for the steady state performances of self-excited induction generators *Intl. Conf. Electrical Engineering and Software Applications (ICEESA) (Tunisia)* p 1–4
[4] Wang Y J and Huang Y S 2009 Analysis of a stand-alone three-phase self-excited induction generator with unbalanced loads using a two-port network model *IET Electric Power Applications* **3** p 445–52
[5] Al-Bahrani A H and Malik N H 1993 Steady-state analysis of parallel-operated self-excited induction generators *IEE Proceedings C-Generation, Transmission and Distribution* **140**, p 49-55
[6] Haque M H 2008 Demonstration of parallel operation of single-phase and three-phase induction generators for off-grid operation *IEEE Intl. Conf. on Sustainable Energy Technologies (Singapore)* p 40–5
[7] Chakraborty C, Bhadra S N and Chattopadhyay A K 1999 Analysis of parallel-operated self excited induction generators *IEEE Trans. on Energy Conversion* **14** p 209–16

IOP Conf. Series: Materials Science and Engineering **635** (2019) 012022 doi:10.1088/1757-899X/635/1/012022

EMG control of a 3D printed myo electric prosthetic hand

U Farooq[1], U Ghani[1], S A Usama[1] and Neelum Y S[1*]

[1]Department of Mechatronics Engineering, Air University, Islamabad, Pakistan

[1*]neelumyousafsattar@gmail.com

Abstract: Recent technological advances have enabled the prosthetic developers to derive an ideal replacement for the human arm in near future. This research presents design of a 3D Printed Myo Electric Prosthetic Hand that grasp symmetrical objects through Electromyographic signals and is intended for people suffering from Trans-radial Amputation. In the current study, a set of data about three motions that include hand open, close and rest position was been acquired from forearm muscles of a human arm. The computer aided model of the prosthetic hand was developed in SolidWorks® which was later 3D printed using Poly Lactic Acid (PLA).The proposed design was based closely on tendon actuation, related to human hand functionality. Signal acquisition and processing has been done using Myo Armband. Different features were selected such that they can be passed through classification process in order to control the hand motions. Using the Bluetooth transmitter, the filtered data was sent and saved in Arduino Uno® controller. Later, the Support Vector Machine (SVM) was been evaluated as a classifier. The classification accuracy obtained from SVM was 96.7%. The results were found significant ($p<0.01$) for twelve able bodied subjects. EMG based grasp control was implemented with successful testing of designed prosthetic hand for three different motions. The practical effectiveness of the Myoelectric Hand was demonstrated by grasping household objects such battery charger, wallet and water bottle.

1. Introduction

During the previous decade, great efforts has been done to improve the human-machine interfaces (HMI) and generating them in a more intuitive way. In order to accomplish HMI, an accurate body signal interpretation is required. Currently, the myoelectric prosthesis is the most advanced commercially available powered trans-radial prosthetic device [1]. The surface electromyogram (EMG) signals of amputee's residual muscles are used to provide the input signal and control the prosthetic device [1-3]. Numerous researches have revealed that electromyographic signals (EMG) are one the most feasible option for this application, as it provides very accurate information regarding limb motions [4][1]. The use of electromyographic signals has been spread out by the medicine field, implementing it on prosthetic devices for amputee[5], for patients who have suffered from limb paralysis[6]. In reference to [8], the measurements acquired from the Inertial Measurement Unit (IMU) worn at the wrist, and the Electromyogram (EMG) of muscles in the forearm were fused together in order to infer hand and finger movements. Furthermore, 12 gestures were defined as a set and Support Vector Machine (SVM) algorithm was implemented to classify and separate activity within the defined gesture classes.

Soon, machines are supposed to be making decisions in any arbitrary environment. They will be able to think and act like humans, which is out of practice in today's human. Development in research sector has been evolved and new trails are emerging. One of them is Machine-learning technique.

Under such a topic comes a bunch of ideas that can be implemented on a machine so that it can learn and behave better than it was in the past[9].

In order to distinguish the data from any daily life problem, classifiers are used. Classifiers cannot work standing alone. There is a need to feed right data into right classifier. Typically, a classifier with several parameters is flexible, but there are also exceptions. Feature selection followed by extraction plays a vital role in performance of any classification technique. For a smaller number of classes, Support Vector Machine (SVM) is preferred in general practice. A classifier that minimizes the sum of training error and a term that is a function of the flexibility of the classifier is thought to be a good classifier. SVM offers relatively easy training. Unlike neural networks, there is no local optimal present. Dimension scaling is carried out significantly; such that it gives better results, .One can easily control classifier complexity.

Raw data like can be used as input to SVM, instead of feature vectors. Virtual environments, as Solid works® provides tools so that a there is no need to manufacture a physical device rather than the feasibility of any design can be tested and implemented, virtually off course.[10] The computer-aided models are been designed in such environment to check how a device would work in reality.

Therefore, this paper proposes design and EMG based grasp control of trans-radial prosthetic device. Next section of the paper proposes the design and the mechanism of the trans-radial prosthetic arm. In section II, EMG Signal Acquisition and processing is presented. In section III, classification results and hardware results are discussed. The final section presents the conclusion

2. Computer Aided Design and Torque Calculations

2.1 Computer Aided Design

The proposed model of the prosthetic hand was designed using Solidworks®. In the proposed design, all fingers have revolute joints and works on the principle of Joint Coupling Mechanism. This mechanism helps to achieve synchronized motion of fingers through lesser number of actuators[7]. The proposed design is represented in Figure 1.

2.2 Design Calculations

The proposed calculations in this section are valid for symmetric objects frequently used in daily life. To find the mathematical relations, it has been supposed that
If '*x*' is the mass of the object, then;

$$\mathrm{W} = \mathrm{x} \times g \quad (1)$$

and;

$$M = \frac{W}{\mu} \quad (2)$$

Where, $\mu = 0.87$, $g = 9.81\frac{m}{s^2}$, μ= Coefficient of rubber,
M= Required value of force for one finger.
For two points of contact,

$$P = \frac{M}{2}$$

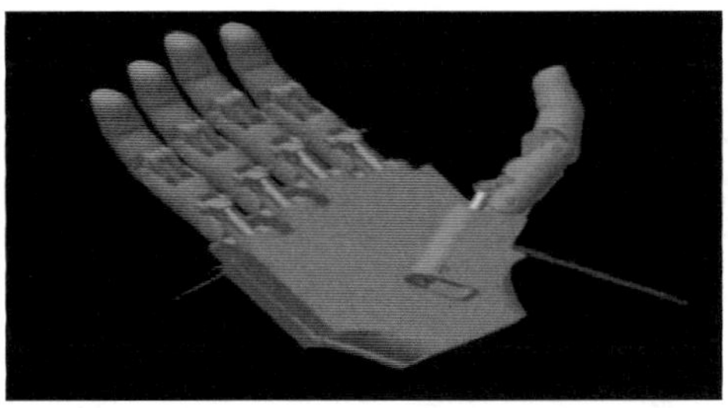

Figure 1. CAD Model of Myoelectric Hand

The calculated values for required force while lifting different weights are represented in Table.1 as

Table :1 Calculated Force for Different Weights

Sr. No.	Mass in Kg	Required Force in N	Constant Point Force in N	Torque in Nm
1	0.7	7.8	1.57	0.2041
2	0.5	5.63	1.12	0.1456
3	0.3	3.38	0.67	0.0871
4	0.2	2.25	0.45	0.0585

3. Signal Extraction and Processing

Electromyography is a technique used to evaluate and record the electrical activity produced by skeletal muscles upon activity. Thalamic Labs product MYO Arm Band has been used to acquire EMG signals. It is an application-based device which can receive surface EMG data at about 200Hz.[11] It consists of 9 axes inertial IMU which contains a gyroscope, accelerometer and magnetometer which are of three axes each respectively as illustrated in figure 3.[5] Working steps of the proposed system is shown in figure 2.

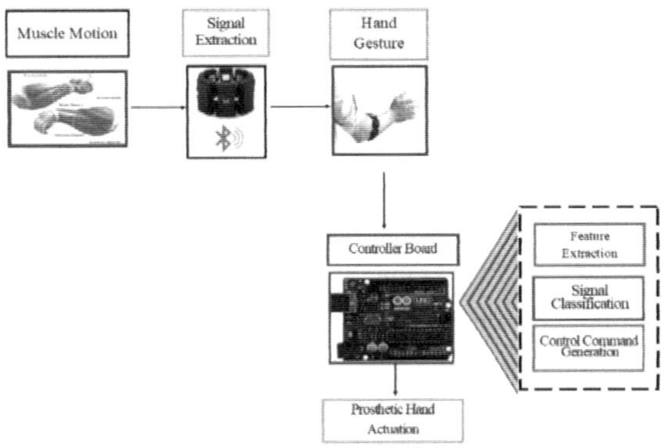

Figure 2. Flow Diagram

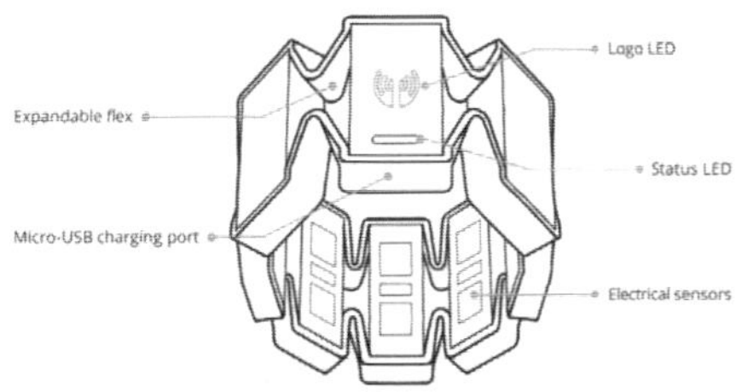

Figure 3. Myo Armband Vectoral Representation

According to [9], Myo provides two types of data for an application: spatial and gestural data. The special data notifies about the orientation and movement of subject's arm. [12]Whereas, the gestural data represents the subject hand motion. The EMG signals were acquired through MYO Capture and then were used to extract features from the signals.

6 males and 4 females (age: 20 ±6) participated in experimentation for data collection. Able-bodied, right-handed, and reported to be 100% functional working hands subjects were chosen. The average wrist size of the 10 participants were found to be 16.3 ±1.1 cm with forearm size of 24.2 ±1.4 cm. The experimental model is shown in Table 2

Table 2 Experimental Model

Total Subjects (9 males & 3 females)	12
Initial Rest Time	30 sec
No. of trials (per subject)	5
Resting Time	3 min.
Time for signal extraction	35 sec

3.1. Signal Investigation

The extracted signals were analyzed for further processing in MATLAB®. Figure 4 symbolizes raw EMG signals extracted from able bodied individuals. It also represents electrodes activity upon forearm muscle action.

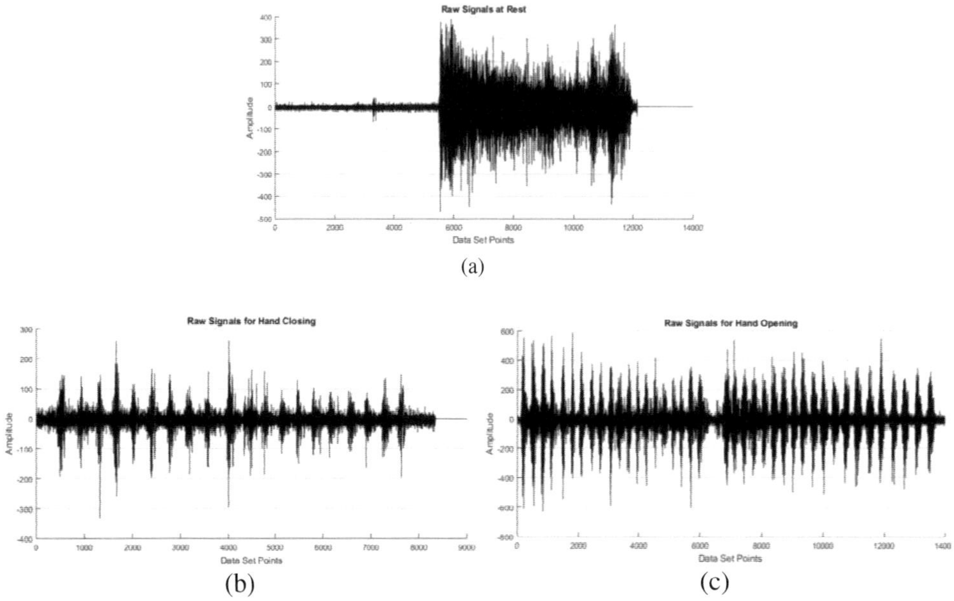

Figure 4. Raw Electromyographic signals of an able-bodied individual for a) Hand close b) Hand open and c) rest position

3.2. Feature Selection

Features measure characteristics from input data, and thus plays an important role when it comes to pattern recognition system design. In this study, feature extraction is carried out on the reference(input) EMG signals to reduce the data dimensionality such that the signal patterns which help to distinguish between the gesture classes were undisturbed. In order to portray an object to be recognized by measurements whose values are alike for objects in same category, and very different for objects in different categories, these characteristics are to be separated [13]. The feature set included MAV(Mean Absolute Value), VAR(Variance), WL(Waveform Length), Kurtosis and Peak. Results in Figure 5. showed that peaks & wave length feature showed the most accurate behavior.

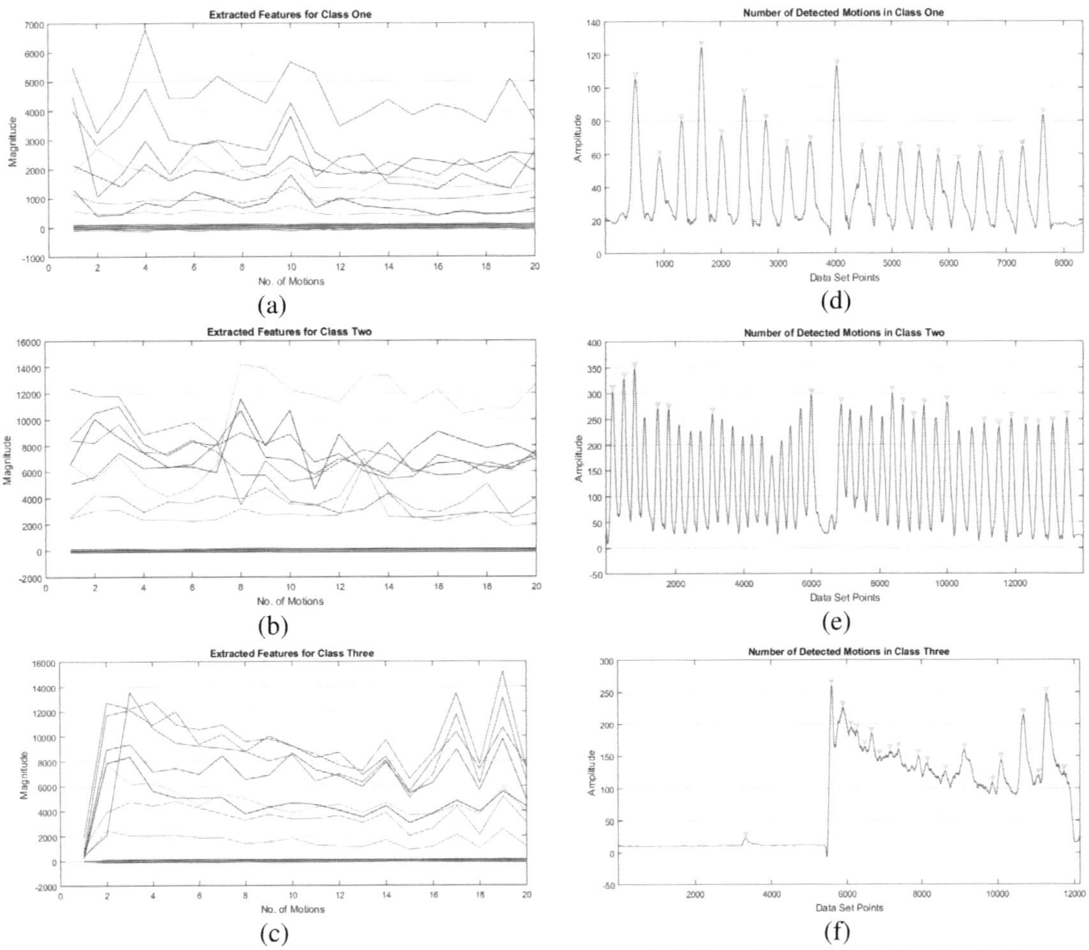

Figure. 5 Extracted features for a) Class A b) Class B c) Class C d) Turning Points or Detected Peaks Showing the Frequency as the Motion is Performed for Class A e) Class B f) Class C

3.3. Support Vector Machine (SVM)

Support Vector Machine (SVM) is an easy understanding classifier. A classifier offers linear algorithms in the output or the feature capacity that they are equivalent to a non-linear algorithm as compared to the input. Standard linear algorithms generally characterized to their non-linear form by analyzing its feature space. SVM is a useful alternative to neural networks. The concept of this classifier is that support vectors are the samples closest to the separating the raw data or featured data. They are the most difficult patterns to classify.[14]

For a smaller number of classes, Support Vector Machine (SVM) is preferred in general practice. A classifier that minimizes the sum of training error and a term that is a function of the flexibility of the classifier is thought to be a good classifier. SVM offers relatively easy training. Unlike neural networks, there is no local optimal present. Dimension scaling is carried out significantly; such that it gives better results, .One can easily control classifier complexity. Raw data like can be used as input to SVM, instead of feature vectors.

Current algorithms for defining the SVM classifier include subgrade gradient and coordinated decline. Both practices have established to offer major advantages over the old-fashioned tactic when dealing with large, uncommon data sets. Subgrade approaches are particularly efficient when there are several training examples and coordinate descent plays well in the ground when the feature possibility dimension is high.

4. Experimental Results

The functional effectiveness of the anticipated Myoelectric Hand was demonstrated by grasping a variety of household objects. The prosthetic device was able to grasp objects such as disposable glass, wallet, PEPSI can, smart phones & water bottles etc. These actions required slight or no assistance at all. Figure 6 represents the 3D printed prototype of the device that was used to check the feasibility of the proposed CAD design.

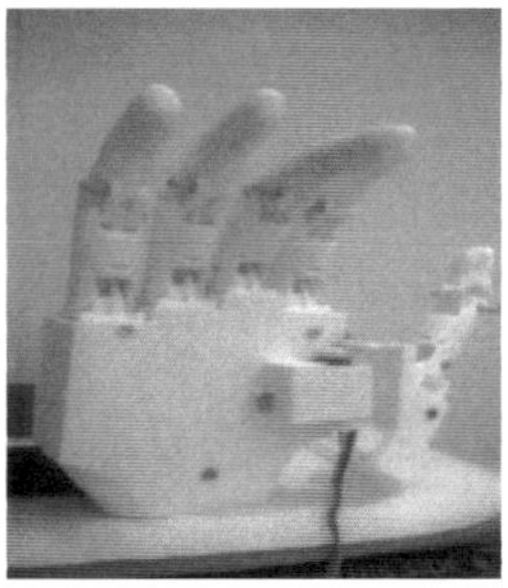

Figure 6. Prototype of the hand during construction phase

The accuracy achieved from SVM is 96.7%. It depicts that the classifier is not doing any mistake as for differentiating the signals coming from human muscle hence resulting in reduced error or no error at all. The real time procedures are feasible with this much accuracy and the system can be implemented on any patient.

The primary task was to move the fingers according to the EMG sensed data that was been achieved successfully. However, there were some problems present in the finger movement due to wrong classification of EMG signals. The Myo-electric hand gripped the following objects without damaging them,

a. Disposable glass
b. Pepsi bottle
c. Wallet
d. Pepsi Can
e. Air-Conditioner remote
f. Smartphone

All these objects were gripped using prismatic power grip. In prismatic power grip, force was been applied between the fingers and palm that in contrast with circular pinch grip is better.

The results for holding real life objects are shown in Figure 7.

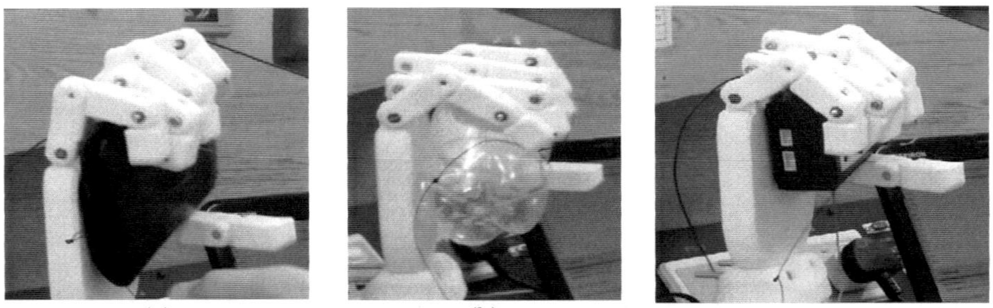

(a) (b) (c)

Figure 7. The prototype is grasping (a) Charger (b) Wallet (c) Hand Holding Water Bottle

5. Conclusion

The results obtained from this research work shows that the day is not far when we will have a complete functional artificial hand. The user must bring the Myoelectric Hand near the object and with the control commands generated based on the EMG signal the hand will start opening or closing based on the intentions coming from the muscle. The controller is responsible for classification of real time signals to actuate the motors. As the accuracy is more than 95%, imparting that it is a rare chance for the hand to behave abnormally. The hand will remain in the desired position as organized from the controller.

References

[1] T. Pistohl, D. Joshi, G. Ganesh, A. Jackson, and K. Nazarpour, "Artificial proprioceptive feedback for myoelectric control," *IEEE Transactions on Neural Systems and Rehabilitation Engineering,* vol. 23, pp. 498-507, 2015.

[2] D. Madusanka, L. Wijayasingha, R. Gopura, Y. Amarasinghe, and G. Mann, "A review on hybrid myoelectric control systems for upper limb prosthesis," in *Moratuwa Engineering Research Conference (MERCon), 2015*, 2015, pp. 136-141.

[3] A. L. Fougner, Ø. Stavdahl, and P. J. Kyberd, "System training and assessment in simultaneous proportional myoelectric prosthesis control," *Journal of neuroengineering and rehabilitation,* vol. 11, p. 75, 2014.

[4] M. A. Oskoei and H. Hu, "Myoelectric control systems—A survey," *Biomedical Signal Processing and Control,* vol. 2, pp. 275-294, 2007.

[5] J. Ding, R.-Z. Lin, and Z.-Y. Lin, "Service robot system with integration of wearable Myo armband for specialized hand gesture human–computer interfaces for people with disabilities with mobility problems," *Computers & Electrical Engineering,* 2018.

[6] J. Ma, N. V. Thakor, and F. Matsuno, "Hand and wrist movement control of myoelectric prosthesis based on synergy," *IEEE Transactions on Human-Machine Systems,* vol. 45, pp. 74-83, 2015.

[7] F. Cordella, A. L. Ciancio, R. Sacchetti, A. Davalli, A. G. Cutti, E. Guglielmelli, *et al.*, "Literature review on needs of upper limb prosthesis users," *Frontiers in neuroscience,* vol. 10, 2016.

[8] M. Georgi, C. Amma, and T. Schultz, "Recognizing Hand and Finger Gestures with IMU based Motion and EMG based Muscle Activity Sensing," in *Biosignals*, 2015, pp. 99-108.

[9] W. Guo, X. Sheng, H. Liu, and X. Zhu, "Toward an Enhanced Human–Machine Interface for Upper-Limb Prosthesis Control With Combined EMG and NIRS Signals," *IEEE Transactions on Human-Machine Systems,* 2017.

[10] S. S. Esfahlani, B. Muresan, A. Sanaei, and G. Wilson, "Validity of the Kinect and Myo armband in a serious game for assessing upper limb movement," *Entertainment Computing,* 2018.

[11] V. Becker, P. Oldrati, L. Barrios, and G. Sörös, "TouchSense: Classifying and Measuring the Force of Finger Touches with an Electromyography Armband," 2018.

[12] T. LabsTM, "Myo SDK Manual: Getting Started," *Consulted November,* vol. 4, 2016.

[13] A. Phinyomark and E. Scheme, "A feature extraction issue for myoelectric control based on wearable EMG sensors," in *Sensors Applications Symposium (SAS), 2018 IEEE*, 2018, pp. 1-6.

[14] A. Phinyomark, R. N Khushaba, and E. Scheme, "Feature Extraction and Selection for Myoelectric Control Based on Wearable EMG Sensors," *Sensors,* vol. 18, p. 1615, 2018.

IOP Conf. Series: Materials Science and Engineering **635** (2019) 012023 doi:10.1088/1757-899X/635/1/012023

Development of an Automated System for Three-Dimensional J-Integral Analysis

Joon-Seong Lee

Dept. of Mechanical System Engineering, Kyonggi University,
154-42 Gwanggyosan-ro, Yeongtong-gu, Suwon, Gyeonggi-do, 16227 Korea

jslee1@kyonggi.ac.kr

Abstract. By integrating the 3D solid modeler with the universal FEM code, an automatic nonlinear analysis system for 3D crack problems has been developed. A geometry model, that is, a solid containing one or several 3D cracks is defined. Multiple distributions of local node density are selected and automatically overlapped with each other over the geometry model using fuzzy knowledge processing. Nodes are created by the bucketing method, and tetrahedral square tetrahedral solid elements are generated by the Delaunay triangulation technique. A complete finite element (FE) model is created and stress analysis is performed. In this system, analysts can significantly reduce the burden of introducing 3D cracks into FE models and estimating fracture mechanics parameters. This paper describes a methodology for realizing such a function and demonstrates the validity of this system.

1. Introduction

Natural cracks often found in structures usually appear as cracks with three-dimensional properties called surface cracks. In order to predict the growth rate and fracture strength of surface cracks, it is necessary to analyze the stress intensity factor K of surface cracks accurately. When the finite element method is applied to the analysis of fracture mechanics, the difficulty of the numerical analysis and the complexity of the experimental parameters have not reached the level that can be applied to everyday design in view of the importance of practical problems. The previous generation of mesh generation techniques, which had to be manually dependent, have been developed by several researchers in recent years [1-3]. However, many of the existing techniques for element generation have been reported to be limited when applied to real problems, and are not adequately supported when applying the 3-D cracking problem [4]. Three-dimensional crack analysis is likely to be a large-scale analysis and requires a tremendous amount of calculator power. In addition, special element partitioning is required near the crack tip, which is a stress-specific element, and therefore it takes a great deal of time and effort to divide the element for three-dimensional analysis and to prepare input data. Therefore, there is a need for an algorithm that automatically and rapidly generates a large-scale finite element model with minimal manipulation. The authors have developed the automatic generation system of elements using fuzzy theory from the information of shape, load, and boundary conditions of structures by using 3D solid modeler with the aim of efficiency of large scale analysis so far [5,6].

In addition, a system for automatically creating and analyzing input files of general-purpose finite element analysis codes has been developed and an automatic K analysis system for three-dimensional cracks of structures has been developed [7]. However, the conventional system has a limitation on the shape of the analytical model with respect to the symmetrically existing cracks in the structure. Also, it

IOP Conf. Series: Materials Science and Engineering **635** (2019) 012023 doi:10.1088/1757-899X/635/1/012023

was a system constructed only for K value as the fracture mechanics parameter. Therefore, in this study, the finite element model is automatically generated with minimal manipulation even when there is a three-dimensional crack at an arbitrary position other than the symmetric plane, and then the J integral, which is a nonlinear fracture mechanics parameter of the crack, can be calculated.

2. Outline of system

It is possible to fully and automatically generate a large number of tens or hundreds of thousands of finite element models with controlled element sizes by performing simple and easy manipulation of geometric models in complex geometrical actual structural devices. The process of this system is to define the shape (geometry model), (b) to define and register the shape of the crack, (c) to specify the boundary conditions and material properties for the geometry model, (d) designation of node density, (e) generation of node, (f) generation of element and smoothing, (g) boundary conditions and material property values to elements, (h) FE analysis, (i) calculation of fracture mechanics parameters (J integral calculations). Among them, (a) to (d) are interactive tasks that the user directly performs on the geometric model, and all others are tasks that are automatically processed.

2.1. Geometric modeling

The geometric model is based on the geometric model definition function of Designbase [8], which is a universal CAD system capable of expressing free curved surfaces such as glass Bezier curved surface and glass Gregory patch, and has abundant library functions for shape processing. In the subsequent processing, necessary information on the shape data can be obtained by using the library function.

2.2. Define the shape of 3D crack

In order to define the shape of a three-dimensional crack at an arbitrary position, the user needs to input the following information.

(1) The center coordinates (x, y, z) of the ellipsoid representing the 3D crack
(2) Radius of the ellipsoid (rx, ry, rz)
(3) Parameters required to create an ellipse from an ellipsoid

Enter the above information for the number of cracks you want to define. A more detailed explanation is given in section 3.

2.3. Enter B.C and material properties into geometric model

Specify vertices, ridges, and loops that make up the geometric model by clicking with the mouse. Also input the type and numerical value of the boundary condition to be added to the component of the specified geometry model.

2.4. Designation of node density

The node density is a function of a) distribution corresponding to local stress concentration, b) distribution that uniformly divides within the finite domain, and c) distribution that uniformly divides the entire area. As a result, the user chooses them according to the object to be analyzed, and if the position is specified, the fuzzy theory automatically calculates the distribution of the nodal density over the entire region of the geometric model [9].

2.5. Generation of node and element

After obtaining the distribution of the density over the entire analysis area, the nodal point is automatically generated using the bucket method [10], which is one of the computational geometric methods. This process uses nodes in the geometry library of the Designbase because it also creates joints on the surface and inside of the geometry model. The generated nodes are well-known to produce a tetrahedron element at high speed using Delaunay's triangulation method [11,12]. When using the Delaunay method, the concave geometry model will generate an element outside the shape as well, so check the inner / outer determination as the center of the element to remove the externally generated element. In addition, a distorted element may sometimes be generated in the vicinity of the junction between the node pattern and the boundary, so that the element shape is modified by introducing the Laplacian smoothing technique [13].

2.6. Boundary condition and property value to element

In this system, the original shape elements (vertices, ridges, and planes) to which the created elements and the nodes belong to a data structure that preserves them as information. Therefore, boundary conditions and material properties directly assigned to the shape model by the user in the work of 2.3 are automatically assigned to the vertices (nodes), sides, faces and inside of the element automatically after the element is created. As a result, the finite element model (element + material property + boundary condition), which is the input file of the finite element analysis code, is automatically generated.

This automatic analysis system can be applied to any FE analysis system in principle, but the current version is able to output data corresponding to the tetrahedron primary and secondary elements used in the general purpose analysis code MARC [14].

3. Introduction of 3D cracks

Assuming that the three-dimensional crack plane is a semi-elliptical shape parallel to the x-y plane, the present system uses the Delaunay method as an element generation method. However, since the Delaunay method is arbitrarily divided by the distribution of the nodes where the elements are generated, it is necessary to adjust the nodal distributions in advance so as not to connect the nodes located above and below the crack plane. Therefore, we propose a cracking method that cuts off the ellipsoid by focusing on the similarity of the crack topology as the ellipsoid. Specifically, when defining the shape of the structure, the ellipsoid is cut at the position where the crack is to be inserted so that the upper and lower surfaces of the crack are not connected. After the element is generated as shown in Fig. 1, the ellipsoid shifted in the z-direction of the nodal point is collapsed to realize the expression of the cracked portion. Where rx is the major radius of the crack, ry is the depth of the crack, and rz is the radius of the ellipsoid in the z direction. It is also necessary to generate an ellipse from the ellipsoid input in Section 2.2 and h is a parameter indicating a range for moving the node.

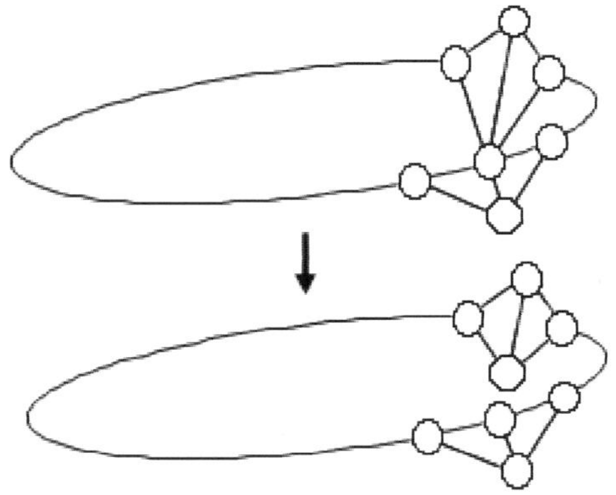

Figure 1. Mapping of an ellipsoid

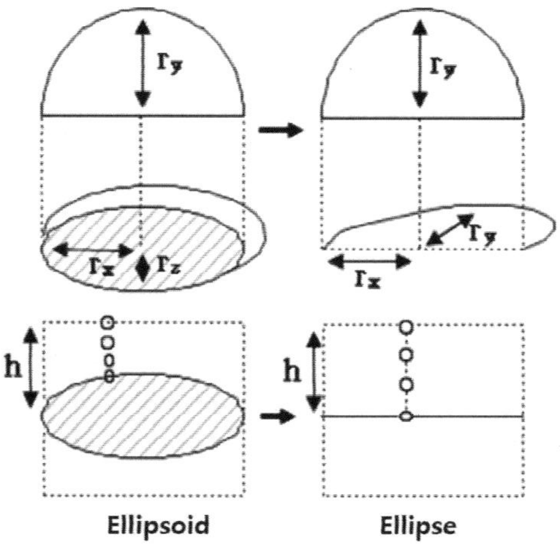

Figure 2. Example of mis-match elements

When defining the shape using the design base, the user can express the desired crack by inputting only the location of the crack center and the rx, ry, rz, and h above.

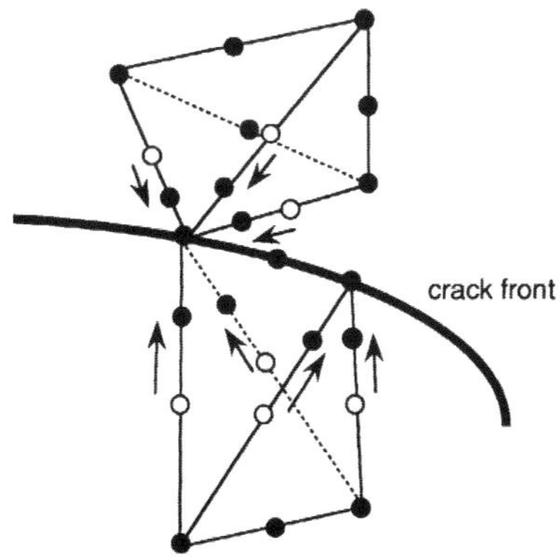

Figure 3. Conversion of quadratic tetrahedral elements along crack front into singular elements

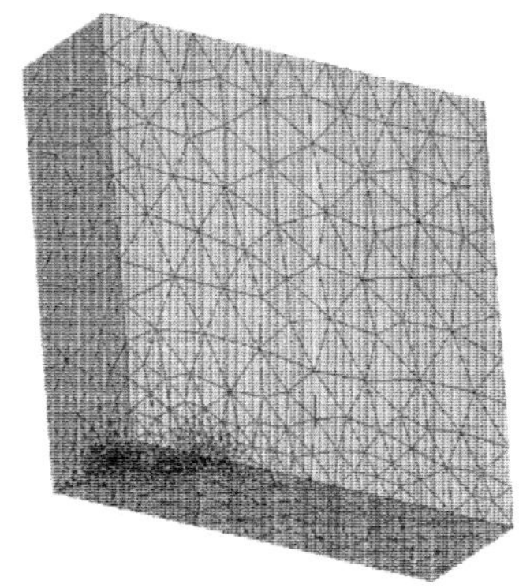

Figure 4. A typical mesh with a surface crack

After inputting the parameters, the cracks are automatically generated, and the following three processes are performed in the automatic analysis system. First, the point on the surface of the ellipsoid is shown in Fig. 2. In some cases, it may be combined with an inappropriate element. This is because, as described in Section 2.5, the elements inside and outside are determined by the inner and outer determinations of the element center, but figure 2 shows the state where the element center remains inside the model. In this case, we introduced a node addition algorithm [9] that eliminates this coupling before collapsing the ellipsoid. Secondly, along with the node movement, the element may be deformed so that the volume is close to zero or minus. In this case, after the ellipsoid is converted into an ellipse, the volume is restored by adding one node vertically so as not to deviate from the crack shape. Finally, Barsoum's singular element [15] is placed at the tip of the crack. That is move the position of the intermediate node to the crack tip side by 1/4 as shown in Fig. 3.

By introducing the process as described above, it is possible to analyze even when the three-dimensional crack is not symmetrically or on a symmetric plane. This technique is applicable to the cracks in the symmetric plane that were analyzed in the previous system, and the inherent cracks can also be analyzed. Fig. 4 shows an example of element partitioning for a 1/4 model of semi-elliptical surface cracks symmetrically to the plate. Also, Fig. 5 shows an example of element partitioning of a flat plate with three surface cracks.

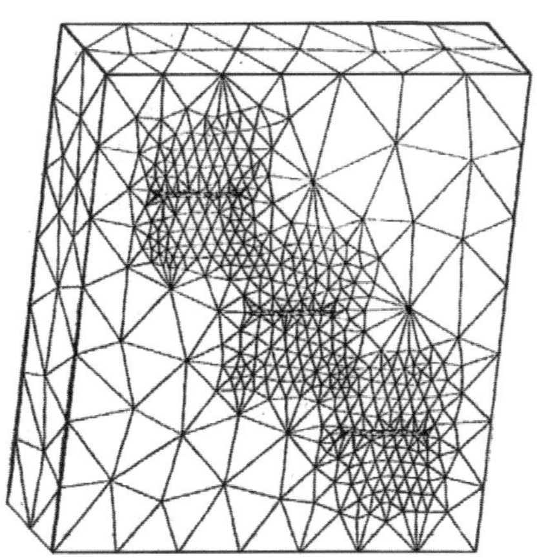

Figure 5. A typical mesh with several cracks

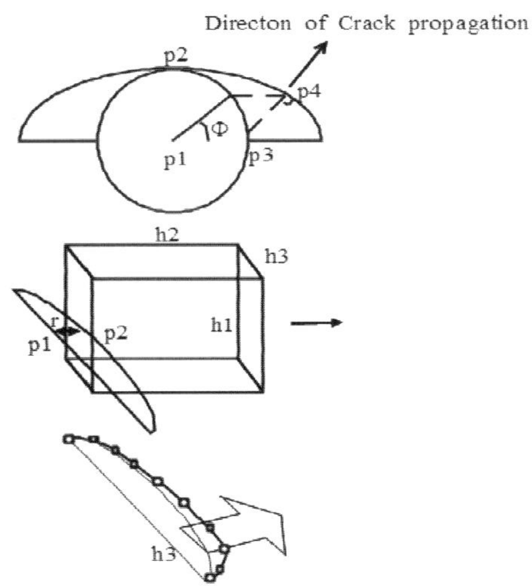

Figure 6. Evaluation of J-integral value

4. Evaluation of J-integral analysis

In this system, the virtual crack propagation method which can be interpreted by MARC and the Lorenzi method which is the calculation method of extension J integral can be used as the J integral calculation method. The user selects the calculation method of the J integral for a plurality of cracks, angle Φ is shown in Fig. 6, and the J-integral computes the crack virtually in the normal direction of the point p4.

Therefore, we can see that J integral is a function of Φ. The user also inputs parameters r, h1, h2 and h3 as shown in the figure. Where r is the ratio of the crack depth to the distance from p1 to the path. The effect of stiffness change can be achieved by moving the node of a certain region. Therefore, in the system, the value of the J integral is calculated by obtaining the change of the strain energy generated by micro-distance movement of the node in the specified rectangular parallelepiped and dividing by the advancement area of the crack represented by the travel distance. The input of these parameters is performed when defining the shape using the Designbase.

However, care must be taken because the number of movements of the nodal point of the crack varies depending on the h3 value. This is because the amount of change of the strain energy and the advancement area of the crack vary greatly. It is better to move two or more joints than one joint movement. However, if too many joints are moved, the J integral at the specified position may not be obtained correctly due to the angle dependency.

In order to verify the effectiveness of the system, it is necessary to analyze the case where semi-elliptical surface cracks with the aspect ratios a / c = 0.6 and a / t = 0.2 exist on the plate under tension load (a) respectively. The finite element model used here is shown in Fig. 4, the number of elements used is 2,980, and the number of nodes is 5,840. The J integral is related to K in the elastic analysis, so it is compared with the non-dimensionalized value obtained by Raju-Newman's equation [16]. The calculations were made by varying the nodal shift by narrow pass for 1 node movement and broad pass over 2 nodes. Fig. 7 shows the results of the analysis. It is found that the values of narrow pass and broad pass are similar.

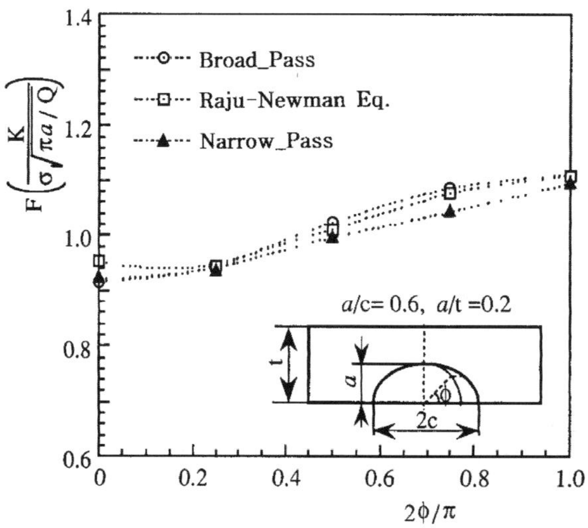

Figure 7. Comparison of present stress intensity factor with Raju-Newman equation

5. Conclusions

In this paper, we constructed a system that enables the analyst to easily find the J integral value for the three - dimensional cracks with minimal interactive manipulation. In particular, we have developed an automatic system that performs a series of input to output operations for nonlinear analysis by adding the introduction function and the analysis function of a three-dimensional crack existing in an arbitrary position. In order to verify the effectiveness of this system, the analysis of the plate with semi-elliptical surface cracks was performed and compared with the results of Raju-Newman equation.

6. Acknowledgments
This work was supported by the GRRC program of Gyeonggi province. [2018-0261, Research on Innovative Intelligent Manufacturing system]

7. References
[1] Sezer, L. and Zeid, I. 1991 Automatic Quadrilateral/Triangular Free-Form Mesh Generation For Planar Region, International Journal for Numerical Methods in Engineering, Vol. 32, pp 1441-1438
[2] Pourazady, M. and Radhakrishnan, M. 1991 Optimiza- tion of a Triangular Mesh, Computers and Structures, Vol. 40, No. 3, pp. 795-804
[3] Talbert, J. A. and Parkinson, A. R. 1990 Development of an Automatic, Two-Dimensional Finite Element Mesh Generator Using Quadrilateral Elements and Bezier Curve Boundary Definition, International Journal of Numerical Methods in Engineering, Vol. 29, pp. 1551-1567
[4] Sugioka, K., Lee, J. S., Yoshimura, S. and Yagawa, G., 2004 A Probabilistic Fracture Mechanics Analysis of Multiple Surface Cracks Considering their Interaction Effects, Proceedings of the '94 Annual Meeting of JSME/MMD, No. 940-37, p 257
[5] J.S. Lee, Y. Genki and M.W. Park 2006 Automatic Mesh Generation of Three-dimensional Structures, Transactions of CAD/CAM Engineers, Vol. 1 , pp. 65-75
[6] Yoshimura, S., Lee, J. S. and Yagawa, G. 2003 FEM Modeler for 3D Solid Geometry with Free-Form Surface , The 6th Computational Mechanics Conference, No. 930-71, pp. 505
[7] Lee, J. S., Yoshimura, S. and Yagawa, G. 2005 Automation of Analysis for Stress Intensity Factors of Three-Dimensional Cracks, Transactions of the Japan Society of Mechanical Engineers A, Vol. 61-587, pp. 1580-1586
[8] Chiyokura, H.2008 Solid Modeling with Designbase : Theory and Implementation, Addition - Wesley
[9] Lee, J. S., 1995 Automated CAE System for Three- Dimensional Complex Geometry, The Doctoral Thesis, The University of Tokyo
[10] Asano, T. 1995 Practical Use of Bucketing Techniques in Computational Geometry, Computational Geometry, North-Holland, pp. 153-195
[11] Watson, D. F. 2008 Computing the n-Dimensional Delaunay Tessellation with Application to Voronoi Polytopes, The Computer Journal, Vol. 24, pp. 162-172
[12] Sloan, S. W., 1997 A Fast Algorithm for Constructing Delaunay Triangulation in the Plane, Advances in Engineering Software, Vol. 9, pp. 34-55
[13] Cavendish, J. C. 1994 Automatic Triangulation of Arbitrary Planar Domains for the Finite Element Method, International Journal of Numerical Methods in Engineering, Vol. 8, pp. 679-696
[14] MARC Manual k7.2, MARC Analysis Research Corporation, Vol. A-E .
[15] Barsoum, R. S. 1984 Application of Quadratic Isoparametric Finite Elements in Linear Fracture Mechanics, International Journal of Fracture, Vol. 10., pp. 603-605
[16] Newman, J. C. and Raju, I. S. 1982 An Empirical Stress-Intensity Factor Equation for the Surface Crack, Engineering Fracture Mechanics, Vol. 15, pp. 185-192

IOP Conf. Series: Materials Science and Engineering **635** (2019) 012024 doi:10.1088/1757-899X/635/1/012024

Decomposition of a complex dynamic system with a network architecture

V A Shevchenko and M Yu Medvedev

Southern Federal University, 44 Nekrasovskiy, Taganrog, 347922, Russian Federation

E-mail: droogg@mail.ru

Abstract. The paper presents the method of decomposition of a complex dynamic network based on the diacoptic approach. The proposed method is aimed at identifying independent network segments, the removal of which from the overall structure will not affect the overall system performance. Segments are formed as simple paths in the graph model of the system with the smallest weight of incident edges and checking the condition for maintaining the balance of the product flow as applied to the transportation problem. In the general case, the presented decomposition method makes it possible to divide a complex technical object with the network architecture into subsystems, simplifying the process of analysis and control. In the case when the general task of control and the dynamics of interaction between subsystems allows the system to be divided into unrelated segments, or to limit the interaction along certain lines, the method allows reducing the load on the transmitting elements of the system.

1. Introduction

Most real technical systems are multiply connected complex dynamic objects consisting of a large number of interconnected components. For example, a network of positioned orbital satellites [1, 2], a group of mobile objects [3], networks of sensor systems [4], power system [5], distributed computer networks [6], paper machines [7], gas turbine engines [8], etc. Traditional decomposition methods are applicable to dynamical systems of small dimension. Analysis of a complex dynamic network by the traditional decomposition methods is a difficult task and often comes down to finding weak links. This can lead to the division of the system not into subsystems, but into separate components. The most preferred methods for decomposing such systems are topological methods based on diakoptics [9]. The introduction of conditions for maintaining the balance of the weights of incident edges within the subsystems will allow the network to be divided into independent segments. The functioning of these segments will be focused on the overall task of the dynamic system. Since within the subsystems (segments) the balance of the edge weights is maintained, such subsystems can be excluded from the network without affecting the system's performance. Further analysis and control of a separated dynamic system is possible by known methods [10, 11].

The paper considers the problem of decomposition of a complex dynamic system with a network architecture. The graphical model of the dynamic system is represented as a bipartite graph, where the set of vertices of the graph M is divided into two subsets X - transmitting elements and Y - consuming elements of the system, so that $X \subseteq M$ and $Y \subseteq M$, $X \cup Y = M$, $X \cap Y = M_{XY}$, where M_{XY} - a set of elements that are both transmitters and consumers depending on the state and operating conditions. The relationship between the elements of the system from X and Y is described by the product stream transferred from Xi to Yj. The product can be power in the power grid, gas flow in gas distribution systems, etc.

2. Decomposition method

The multi-connected model of the system is a bipartite graph, on the one hand of which are transmitting elements X_i, on the other hand consuming elements Y_j, each vertex of the graph has several incident edges, which reflects possible channels for transporting some product (Figure 1). The task of decomposition is to identify network sections, the allocation of which from the overall system will not affect the overall performance and the ability to achieve the overall goal of the system. This statement of the problem allows to define segments with minimal weight of incident edges. If the weight is the value of the cost of passing the resulting path, it is possible to minimize the load on the transmitting elements of the system. The most preferable channel of product transportation can be considered a way that can transfer the largest amount of product at the lowest cost. Thus, the value of losses on the way W_{ij} is put in correspondence to each edge, and the value of the produced P_{gi} or consumed P_{pj} product is put in correspondence to each vertex.

It is necessary to single out a graph in which each consumer will be associated with two transmission components capable of meeting his needs for a product with a certain supply of R. To determine these transmitters is needed to meet the minimum expenditures for transportation of the product. To do this, we construct the reduced adjacency matrix of the graph for the vertices of the fraction Y_j, where the adjacency indicator will be the weight of the incident edge to adjacent vertices:

$$A = \begin{array}{cc} & X_i \\ Y_j & \begin{bmatrix} W_{ij} \end{bmatrix} \end{array} \tag{1}$$

In case the vertices are not adjacent $W_{ij}=0$. For the considered graph (Figure 1) the reduced adjacency matrix will take the form:

$$A = \begin{array}{cc} & \begin{array}{ccccc} X_1 & X_2 & X_3 & X_4 & X_5 \end{array} \\ \begin{array}{c} Y_1 \\ Y_2 \\ Y_3 \\ Y_4 \end{array} & \begin{bmatrix} W_{11} & W_{21} & 0 & W_{41} & 0 \\ W_{12} & W_{22} & W_{32} & 0 & W_{52} \\ 0 & 0 & 0 & W_{43} & W_{53} \\ 0 & W_{24} & W_{34} & W_{44} & W_{54} \end{bmatrix} \end{array} \tag{2}$$

We define an A_{m2} matrix that will contain two adjacent vertices to Y_j (each element of the set Y) with the minimum weight of incident edges, so that the condition is satisfied:

$$A_{m2} = \left[Z^{*}_{Y\times X} \right] + \left[Z^{**}_{Y\times X} \right] \tag{3}$$

where $Z^*(n,X)=A(n,X)\rightarrow min$, $Z^{**}(n,X)=A^*(n,Y)\rightarrow min$, $A^*=A-Z^*_{Y\times X}$, $n=1,..,Y$, $Z^*_{Y\times X}$ – a zero matrix with a nonzero element $Z^*(n,X)$ in each row that is equal to the minimum element in each row of the matrix $A_{Y\times X}$.

We introduce conditions for the weights of the edges: Y_1: $W_{11}<W_{21}<W_{41}$; Y_2: $W_{22}<W_{32}<W_{12}<W_{52}$; Y_3: $W_{43}<W_{53}$; Y_4: $W_{54}<W_{44}<W_{34}<W_{24}$. Based on these conditions, we define the matrix A_{m2}:

$$A_{m2} = \begin{bmatrix} W_{11} & W_{21} & 0 & 0 & 0 \\ 0 & W_{22} & W_{32} & 0 & 0 \\ 0 & 0 & 0 & W_{43} & W_{53} \\ 0 & 0 & 0 & W_{44} & W_{54} \end{bmatrix} \tag{4}$$

Thus, two unrelated graphs are obtained (Figure 2).

To isolate the resulting graphs into separate network segments, it is necessary that two conditions be met:

$$\begin{cases} P_{pj} \le (1-R)\cdot \sum P_{gi},\ i \in A_{m2}(\mathrm{Y}_j,:),\ j \in A^*_{m2}(\mathrm{Y}_j,\mathrm{X}_i); \\ \sum P_{pj} \le (1-R)\cdot \sum P_{gi},\ i \in A_{m2}(\mathrm{Y}_j,:),\ j \in A^*_{m2}(\mathrm{Y}_j,\mathrm{X}_i). \end{cases} \tag{5}$$

where $A_{m2}(\mathrm{Y}_j,:)$ – set of connected vertices to a vertex Y_j from A_{m2} matrix; $A^*_{m2}(\mathrm{Y}_j,\mathrm{X}_i)$ – a set of vertices formed from simple paths obtained from A_{m2} matrix; P_{pj} – product quantity required by the j-th user; P_{gi} – the amount of product that can be produced by the i-th transmitting component; R – the level of the required reserve of product up to a maximum transmitter performance.

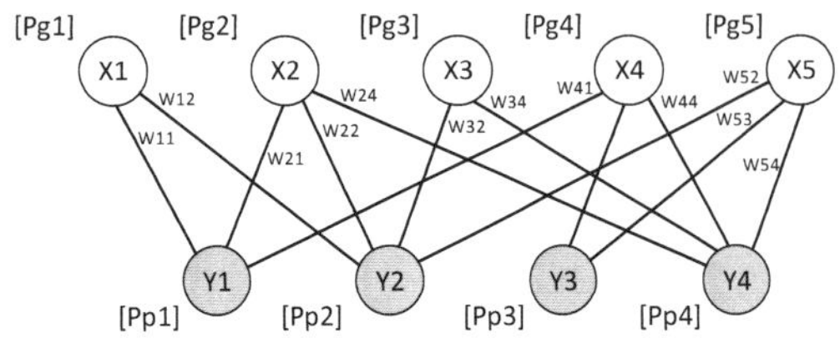

Figure 1. Segment of a dynamic system

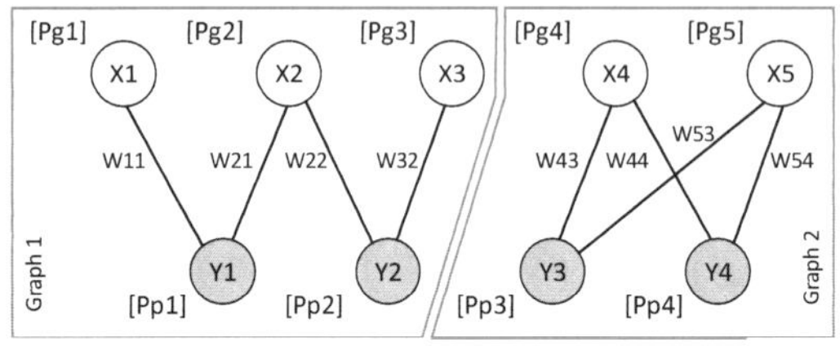

Figure 2. Decomposition of the dynamic network (1)

Check the obtained graphs (Figure 2) for compliance with the conditions (5):

$$Graph1: \begin{cases} P_{p1} \le 0.8\cdot(P_{g1}+P_{g2}); & (6.1.a) \\ P_{p2} \le 0.8\cdot(P_{g2}+P_{g3}); & (6.1.b) \\ P_{p1}+P_{p2} \le 0.8\cdot(P_{g1}+P_{g2}+P_{g3}). & (6.2) \end{cases} ;Graph2: \begin{cases} P_{p3} \le 0.8\cdot(P_{g4}+P_{g5}); & (6.1.a) \\ P_{p4} \le 0.8\cdot(P_{g4}+P_{g5}); & (6.1.b) \\ P_{p3}+P_{p4} \le 0.8\cdot(P_{g4}+P_{g5}). & (6.2) \end{cases}$$

In case of complete non-fulfillment of inequality (6.1), and (6.2) is not fulfilled, but in case of non-fulfillment (6.2) inequality (6.1) can be partially fulfilled (for example, inequality (6.1.a) - true, and (6.1.b) – wrong). It is possible that (6.2) is true and (6.1) is partially true. If all inequalities are satisfied, we can assume that the segments of the system are found. In the case where (6.1) is partially satisfied, we determine the vertices from the set Y for which (6.1) is incorrect and select the next adjacent vertex from the set X with the lowest weight of the incident edge, and form a matrix A_{m3}:

$$A_{m3} = A_{m2} + \left[Z^{*_3}_{Y\times X} \right] \tag{7}$$

in case $P_{pj}>R\cdot\Sigma P_{gi}$, $\Sigma P_{pj}\le R\cdot\Sigma P_{gi}$, where $Z^{*2}(Y_j,X)=A^{*2}(Y_j,X)\rightarrow min$, $A^{*2}=A-A_{m2}$.

Suppose that the inequality (6.1.b) for graph 1 is not executed, in this case we define the matrix $Z^{*3}_{Y\times X}$ and A_{m3}, and construct graphs 1 and 2 (Figure 3) according to the obtained A_{m3} matrix:

$$Z^{*_3}_{Y\times X} = \begin{bmatrix} 0 & 0 & 0 & 0 & 0 \\ W_{12} & 0 & 0 & 0 & 0 \\ 0 & 0 & 0 & 0 & 0 \\ 0 & 0 & 0 & 0 & 0 \end{bmatrix};\ A_{m3} = \begin{bmatrix} W_{11} & W_{21} & 0 & 0 & 0 \\ W_{12} & W_{22} & W_{32} & 0 & 0 \\ 0 & 0 & 0 & W_{43} & W_{53} \\ 0 & 0 & 0 & W_{44} & W_{54} \end{bmatrix}.$$

If partially executed (6.1) and not executed (6.2), it is necessary to look for an adjacent vertex not belonging to the set $A^*_{m2}(Y_j,X_i)$.

$$A_{m3} = A_{m2} + \left[Z^{*_3}_{Y\times X} \right] \tag{8}$$

in case $P_{pj}>R\cdot\Sigma P_{gi}$, $\Sigma P_{pj}>R\cdot\Sigma P_{gi}$, where $Z^{*3}(Y_j,X^*)=A^{*2}(Y_j,X^*)\rightarrow min$, $A^{*2}=A-A_{m2}$, $X^* \in (X \setminus A^*_{m2}(\mathrm{Y}_j,\mathrm{X}_i))$.

Suppose that (6.1.b) and (6.2) for graph 1 is not performed, in this case we define Z^{*3} matrix and A_{m3} and construct graphs 1 and 2 (Figure 4) according to the obtained matrix A_{m3}:

$$Z^{*_3}_{Y\times X} = \begin{bmatrix} 0 & 0 & 0 & 0 & 0 \\ 0 & 0 & 0 & 0 & W_{52} \\ 0 & 0 & 0 & 0 & 0 \\ 0 & 0 & 0 & 0 & 0 \end{bmatrix}; \; A_{m3} = \begin{bmatrix} W_{11} & W_{21} & 0 & 0 & 0 \\ 0 & W_{22} & W_{32} & 0 & W_{52} \\ 0 & 0 & 0 & W_{43} & W_{53} \\ 0 & 0 & 0 & W_{44} & W_{54} \end{bmatrix}.$$

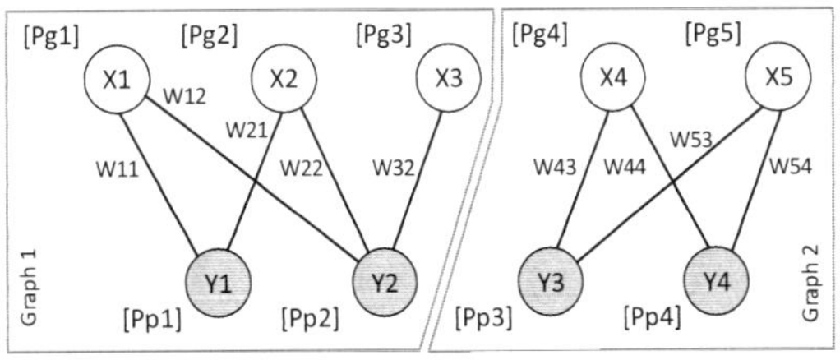

Figure 3. Decomposition of the dynamic network (2)

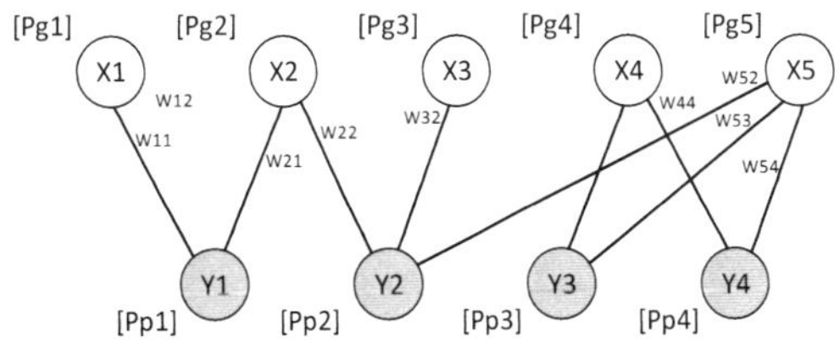

Figure 4. Decomposition of the dynamic network (3)

For graph 1 (Figure 3), the adjacent vertex was found from the set of vertices of graph 2, forming a single simple path combining both graphs (Figure 4). Thus, the set $A^*_{m2}(Y_j,X_i)$ of graph 1 is replenished with a similar set of graph 2, and this segment becomes one and is highlighted when performing (5). If inequalities (5) are not satisfied, a return to the search for the next adjacent vertex occurs (7) or (8), depending on which inequality is not satisfied.

In the general case, the problem of finding adjacent vertices when inequalities (5) are not fulfilled is written in the form:

$$\begin{cases} A_{mk} = A_{m(k-1)} + \left[Z^{*_k}_{Y\times X} \right], \; in\,case\, P_{pj} > 0.8 \cdot \sum P_{gi}, \sum P_{pj} \le 0.8 \cdot \sum P_{gi}, \; \mathrm{k} > 2, \\ Z^{*_k}(\mathrm{Y}_j, \mathrm{X}) = A^{*(\mathrm{k}-1)}(\mathrm{Y}_j, \mathrm{X}) \to \min, A^{*(\mathrm{k}-1)} = A - A_{m(\mathrm{k}-1)}. \\ A_{mk} = A_{m(\mathrm{k}-1)} + \left[Z^{*_k}_{Y\times X} \right], in\,case\, P_{pj} > 0.8 \cdot \sum P_{gi}, \sum P_{pj} > 0.8 \cdot \sum P_{gi}, \; \mathrm{k} > 2, \\ Z^{*_k}(\mathrm{Y}_j, \mathrm{X}^*) = A^{*(\mathrm{k}-1)}(\mathrm{Y}_j, \mathrm{X}^*) \to \min, A^{*(\mathrm{k}-1)} = A - A_{m2}, \; X^* \in (X \setminus A^*_{m2}(\mathrm{Y}_j, \mathrm{X}_i)). \end{cases} \tag{9}$$

where k – the number of iterations of the search cycle for adjacent vertices, starting with k=3 at the first access to the system of inequalities (5) (since the search for the third adjacent vertex from X for the vertex from Y).

At the end of the cycle of search for adjacent vertices and the formation of the segment will be obtained matrix A_{mk}. A_{mk} – adjacency matrix for a network segment filled with interconnected vertices and edges between which the product is transferred. Let's define an adjacency matrix for vertices connected but not involved in the product transportation process (the connection exists but will be involved in emergency situations):

$$A_{ne} = A - A_{mk} \tag{10}$$

For the segments found in fig. 3, we define the A_{ne}, A_{mk} matrices and construct the resulting segments taking into account the temporarily unused edges (Figure 5):

$$A_{mk} = \begin{bmatrix} W_{11} & W_{21} & 0 & 0 & 0 \\ W_{12} & W_{22} & W_{32} & 0 & 0 \\ 0 & 0 & 0 & W_{43} & W_{53} \\ 0 & 0 & 0 & W_{44} & W_{54} \end{bmatrix}, \; k = 3 \; ; \; A_{ne} = \begin{bmatrix} 0 & 0 & 0 & W_{41} & 0 \\ 0 & 0 & 0 & 0 & W_{52} \\ 0 & 0 & 0 & 0 & 0 \\ 0 & W_{24} & W_{34} & 0 & 0 \end{bmatrix}.$$

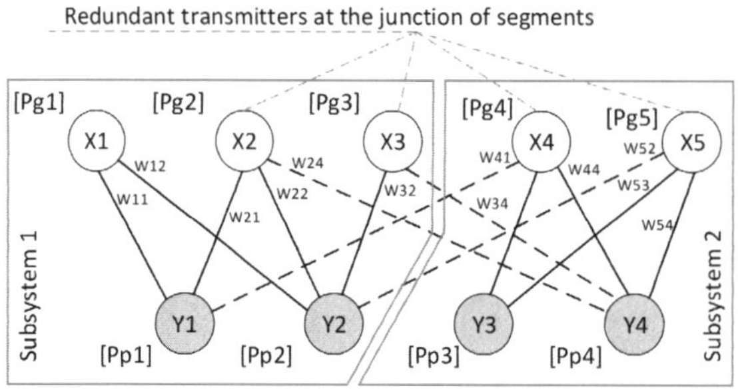

Figure 5. Decomposition of the dynamic network (4)

3. Modelling the decomposition method on the example of the grid

Consider the operation of the algorithm on the example of the test scheme IEEE 118-node power system in graphical representation [12] (Figure 6). In the graphical model the black dots represent the nodes load, white dots – nodes generators. When modeling the decomposition algorithm for each edge of the graph, the value of the active resistance of the line was set in accordance. Information about lines, load, generators, mode parameters are presented in [13]. The result of the network decomposition algorithm is shown in Figure 7, the requirements for the residual power reserve were not introduced.

Based on the simulation results, the decomposition algorithm divided the original network into five segments: 1st segment consumes 458 MW and has a power reserve of 77 MW; 2nd segment consumes 462 MW and has a power reserve of 72 MW; 3rd segment consumes 1,859 MW and has a power reserve of 23 MW; 4th segment consumes 600 MW and has a power reserve of 11 MW; 5th segment consumes 800 MW and has a power reserve of 5 MW. When calculating the power balance, the maximum and total power consumption inside each segment and the maximum power of energy sources were taken into account.

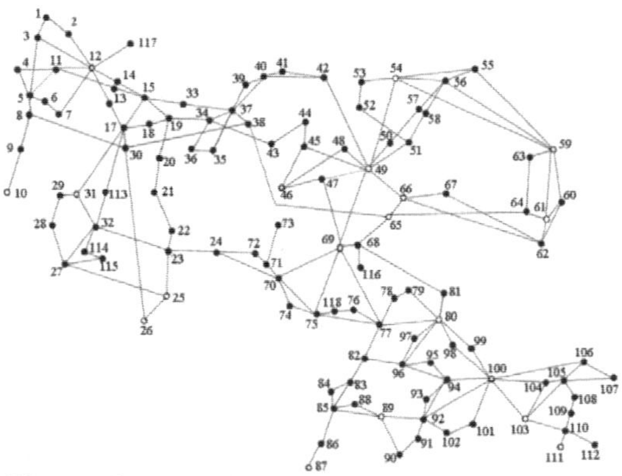

Figure 6. A graphical model of the grid

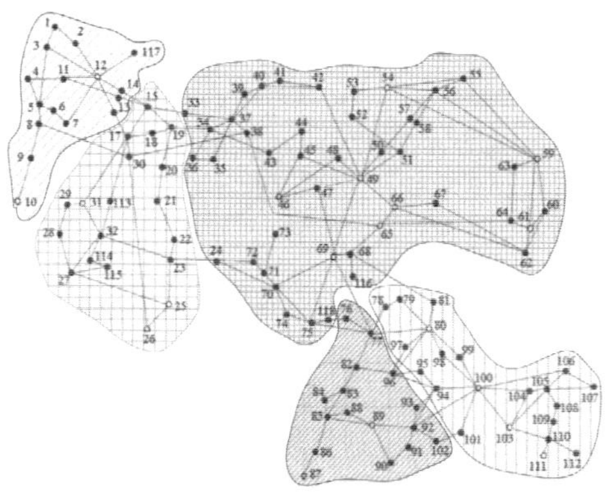

Figure 7. Result of the grid decomposition algorithm

We will analyze the energy losses in the redistribution of power flows. Suppose that the power system is operating normally,and intentional disconnection of inter-segment links will not affect the parameters of the mode, and the distribution of power flows on other lines maintains the balance of power.

Disabling the link between nodes 14-15 limits the flow of active power to 4.18 MW. Analysis of the mode parameters shows that this power is transferred from the 12th generator node, the power flow path is from node 12 to node 15: 12-14-15. If the supply of power is limited along this path, there is a shortage of power in node 15, which can be replenished from the 26th node-generator of segment 2. In the analysis of active power losses, we will operate with the active resistance of the line. For the line 12-14-15: R = 14 Ohm, for the line: 26-30-17-15: 12 Ohm. Based on the traditional ratio for calculating losses in power lines [14], we obtain that when transferring power through line 26-30-17-

15, losses will decrease by 14.29% (from 0.05 MW to 0.043 MW) compared to transfer through line 12-14 -15. Further analysis of the operating modes of the power grid segments indicates that this power does not participate in the formation of the power balance within the 2nd segment and is transferred further along the line 15-33 to the 3rd segment. Taking into account that the capacity of each segment is sufficient, the power flow along the line 12-15 can be limited without damage to the power system. Thus, it is necessary to cover the resulting shortage of power in the 33rd node. On lines 15-33 transmitted power 7.22 MW, of which 4.21 and 0.74 MW was transported along the lines 12-14-15 and 10-9-8-5-4-11-13-15, respectively. The resistance of the 12-33 line will be 22.2 Ohms, and the 10-33 line will be 57.3 Ohms, while the replacement line 65-38-37-33 has a resistance of 18.6 Ohms. Thus, the loss of line 65-33 in the transmission of 4.95 MW will be 0.012 MW, and total losses along the lines 10-33 and 12-33 will be 0.0401 MW. Reduced power loss will be 70%.

Analysis of power flows along the lines connecting the segments of the grid, compared with the power losses of the replacement lines within the entire divided power system, shows a decrease in power losses of 57%. Depending on the network topology and parameters of its mode, the calculated reduction in power losses along the replacement lines connecting the segments may vary.

4. Conclusion

The article proposes a method for decomposing a complex dynamic system with a network architecture. The method belongs to the class of topological methods and is based on the diakoptic approach to the study of complex dynamic systems in parts. The paper introduces a number of conditions that reflect the preservation of the product flow balance between the components. Thus, segments are formed, the removal of which from the general structure will not affect the overall performance of the system. If it is possible to physically divide the system into segments, the method reduces the load on the transmitting elements on the destroyed communication lines to 55-60%. Otherwise, the interaction between the segments is taken into account by means of analysis of the new graph, where the segments act as elements of the system.

References

[1] R. Ghadami, B. Shafai, *IEEE Transactions on Automatic Control*, **58**, 1, 258-264 (2013).
[2] P. Massioni, M. Verhaegen, *IEEE Transactions on Automatic Control*, **54**, 1, 124-135 (2009).
[3] V. Pshikhopov, M. Medvedev, *5th International Conference on Control, Decision and Information Technologies*, 452-457 (2018).
[4] I. Akyildiz, W. Su, Y. Sankarasubramniam, E. Cayirci, *IEEE Commun. Mag.*, **8**, 102-114 (2002).
[5] E. Camponogara, D. Jia, B. Krogh, S. Talukdar, *IEEE Control Systems Magazine*, **22**, 1, 44-52 (2002).
[6] G. Cafaro, P. Pugliese, F. Vacca, *International Journal of Electrical Power and Energy Systems,* **6**, 131–138 (1984).
[7] P. Massioni, M. Verhaegen, *IEEE Transactions on Automatic Control*, **54**, 1, 124-135 (2009).
[8] Choi, J.W., Sung, H.-G., *International Journal of Aeronautical and Space Sciences*, 15, 4, 434 – 443 (2014).
[9] F.M. Uriarte, *Electric Power Systems Research*, 88, 146 – 150 (2012).
[10] V. Shevchenko, V. Pshikhopov, M. Medvedev, *MATEC Web of Conferences, 3rd International Conference on Mechatronics and Mechanical Engineering*, 126345 (2017).
[11] V. Pshikhopov, M. Medvedev, *Path Planning for Vehicles Operating in Uncertain 2D Environments*, 1-23 (2017).
[12] K. Sun, D.-Z. Zheng, Q. Lu, *IEEE Transactions on Power Systems*, **18(2)**, 912-923 (2003).
[13] IIT, "Index of Data Illinois Institute of Technology." URL: motor.ece.iit.edu/data.
[14] P. Lezhnuk, O. Rubanenko, I. Hunko, Lecture Notes in Mechanical Engineering, **F2**, 289-298 (2019).

Acknowledgments

IOP Conf. Series: Materials Science and Engineering **635** (2019) 012024 doi:10.1088/1757-899X/635/1/012024

This work was supported by Russia's Southern Federal University grant №ВнГр-07/2017-15: Development of theoretical foundations and creation methods of intelligent multivariable control systems for generating processes, transportation, distribution and consumption of energy. The reported study was funded by RFBR according to the research project № 16-08-00013 A.

IOP Conf. Series: Materials Science and Engineering **635** (2019) 012025 doi:10.1088/1757-899X/635/1/012025

An Accuracy and Repeatability of a Robot made with V-Slot Extrusion with built-in Linear Rails

R Chancharoen, P Veerakiatikit, L Kriathkungwalkai, P Daraseneeyakul, T Loetchaipitak and M Prayongrat

Department of Mechanical Engineering, Faculty of Engineering, Chulalongkorn University, Bangkok 10330, Thailand

Email: Ratchatin.C@chula.ac.th, maimai.p@student.chula.ac.th, Lalit.k@student.chula.ac.th, paporn.da@student.chula.ac.th, paphimon.v@student.chula.ac.th and tapakorn.L@student.chula.ac.th

Abstract. Recently, aluminum extrusion with a built-in V-rail linear bearing draws increasing attention, especially among the hobbyist and makers, and is applied into many of their projects. The aluminum extrusion is an open source, light weight, and coming with extensive accessory. This significantly reduces cost and gives flexibility to their projects. Its performance is reasonable and adequate for operations among hobbyist and university's projects. However, the precise accuracy of this kind of robot is usually unavailable. Accordingly, the purpose of this paper is to investigate and analyse performance of the V-Slot extrusion with built-in linear rails robot in terms of accuracy and repeatability. The tools, in this investigation, can be obtained within the ecosystem and the process can be done by the user. This performance result will help general users to define the scope of the operation that can be achieved with reasonable result when using this robot. The methodology, in this experiment, is Trajectory analysis approach that commands the robot to move in both linear and circular, with various travel distances, in a single run, and determine the defects. Effects of travel speed are also investigated. From experimental result, the accuracy of the experimental robot is 0.45 mm in the entire range of motion. This methodology can be applied to other machines with their accuracy not better than 0.1 millimetre due to the precision of tools that are used.

1. Introduction

Since 1980, industrial robots were designed with high accuracy by using high-grade materials [1], intricate assembly process, and advanced controller. They are comparatively expensive among automation devices. By the technological advancement, robot's performance is enhanced, while its cost continues to decrease. As the robot's cost to perform ratio reduces, the robots have been widely introduced into new applications. Recently, aluminum extrusion with a built-in V-rail linear bearing catches more attention, especially among the hobbyist and makers, as it can be seen in many activities; 3D printer (1), Farmbot (2), Braille Printer in Figure 1. The aluminum extrusion is an open source, light weight, low cost, and coming with extensive accessory. Its ecosystem is very interesting. This gives flexibility to both the design and how we source parts, which can be obtained by various vendors. The performance of this robots also appropriates for daily activities which only sub-millimetre precision is adequate. Though, for this robot, its resolution is usually given by provider, its accuracy and repeatability are usually missing and cannot be exactly identified since the robot was fabricated under ecological communication. The international standard ISO 9238 [2], that is used to determine the accuracy and repeatability of an industrial robot [3], requires precision, expensive tools and time-consuming process, is not applicable for the user in aluminum extrusion's ecosystem. Furthermore, the accuracy and repeatability are for the robots, not the task. This paper focuses on the investigation of accuracy and repeatability of aluminum extrusion robots while it is performing on specific tasks – gripping and moving-- by using trajectory analysis approach. Trajectory analysis approach is designed to be a simplified method that requires only ecological equipment and lessens performing time, which is accessible by user. Trajectory analysis approach is the accuracy and repeatability investigating method focusing on the shifting of value between desired trajectory and actual performing trajectory of the robot.

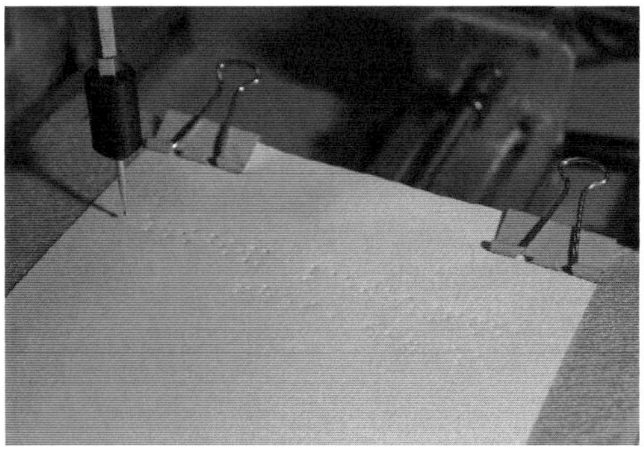

(a) Braille Printed on Paper

(b) Demonstration

Figure 1. Braille Printer

2. The Cartesian robotic system

The robot used in this experiment is fabricated by Thai StartUp company "Easy Builds". The components of robot shown in Figure 2 are built with 3030 V-Slot aluminium Extrusion profile, a feed screw and rollers. The thread of the screw is 1 mm. Each axis is driven by NEMA stepper motor with micro stepping drive and Arduino board that can give the resulting velocity up to 50 mm/s for X and Y axes and 10 mm/s for the Z axis with 0.0025 mm resolution. The workable space is 150×150×150 mm^3. The controller board is RUMBO 3D printer control board with Repetier firmware.

3. Design of experiment

To investigate the accuracy and repeatability of the experimental Cartesian robot in moving and gripping task, trajectory analysis approach has been assigned. The Z-axis of this experimental robot is in vertical direction against gravity field while X- and Y-axis are in horizontal plane. Thus, the characteristic of motion robot's arm in XY-plane will be differ from the motion in Z-axis. Furthermore, the G-code [4], which is the ISO standard language used to control the machine, commands the motion with linear or circular trajectory. It is noted that G01 is for linear movement, G02 and G03 for clockwise and counter clockwise circular curve respectively, G04 for time delay which could stop the motion of all axes for a given amount of time. For the accuracy investigation, we command the robot to move in circular motion with various radius, e.g. 10,20,30,40,50,60 mm and radial motion with various directions. E.g. and each set up of the robot is commanded with two velocity, high and low velocity. In this designed moving pattern, both linear and circular trajectories are investigated in a single run. For the repeatability test, we command the robot to move repeatedly in circular motion for 30 mm in an entire workspace in horizontal plane. In the task, the robot grips a pen in its hand and draws lines on the paper. A 0.5 mm felt-tip pen is used. For the image processing process, scanning ink trace graphical is result in a resolution of 300 dots per inch (about 0.085 mm resolution) to measure all necessary dimensions in sub-millimetre tasks; the circle's radius, length of the line. Moreover, for an accuracy investigation, constructing the circumferential band that cover the ink trace to represent the minimum and maximum radius of the ink trajectory; the width of the band will represent the accuracy. The repeatability of the robot can be obtained by the maximum deviation between desired trajectories and ink trajectories. The effect of length and velocity are investigated.

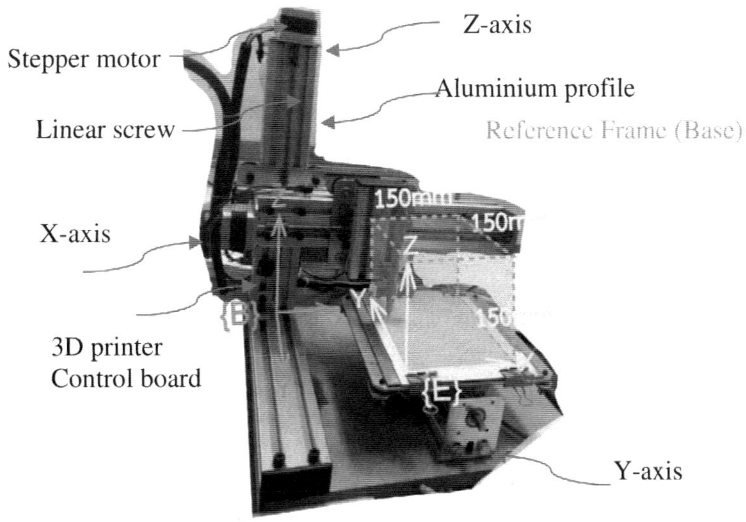

Figure 2. Cartesian Robot made with aluminium extrusion

4. Experimental result

The experimental accuracy is demonstrated in Figure 3. From the experiment, in the work space of 150x150x150 m^3, the accuracy, as analysed in Table 1, of this robot is 0.45 millimetre under the entire range of travel velocity. The accuracy linearly degrades as the robot's travel velocity increases. If the travel velocity is too high (2400 mm/min in the repeatability test, as in Figure 4 and summarized in Table 2), the repeatability degrades quickly. The tools, that are used in the proposed technique, give the variation of result around 0.1 millimetre which could be used to examine the accuracy at about 0.5-millimetre resolution. This is sufficient for this type of robot that its accuracy is about 0.5 millimetre and for our case that the resulting accuracy is only 0.45 millimetre. The effect of travel's distance also linearly degrades the accuracy. This could be corrected at the step/millimetre in the controller's firmware. The accuracy is better when the robot travels along its hardware's axis, in X-direction in this case, and slightly poorer in the incline direction. The linear trajectory gives better accuracy. These preliminary results may be used to design the task's trajectory for a better accuracy.

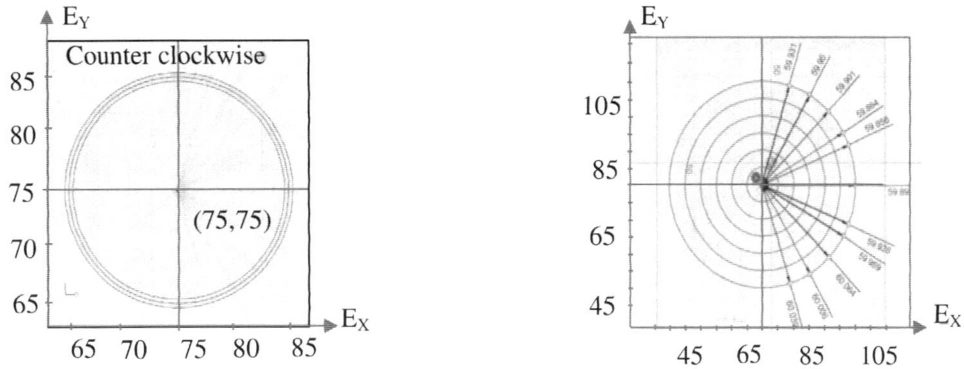

(a) Circular trajectory (b) Linear trajectory

Figure 3. Accuracy analysis in the horizontal plane

IOP Conf. Series: Materials Science and Engineering **635** (2019) 012025 doi:10.1088/1757-899X/635/1/012025

Table 1. Accuracy analysis

(a) circular trajectory (G02 G03 command)							(b) linear trajectory (G01 command)		
velocity	Circle label	Commanded distance (mm)	Band radius (mm) Inner band	Band radius (mm) Outer band	accuracy (mm)	Max accuracy (mm)	Degree (°)	Radius got (mm)	Accuracy (mm)
300 mm/min	A	10	9.8718	10.128	0.128		73	60.356	0.356
	B	20	19.835	20.165	0.165		60	60.286	0.286
	C	30	29.775	30.225	0.225		45	60.318	0.318
	D	40	39.703	40.297	0.297		30	60.178	0.178
	E	50	49.758	50.242	0.242		22.5	60.154	0.154
	F	60	59.593	60.407	0.407	0.407	0	60.075	0.075
							-22.5	60.135	0.135
							-30	60.262	0.262
							-45	60.283	0.283
							-60	60.142	0.142
							-73	60.262	0.262
600 mm/min	A	10	9.9088	10.091	0.091		73	59.931	0.069
	B	20	19.873	20.127	0.127		60	59.95	0.05
	C	30	29.798	30.202	0.202	0.251	45	59.901	0.099
	D	40	39.841	40.159	0.159		30	59.884	0.116
	E	50	49.826	50.174	0.174		22.5	59.856	0.144
	F	60	59.749	60.251	0.251		0	59.89	0.11
							-22.5	59.928	0.072
							-30	59.959	0.041
							-45	60.064	0.064
							-60	60.008	0.008
							-73	60.039	0.039
600 mm/min time delay 2s	A	10	9.8418	10.158	0.158		73	60.278	0.278
	B	20	19.877	20.123	0.123		60	60.227	0.227
	C	30	29.854	30.146	0.146	0.302	45	60.208	0.208
	D	40	39.777	40.223	0.223		30	60.148	0.148
	E	50	49.784	50.216	0.216		22.5	60.184	0.184
	F	60	59.698	60.302	0.302				

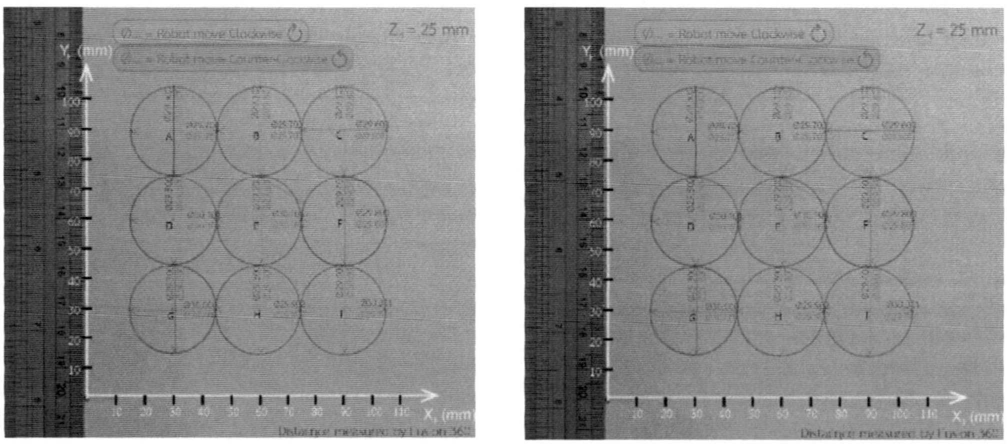

(a)Velocity 2400 mm/min (b)Velocity 200 mm/min

Figure 4. Repeatability test

Table 2. Repeatability test

Circle label	Diameter of circle [mm]			
	Velocity 200 mm/min		Velocity 2400 mm/min	
	X	Y	X	Y
A	29.900	29.601	29.704	29.300
	29.900	29.601	29.704	29.900
B	29.900	29.701	30.101	29.500
	29.900	29.701	30.101	30.300
C	30.001	29.601	30.000	29.300
	30.001	29.601	30.000	29.800
D	29.604	30.000	29.703	30.601
	29.604	30.000	29.703	29.102
E	30.001	30.003	30.103	29.200
	30.001	30.003	30.103	30.300
F	29.900	30.001	29.902	29.300
	29.900	30.001	29.902	30.400
G	29.804	29.702	29.600	29.100
	29.804	29.702	29.600	29.800
H	30.004	29.800	29.800	29.401
	30.004	29.800	29.800	30.301
I	29.902	29.900	30.001	29.400
	29.902	29.900	30.001	30.000
Repeatability	0.296		0.753	

5. Conclusion

In this investigation, the robot with V-slot extrusion with 150x150x150 m^3 workspace, give the resulting task accuracy of 0.45 millimetre under the entire range of motion. This task accuracy is a lot lower than 0.0025 mm resolution of the robot actuation system. However, this performance is enough to accomplish most of operation in hobbyist's works and university's projects which are only sub-millimetre tasks. The travel's distance and speed also affect the accuracy of the robot. If the circular trajectory is too big, the robot may degrade its accuracy quickly. The effect of travel speed is observed in the repeatability test as we test the robot with lower and higher velocity, compared to in the accuracy test. The repeatability degrades as the robot moves faster.

6. Notes

(1) "3-in-1 3D printer". https://www.snapmaker.com (accessed 25 November 2018)
(2) "FarmBot". https://farm.bot (accessed 28 November 2018)
(3) "OX CNC machine". https://openbuilds.com (accessed 28 November 2018)

7. References

[1] Johanna, Wallén. "The History of the Industrial Robot" 1400-3902; 2853 (2008), Linkoping University, The Institute of Technology (accessed 8 December 2018)
[2] International Organization for Standardization 1998 *Manipulating industrial robots - Performance criteria and related test methods 2nd edition*
[3] Şirinterlikçi A, Bird A, Harris A and Kweder K 2009 Repeatability and Accuracy of an Industrial Robot: Laboratory Experience for a Design of Experiments Course, the *Technology Interface Journal*
[4] Daniel N 2016 *G-Code to RAPID translator for Robot-Studio*

Acknowledgments

This work was a part of the intelligent 3D printing for development project, which was supported by the Development into the 2nd Century Fund, Chulalongkorn University.

IOP Conf. Series: Materials Science and Engineering **635** (2019) 012026 doi:10.1088/1757-899X/635/1/012026

Features of the complex power unit for mobile robotic

V A Shevchenko and V A Kostjukov

Southern Federal University, 44 Nekrasovskiy, Taganrog, 347922, Russian Federation

E-mail: wkost-einheit@yandex.ru

Abstract. This article discusses the prospects for the use of power units based on renewable energy sources for additional and emergency power supply of surface robotic systems. This kind of complex power unit can be built on the basis of renewable wind and solar energy and generate at least 10-15% of all the electrical energy required for a vessel. Considered one of the main problems of designing this type of complex power unit - the creation of a promising wind power systems (windmills). The windmill under development should meet a number of necessary criteria regarding reliability, power, noise level, and design constraints of the surface platform itself. The rationale for the design of the windmill is given. Further, the design of the windmill is optimized by the criteria of aerodynamic power, taking into account these limitations. The aerodynamic comparison shows the superiority of the proposed design of a windmill in relation to analogues on all the most important quality criteria. The features of the implementation of the electrical connection scheme of a windmill, solar panels and a diesel generator set into a single complex power unit are considered. For a robotized ship with a given power of the propulsion system, the complex power unit has been developed, which allows generating at least 10% of the power consumption.

1. Introduction

Currently, there is a large group of offshore facilities, stationary and mobile, including robotic complexes, which need auxiliary autonomous energy sources. Diesel generators used as such energy sources are quite expensive. Also, the fuel that is used by such units is quite expensive. In addition, these units are highly polluting.

Many international companies are developing and introducing alternative energy sources on offshore surface platforms to reduce their overall fuel consumption. [1,2].

One of the approaches to significantly reduce the consumption of conventional fuel (more than 10%) is the use of a complex power unit consisting of a windmill and solar panels.

The key challenges are: selection of a windmill type suitable for installation on a fixed/mobile platform as an element of a complex power unit; development of optimal configuration of solar panels and windmill; selection of types and relevant parameters of solar panels and a windmill, allowing to optimize the operation of a complex power unit by the criterion of the maximum power generated under severe restrictions imposed by the requirements of safety and reliability of operation of all ship systems.

2. Problem statement

It is necessary to develop a comprehensive power unit based on renewable energy sources. The unit should produce at least 10% of the nominal power $P_{\text{engine}} = 10\ kW$ of propulsion boats of small displacement ($L_{max} = 10$ m long).

We take the standard system of solar panels, so we will further focus on the necessary requirements for the design and features of the windmill.

From the requirements of safety and efficiency of the functioning of a surface ship follow certain restrictions imposed on the used design of the windmill:

1) compactness. Installing a windmill on the ship should not greatly increase the windage due to the flow of wind;

2) exceed power output. The windmill should surpass all similar designs at the same wind speeds;

3) reduced noise level. The windmill should have a lower noise level compared with the considered analogues;

4) sufficient structural strength;

5) independence from the direction of the air flow relative to its moving part;

6) efficient rotor stop system when critical speed is reached;

7) simple assembly, disassembly and repair.

The requirements of 1, 3, 4 and 5 classic horizontal-axis windmills do not satisfy. Requirements 1 and 4 and 7 also do not meet vertical-axial windmills with strongly elevated blades of the movable part. The requirements correspond to the design of vertical-axial windmills, which have a movable part of the rotor with a minimum ratio of the height of the rotor to its diameter. To meet the requirements of 1 and 4, it is advisable to use windmills with special static elements that protect the rotor from damage.

It is necessary to develop a structural electrical circuit of an integrated power unit, taking into account the features of the operation of a windmill that meets the specified requirements, in conjunction with solar panels.

3. Synthesis and justification of the design of a promising windmill, which can be integrated into a complex power unit

In works [3-9] one of the possible designs of a windmill of a vortex type is presented that satisfies all the above conditions. shows the advantages of the proposed windmill in terms of the power generated and the level of noise compared with analogues, including those presented in [7-13].

The proposed design is patented, and its mechanical part includes the following main elements [5, 6, 16]: a special-shaped turbine with eight blades, the upper part of the stator is a bell, the lower part of the stator is a guide structure; the rack-holder of the socket and the rotor; rotor shaft (Figure 1).

The electrical part of the integrated power unit includes: power generator based on synchronous machine, output voltage stabilization system, power converters, energy conversion control system, batteries, backup source - diesel generator.

As a result of the aerodynamic optimization of the shape of the windmill, the design was obtained according to the criterion of the maximum of the aerodynamic moment of the rotor [5-6] (Figure 2). This form is more than 2 times more efficient in power compared to horizontal-axial windmills, and more than 30% superior to some modern types of vertical-axial windmills in power, with other things being equal (Table 1).

Table 1. Values of the useful aerodynamic moment on the moving parts of a windmill

Types of windmills	Wind swept surface area, m^2	Torque at 5 m/s, N*m	Torque at 2.5 m / s, N*m
Proposed windmill	0,70263	1,8	0,87
Horizontal classic windmill	0,70263	0,223	0,071
Vertical spiral windmill	0,70263	0,757	0,316
Vertical windmill with bell	0,70263	1,242	0,577

The principle of operation of the vortex wind power unit is based on the useful aerodynamic interference between the static and rotor parts of the unit and the use of a special twisted shape of the rotor blades [3-7, 14-16] (Figure 2).

Figure 1. Explanation of the principle of operation of the windmill

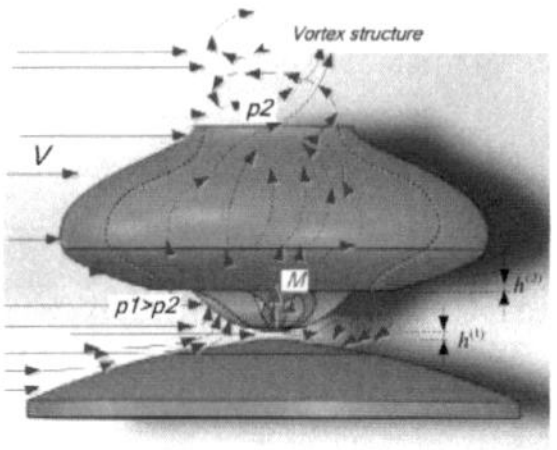

Figure 2. The developed design of the vortex windmill

Figure 3. 3D-model of a small-size boat with a complex power unit

Let us consider the main technical characteristics of the complex power unit being developed with the vortex type windmill proposed above for a boat with a length of L = 10 m.

For wind speed acting on a windmill V = 3 m / s, the average power of a complex power unit is P_{CP} = 2 kW, which is 14% of the power of the propulsion unit P_{PU} = 10 kW. The windmill generates P_{WM} = 0.1 kW, and the solar panel - P_{SP} = 1.3 kW. This ratio is calculated on the basis that a solar panel generates an average power of 190 W per 1 m^2, which corresponds to a solar panel efficiency of 19%. The power of the windmill and the complex power unit with a wind speed of V = 10 m/s are respectively equal to: P_{WM} = 3.7 kW, P_{CP} = 5 kW, which, taking into account the P_{SP}, is already 56% of the power consumption of the propulsion unit.

Dimensions of the windmill are 2.4x1.6x2.4 m, solar panels – 2.2x2.5x2.9 m. The vertical size of the windmill can be reduced by using existing structural elements of the boat as the lower part of the stator (Figure 3).

4. Electromechanical control circuit of complex power unit

Consider a generalized structural electromechanical scheme of a complex power unit including a windmill a block of solar panels and a backup diesel generator (DG).

To combine the windmill, solar panels and DG in a single integrated power unit can be used classical scheme (Figure 3). The circuit consists of a synchronous generator with permanent magnets, a low-pass filter, AC/DC converters, DC/AC inverter, a charge controller and batteries. The solar panel module (SP) uses its own charge controller supplying the battery; the DG can connect to the load using a software-controlled key 2.

When the wind speed is greater than the minimum V_{min} value required to start the windmill and less than the maximum V_{max} value, controlled key 1 should be in position 1 if the battery is already fully charged. In this case, a output voltage from the synchronous generator is rectified, and then it is inverted to produce a three-phase sinusoidal voltage and then it goes to the load. If under the same conditions the battery is not fully charged, the key 1 is transferred to position 2, and the load begins to be powered by the battery through the inverter.

If the speed $V \geq V_{min}$, then controller starts to limit the current from the rectifier and the voltage on the generator increases, the electromechanical moment of resistance to the rotation of the rotor decreases, contributing to an increase in the speed of rotation. To prevent such a situation, two options are provided: a) emergency braking of the propeller; b) switching generator to ballast load. If key 1 was in position 2, then a command is given to the braking devices of the rotor for its forced braking. In case V>Vmax is detected in position 1 of key 1, first the key 1 is moved to position 2, and then the battery is disconnected from the generator and the rotor is braked.

A special feature is the aerodynamic braking of the rotor, for which the variable geometry of the elements of the windmill is applied to regulate its output characteristics [4-6]. This braking is proposed to be carried out by means of controlled pull-up of the rotor, i.e. by regulating the wind-blown surface of the rotor. It uses two controlled parameters of the variable geometry of the elements of the windmill: the distance from the lower part of the rotor to the upper edge of the lower stator $h^{(1)}$, and to the edge of the upper stator $h^{(2)}$, respectively (Figure 3). Geometry changes are carried out through the deviation of actuators from their normal positions for the implementation of a given interaction of wind flow with a windmill. For the implementation of feedback on the management of variable elements of the geometry used sensors deviation. Control signals for actuators are generated in the control pulse generation unit for variable geometry elements.

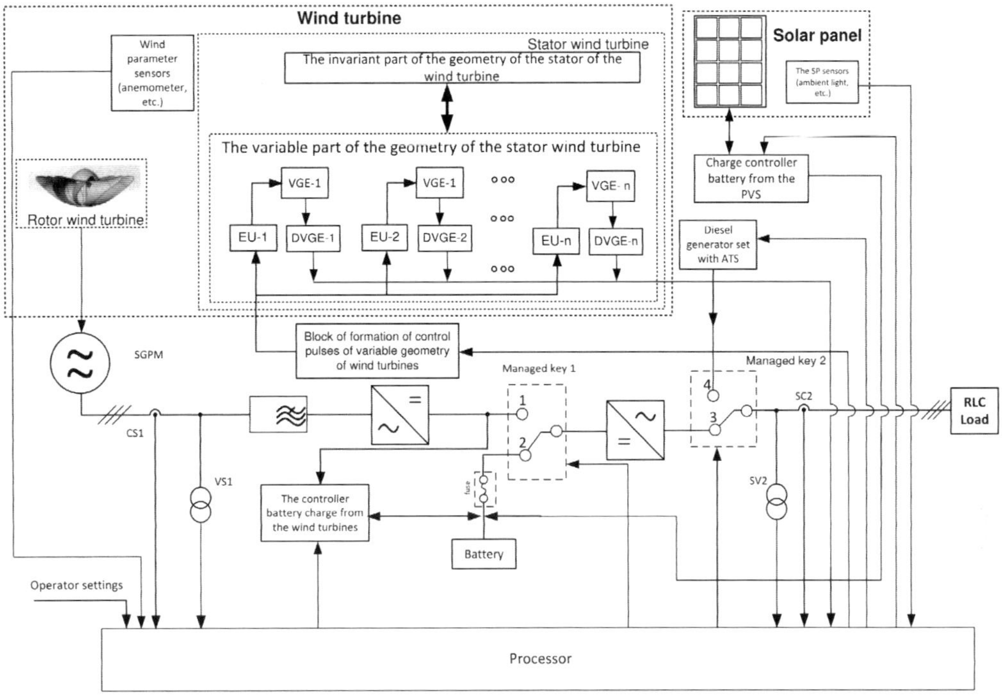

Figure 4. Generalized classical electromechanical circuit of an complex power unit.

The battery must supply the load and, accordingly, be designed for one of two possible modes: 1) supply the consumer after the loss of power from renewable energy sources as long as the capacity of the battery allows; at the same time, the indicated capacity should be as high as possible at a given limit of the cost of the battery; 2) consumer supply before turning on the diesel generator.

Consider the second mode, since it will determine the lower limit of the battery capacity in the general case. Thus, the battery unit should provide power to the load, powered by a complex power unit, during short-term drawdowns of its power (due to unstable primary energy flow of renewable energy), the idle state of the windmill and solar panels for a time before the diesel generator is commissioned. Set the power of the complex power unit $P_{CP}=5\ kW$, suitable wind V=10 m/s.

The time of putting the diesel generator into operation, depending on the power, is from 2 to 15 seconds. Thus, the choice of battery capacity is determined by the time during which the consumer may need power only from batteries. Choose a 5-minute interval with a maximum load power of 5 *kW*. Consider a battery, according to the passport data, the voltage of which will be 12V with a capacity of 200 Ah, providing 5352 W for 5 minutes with a voltage drop of 20%. A deeper discharge of the battery is impractical and can reduce its service life.

The use of additional solar panels requires the connection of an appropriate charge controller. Note that the use of a single charge controller for windmills and solar panels is impossible, since the transistors used in the solar panels charge controller, failure when the voltage on the generator is close to the open circuit voltage at a high rotor speed. This limitation can be overcome by using a hybrid controller for windmill / solar panels [17-19]. The introduction of a hybrid controller allows you to abandon one of the charge controllers and electronically controlled key 1 in the classical scheme.

IOP Conf. Series: Materials Science and Engineering **635** (2019) 012026 doi:10.1088/1757-899X/635/1/012026

The hybrid controller controls additional external charging / discharge currents by automatically limiting them. Thus, the problem of exceeding the permissible charge current in the classical scheme is solved, since the instantaneous currents from the controller of the windmill and the solar panels can not be taken into account simultaneously. In addition, the use of a hybrid controller has a number of advantages: a) application of MPPT technology (maximum power point tracking); b) detection of excess speed of rotation of the windmill rotor (alarm indicating the need to use an aerodynamic brake); c) current and voltage control from two sources (windmill and solar panel); d) application of solar panel protection against reverse, reverse polarity; and battery from idle, reverse polarity, overload, lightning, overcharge; e) dynamic correction of battery charge/discharge parameters.

Depending on the specific hybrid controller, different circuit solutions can be applied. Analysis of serially-produced hybrid controller showed that there are three main options for connections: 1) The windmill and solar panels feed the battery through a hybrid controller, and the battery feeds the load; 2) Windmill and solar panels through a hybrid controller supply battery and load in parallel, and the battery feeds the load directly; 3) Windmill and solar panels through the hybrid controller in parallel feed the battery and the load, and the battery is connected to the load through the hybrid controller.

Since we preferred to use the battery exclusively to maintain the required total capacity of the complex power unit, the third option is the best in terms of ensuring maximum battery control.

Thus, the classical scheme of connection of a windmill, solar panels and a diesel generator is presented in Fig. 4, can be converted to the view shown in Figure 5.

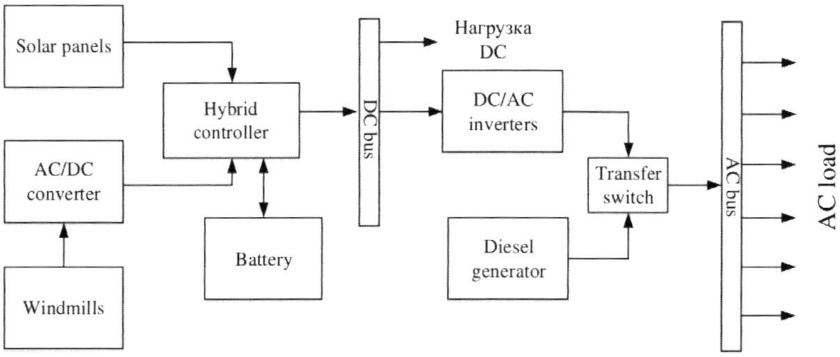

Figure 5. Generalized block diagram of a complex power unit using a hybrid controller

5. Conclusion

In this paper, the substantiation of the design of windmills which can be a structural element of a complex power unit to be used on surface ships of small displacement in order to save fuel, including on robotic platforms.

The analysis of the specific shape of a windmill with a vertical axis of rotation, axisymmetric stator and rotor, special forms of which and their layout are obtained as a result of aerodynamic optimization by the criterion of the maximum torque of the rotor is carried out. The paper shows that the proposed design of windmills meets all the key requirements for ensuring safety, reliability and efficiency, dictated by the use of an complex power unit on a ship with a small displacement including a robotic surface platform. The proposed design significantly exceeds the analogs in the generated power - by at least 30% for the best types of vertical-axial windmills from the comparison list.

For a boat with a small displacement with the power of the propulsion unit P_{CP} = 10 kW, a generalized design of an integrated power plant has been designed, which produces at the accepted minimum wind speed V = 3 m/s not less than 14% of the energy from the P_{PU}.

The structural electric scheme of the complex power unit including solar panels and diesel generator in addition to the windmill is considered; the diesel generator can be derived from the scheme at short-term failures of the generated energy. Also the scheme with use of the hybrid controller is considered, the choice of the main equipment of this scheme is proved.

The commonality of consideration of the electrical part of the complex control system, containing a diesel generator in its composition, allows the use of the proposed circuit solutions for the organization of both the main and backup power supply of medium displacement vessels.

6. References

[1] Innovative Technologies & Solutions for Sustainable Shipping // Eco Marine Power URL: ecomarinepower.com/en/about-us (дата обращения: 16.10.2018).

[2] What is a USV? // ocius URL: https://ocius.com.au/usv/ (дата обращения: 24.10.2018).

[3] V. A. Kostjukov, A. M. Maevskiy, N. K. Poluyanovich and M. N. Dubyago, "Adaptive mechatronic management system of wind-driven power-plant with variable geometry," 2017 18th International Conference of Young Specialists on Micro/Nanotechnologies and Electron Devices (EDM), Erlagol, 2017, pp. 460-464.

[4] M.Yu. Medvedev, V.A. Kostjukov, A.M. Maevskiy, D.D. Pavlenko, Izvestiya YuFU. Tekhnicheskie nauki. 2018. №01. pp. 194-208. [RUS]

[5] Kostjukov V.A., Medvedev M.Yu., Maevskiy A.N., Poluyanovich N.K., Savchenko V.V., Vestnik donskogo gosudarstvennogo tekhnicheskogo universiteta, 2017, 1(88), pp. 85-91. [RUS]

[6] Kostjukov V.A., Medvedev M.Yu., Maevskiy A.N., Poluyanovich N.K., Savchenko V.V., Vestnik donskogo gosudarstvennogo tekhnicheskogo universiteta, 2017, 4(91), pp. 61-68. [RUS]

[7] P. Ying, Y.K. Chen, Y.G. Xu, Y. Tian. Computational and experimental investigations of an omni-flow wind turbine. Applied Energy, Volume 146, pp. 74–83.

[8] Rafał Wróżyński, Mariusz Sojka, Krzysztof Pyszny, Krzysztof Pyszny. The application of GIS and 3D graphic software to visual impact assessment of wind turbines. Renewable Energy, Volume 96, Part A, pp. 625–635.

[9] Qing'an Li, Junsuke Murata, Masayuki Endo, Takao Maeda, Yasunari Kamada. Experimental and numerical investigation of the effect of turbulent inflow on a Horizontal Axis Wind Turbine (part II: Wake characteristics)./ Energy, Volume 113, pp. 1304–1315.

[10] Y.G. Heoa, b, N.J. Choic, K.H. Choib, H.S. Jia, K.C. Kima. CFD study on aerodynamic power output of a 110 kW building augmented wind turbine./Energy and Buildings, Volume 129, pp. 162–173.

[11] L. Wanga, X. Liub, A. Koliosa. Renewable and Sustainable Energy Reviews, Volume 64, pp. 195–210.

[12] Mikhnenkov L.V. Nauchnyy vestnik MGTU, №125, 2008, pp. 22-24.[RUS].

[13] Wenyi Liu. Design and kinetic analysis of wind turbine blade-hub-tower coupled system Renewable Energy, Volume 94, pp. 547–557.

[14] Mats Wahl, «Designing an H-rotor type Wind Turbine for Operation on Amundsen-Scott South Pole Station», 2007, ISSN: 1650-8300, UPTEC ES07 030.

[15] Patent na poleznuyu model' «Ustroystvo preobrazovaniya kineticheskoy energii vetra v mekhanicheskuyu energiyu s ispol'zovaniem nizhney napravlyayushchey struktury» [Device for converting kinetic wind energy into mechanical energy using the lower guide structure]/ Kostjukov V.A., Medvedev M.Ju., Maevskij A.M., Polujanovich N.K., Savchenko V.V., 11.08.2016 г., № 175397. [RUS]

[16] Haskin L.Ja. Uchenye zapiski CAGI. 1993, т.24, №4, pp. 15-17. [RUS]

[17] Nema P., Nema R. K., Rangnekar S. A current and future state of art development of hybrid energy system using wind and PV-solar: A review //Renewable and Sustainable Energy Reviews, 2009, Vol. 13, №. 8, pp. 2096-2103.

[18] Wei Zhou, Chengzhi Lou, Zhongshi Li, Lin Lu, Hongxing Yang, Current status of research on optimum sizing of stand-alone hybrid solar–wind power generation systems, Applied Energy, Volume 87, Issue 2, 2010, pp. 380-389.

[19] Chong W. T. et al. Techno-economic analysis of a wind–solar hybrid renewable energy system with rainwater collection feature for urban high-rise application //Applied Energy, 2011, Vol. 88, №. 11, pp. 4067-4077.

IOP Conf. Series: Materials Science and Engineering **635** (2019) 012026 doi:10.1088/1757-899X/635/1/012026

Acknowledgments
The work is executed at support of RFBR grant No. 18-08-00473 and, from 25/12/2018 on the theme: "Development and research of methods of optimization and management of energy conversion processes in power plants of complex type, including transforming the continuous flow».

IOP Conf. Series: Materials Science and Engineering **635** (2019) 012027 doi:10.1088/1757-899X/635/1/012027

Deformation behaviors of non-oriented silicon steel plate in hot rolling process

Qiang Dong[1], **Huihui Liu**[2]

[1]School of Mechanical Engineering, Shandong Jiaotong University, Jinan 250357, China

[2]Shandong Institute for Product Quality Inspection, Jinan 250102, China

successin2010@163.com

Abstract. In this paper, the flow stress behaviors of non-oriented silicon steel are investigated by hot compressive experiments and a constitutive model is built based on the experimental data. The constitutive model provides material model for FE numerical investigation. The deformation behaviors of the non-oriented silicon steel during hot rolling process are investigated by a 3D FE model. Contact stress and plate shape in different working conditions are presented. The FE results can provide primary data for the precise shaping of non-oriented silicon steels in hot rolling process.

1. Introduction

Silicon steel, also known as electrical steel, is an important metallic functional material, which is used to make core plates of transformer, electromotor and generator. Non-oriented silicon steel is the largest consumed ferromagnetic alloy in the world. The deformation of silicon steel plate has important effects on its magnetic property. Landgraf et al.[1] found that magnetic properties of silicon steels are more sensible to small deformation. Paepe et al.[2] pointed out that hot rolling parameters have significant influence on the magnetic properties of a low silicon steel. Hot rolling is a critical process for silicon steel strip production. The defects produced in hot rolling process have a genetic effect on cold rolling strip, ultimately degrading its magnetic property. The quality of final product in hot rolling process depends mainly on the material properties, thickness and plate shape.[3] At present time, the most common defects in hot rolling process are plate shape defects. The hot strip mill is subject to adverse service conditions when producing silicon steel due to its strict technique. [4] The desired plate shape of silicon steel is difficult to achieve because of its inconclusive high temperature deformation characteristic in rolling process. The quality defects generated from hot rolling process are difficult to eliminate in cold rolling and other processes afterwards. The topic of this work is to investigate deformation behavior of a non-oriented silicon steel in hot rolling process by thermal simulation experiment and finite element (FE) caculation, thereby providing guidance for precise shaping of silicon steel.

2. Experiment

2.1. Experimental procedure

Rolling is a process to reduce the thickness of a steel plate from the initial thickness to a pre-determined final thickness with revolving rolls.[5] The HSM in this work consists of a 2-high roughing mill R1, a

4-high roughing mill R2 and seven 4-high finish rolling stands from F1 to F7 [4]. The work piece is heated up to more than 1200°C in the reheating furnace. When passing through the finishing mill, the plate temperature decreases from 945°C in stand F1 to 860°C in stand F7 with the cooling process. Chemical compositions the investigated non-oriented silicon steel is shown in Table 1.

The obtained sample from rolling mill was processed into dozens of standard cylindrical specimens with the diameter of 8 mm and height of 15 mm. Before hot compressive tests, the start and finish temperatures of phase transformations were determined using a dilatometer based on the principle of the expansion differences between austenite and ferrite. The critical temperatures *Ar*1 and *Ar*3 at the average cooling rate of 9.8°C/s in actual steel rolling process are 940°C and 858°C respectively. Therefore, work piece in almost all the finishing stands is in austenite-ferrite dual-phase region. The hot compressive experiments were conducted on Gleeble 1500 thermo-mechanical simulator in the temperature range of 1050~750 °C with strain rates of 0.05~10 s^{-1}.

Table 1. Chemical composition of the investigated non-oriented silicon steel

Element	C	Si	Mn	Al	Cu	Cr	Ni	Mo
Wt, %	0.001	0.301	0.249	0.053	0.032	0.024	0.012	0.003

2.2. Flow stress behaviors

Figure1(a-c) show stress-strain curves of silicon steel in austenite-ferrite region. The test results show that dynamic recovery plays a leading role in silicon steel hot compressive deformation. The flow stress increases with increasing strain rate, while it increases as well when deformation temperature increases at this temperature range from 900~950°C. Figure1(d) shows stress-temperature curve at different strain rates with the strain of 0.6. The flow stress increases in austenite or ferrite region with the decreasing deformation temperature. The flow stress tendency is on the contrary when the steel is in austenite-ferrite region. The experimental data has great significance for constitutive model and numerical simulation of the non-oriented silicon steel.

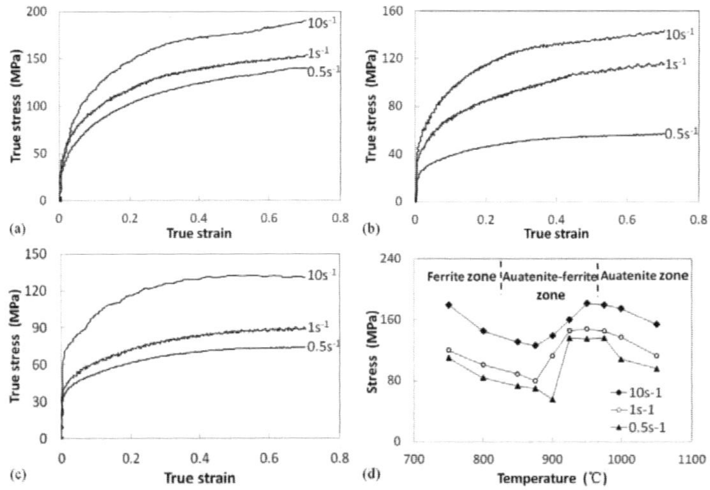

Figure 1. Flow stress behaviors of non-oriented electrical steel. (a) stress-strain curve at 950°C (b) stress-strain curve at 900°C (c) stress-strain curve at 850°C and (d) stress-temperature curve

2.3. Constitutive model

The constitutive model is used to predict mechanical response of materials during deformation in a certain range of temperature and strain rate [6]. Many constitutive models for steel have been proposed. Arrhenius-type model[7,8] is applied extensively recently to describe hot deformation of materials. In this work, the constitutive model of non-oriented silicon steel during hot compression was constructed based on Arrhenius-type model. The established model manifests excellent property to predict the flow stress of non-oriented silicon steel in hot compressive deformation process. The average absolute relative error of this model is 4.98. The constitutive model can provide material property of the silicon steel in numerical modeling with various temperatures and strain rates.

3. Numerical simulation

3.1. Modeling

FE method is widely used in material forming process. Thermodynamics, edge defect formation and microstructure evolution during rolling process were investigated by FE method in previous studies. In order to investigate the deformation behaviors of the silicon steel plate in actual hot rolling process, a three-dimensional (3D) FE model for roll system of the finishing mill is built based on ANSYS/LS-DYNA software package. The modeling parameters are presented in Table 2. The FE modeling is constructed by Ansys parametric design language. To get accurate nodes solutions, mesh refinements are performed around the contact surface, which is shown in Figure 2. The rolls in the model are elastic with the elastic modulus of 210 GPa and Poisson's ratio of 0.3. The material property of steel plate is elastic-plastic. As shown in Table 3, temperature of the steel plate decreases and strain rate elevates during finish rolling, resulting in different material properties. The material model of the steel plate is acquired from the thermal simulation test and the above constitutive model.

Table 2. Geometric parameters of the FE model

Entry	Parameters [mm]
Work roll barrel size, $D_W \times L_W$	⌀820×1880
Work roll neck size, $D_N \times L_N$	⌀480×610
Backup roll barrel size, $D \times L_B$	⌀1600×1550
Backup roll neck size, $D \times L_E$	⌀960×725
Plate thickness, H	40

Table 3. Rolling parameters of non-oriented silicon steel in the finishing mill

Finishing mill NO.	F1	F2	F3	F4	F5	F6	F7
Reduction rate [%]	24.1	36.4	32.4	41.8	35.4	27.3	18.4
Strain rate [1/s]	4.2	10.4	16.9	47.6	89.5	126	143
Temperatures [°C]	945	907	890	874	868	865	863

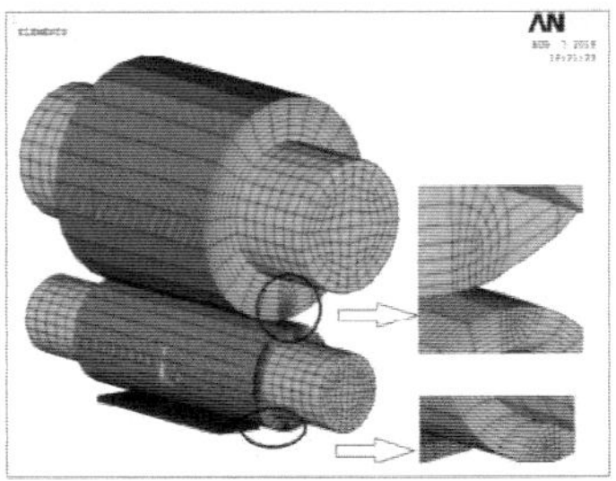

Figure 2. The FE model for the 4-hi finishing mill

3.2. Results

The deformation behaviors of the steel plate in hot rolling process were demonstrated below with the following outputs: contact stress and plate shape. The outputs were obtained from FE calculation with various inputs, such as reduction rate, entrance crown, thickness, width, rolling temperatures and strain rate of the work piece. Reduction rate (R_d) is defined as the ratio between work piece thickness reduction (ΔH) during a rolling pass and the original thickness (H), i.e., $R_d = \Delta H/H$. Entrance crown of the work piece is defined as the thickness difference between the central point of the work piece and the edge along transverse direction, i.e., $C_{st}=H_{cen}-H_{edg}$.

3.2.1. Contact stress

Contact stress distributions between work roll and the plate in transverse direction (plate width direction) with different working conditions are shown in **Fig. 3**. Reduction rate, plate width, rolling temperature and strain rate are the main influence factors of contact stress. The previous studies have revealed that unrealistically large contact stress would be obtained with the elastic-perfectly materials[9]. Actually, when plastic dominates the total deformation, contact stress increases slightly. Stress concentrations locate about 80 mm from the plate side in all the working conditions. The locations of "cat ear" type wear[10] and surface spalling[4] on work roll are coincide with the position of stress concentrations, which indicates that contact stress distributions are closely related to localized wear and fatigue of rolls.

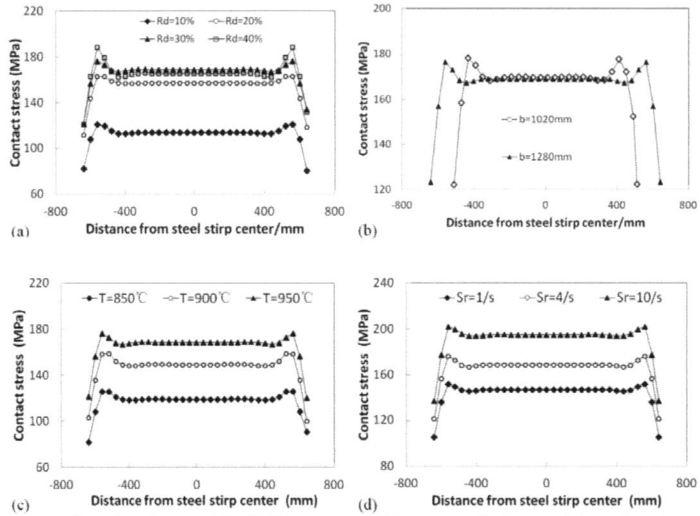

Figure 3. Contact stress distribution with different working conditions: (a) reduction rate (b) plate width (c) deformation temperature and (d) strain rate

3.2.2. Plate shape

Plate shape is defined as the thickness profile of steel plate along transverse direction. Plate shape is the core quality factor of the final product, especially for silicon steel plate. The production of plates with high mechanical yield stress, low thickness and less flatness defects is a major challenge for plate steel.[11] Plate shape is drawn the most attention in hot rolling process. The objective of the shape control in steel rolling process is to minimize the thickness deviation and asymmetrical shape in the transverse direction of the plate.[12] The plate shape is influenced by many rolling parameters. In actual steel production, work roll bending is the main on-line control approach, while the other parameters are mostly presetting ones. Plate shape is sensitive to rolling parameters. Almost all the main rolling parameters would affect plate shape, just as shown in Figure 4. The steel plate will get larger crown with the reduction rate, entrance plate crown and harding level (mainly influenced by deformation temperature and strain rate) in hot deformation (Figure 4 a, b, e, f). Contrarily, thicker and wider plates are liable to get smaller crown, which are demonstrated in Figure 4 c,d.

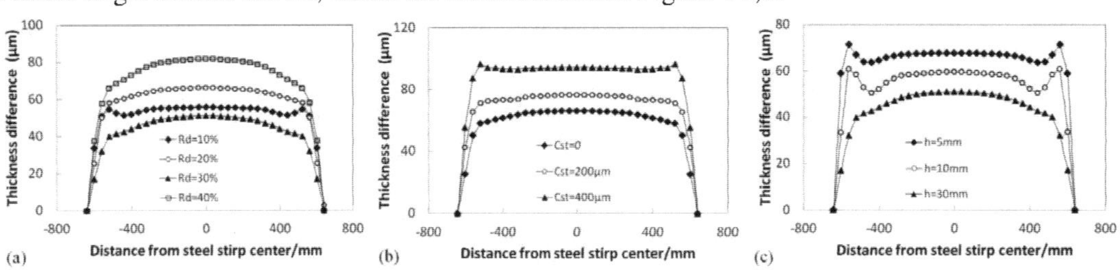

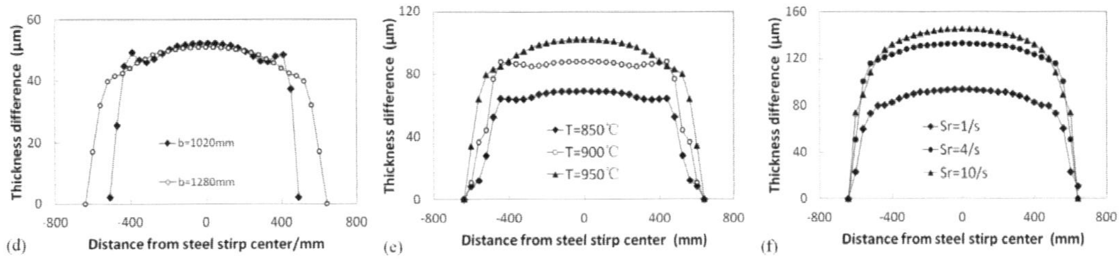

Figure 4. Plate shapes in different working conditions: (a) reduction rate (b) entrance plate crown (c) plate thickness (d) plate width (e) rolling temperatures and (f) strain rate of the steel

4. Discussion and outlook

Thinner and more flat steel sheet is the target of HSM, which made a direct challenge to the rolling strategy of the finish rolling process. A large amount of silicon steels with the minimum thickness of 2 mm and maximum width of 1280 mm are yielded in the HSM studied in this paper. In actual rolling production, the desired plate profile is difficult to achieve and "edge wave" occurs frequently. The knowledge of silicon steel deformation behaviors in rolling process is crucial for its precise shaping.

The deformation behaviors of the silicon steel will provide basis for predicting and controlling the plate profile during rolling process with real-time control systems. Moreover, the optimization and decision of rolling strategy, including the selections of suitable roll profile and adjustment of load sharing in different stands, must base on the steel deformation behaviors in order to improve the plate product quality.

5. Conclusions

The main findings and contributions from this work are as follows:

(1) The flow behaviors of the silicon steel investagated in this work during hot compressive process were investigated by thermal simulation experiments. The silicon steel showed strain hardening and thermal softening effects in single-phase region, while it appeared strain hardening and low temperature softening in dual-phase region.

(2) Stress concentrations exist adjoin the plate sides, which are related to localized roll wear and fatigue. Rolling parameters, plate sizes and material properties have significant influence on plate shape. Larger convexity strip will acquired by larger reduction rate, narrower and thinner plate, higher flow stress of the steel.

Acknowledgment

The authors would like to acknowledge the financial support provided by the Shandong Jiaotong University Doctorial Scientific Research Fund and Climbing Project for Advanced Equipment and Control.

References

[1] Landgraf, F.J.G., et al., Effect of plastic deformation on the magnetic properties of non-oriented electrical steels. Journal of Magnetism and Magnetic Materials, 2000. 215-216: p.94-96.
[2] De Paepe, A., et al., Effect of hot rolling parameters on the magnetic properties of a low-silicon ultra-low-carbon steel. Journal of Magnetism and Magnetic Materials, 1996. 160: p.129 - 130.
[3] Schausberger, F., A. Steinboeck and A. Kugi, Mathematical modeling of the contour evolution of heavy plates in hot rolling. Applied Mathematical Modelling, 2015. 39(15): p.4534-4547.
[4] Dong, Q., et al., Analysis of Spalling in Roughing Mill Backup Rolls of Wide and Thin Strip Hot Rolling Process. steel research international, 2015. 86(2): p.129-136.
[5] Lenard, J.G., Primer on Flat Rolling. 2007: Elsevier Science.
[6] Puchi-Cabrera, E.S., et al., Constitutive description for the design of hot-working operations of a 20MnCr5 steel grade. Materials & Design, 2014. 62: p.255-264.
[7] Zener, C. and J.H. Hollomon, Effect of Strain Rate Upon Plastic Flow of Steel. Journal of Applied Physics, 1944. 15(1): p.22.

[8] Sellars, C. and W. McTegart, On the mechanism of hot deformation. Acta Metallurgica, 1966. 14: p.1136-1138.
[9] Liu, M., Finite element analysis of large contact deformation of an elastic–plastic sinusoidal asperity and a rigid flat. International Journal of Solids and Structures, 2014. 51(21-22): p.3642-3652.
[10] Li, W., et al., Roll shifting strategy with varying stroke and step in hot strip mill. Journal of Central South University, 2012. 19(5): p.1226-1233.
[11] Tran, D.C., N. Tardif and A. Limam, Experimental and numerical modeling of flatness defects in strip cold rolling. International Journal of Solids and Structures, 2015.
[12] HUR, Y. and Y. CHOI, A Fuzzy Shape Control Method for Stainless Steel Strip on Sendzimir Rolling Mill. Journal of Iron and Steel Research, International, 2011. 18(3): p.17-23.

IOP Conf. Series: Materials Science and Engineering **635** (2019) 012028 doi:10.1088/1757-899X/635/1/012028

On the analysis of compressive failure load of single-lap bolted joint of green composites

M Roy Choudhury and K Debnath*

Department of Mechanical Engineering, National Institute of Technology Meghalaya, Shillong – 793 003, India

*debnath.iitr@gmail.com

Abstract. In the present study, bamboo fiber reinforced polylactic acid (PLA) composites have been fabricated by hot compression molding using film stacking method. The compressive failure load behavior of single lap bolted joint of the developed composites has been experimentally investigated. The influence of the geometric parameters of mechanical bolted joint such as width to diameter ratio (*w*/*d*), edge to diameter ratio (*e*/*d*) and bolt fastening torque on the compressive failure load has been statistically analyzed. Taguchi's analysis has been performed to find the optimum level of geometric parameters to obtain higher compressive failure load of the bolted joint. The *w*/*d* and bolt fastening torque were found to be significant parameters influencing the compressive failure load of the bolted joint. The compressive failure load of the bolted joint is more at a higher value of *w*/*d* and bolt fastening torque.

1. Introduction

The application of composites in different fields of engineering has increased because of their high specific strength, stiffness and ease of complex shape fabrication characteristics. In recent years, green composites composed of natural fiber and biodegradable polymer has emerged as an excellent material that replaced many synthetic fiber reinforced composites. The biodegradable characteristic makes these composites environmental friendly and eliminates the waste disposal problem [1]. Polymer composites are fabricated to near-net-shape in the primary manufacturing process. But in order to obtain the intricate structure, several composite components are fabricated separately and then they are joined to obtain the complete structure. However, joining of composites materials is not an easy task. High-stress concentration is developed in the joining area due to the anisotropic nature of the fibrous composites. The localized stresses at the joint are difficult to relive because of the brittle nature of most of the composites and consequently limits the large displacement [2]. There are three basic methods available for joining of polymer composites: adhesive joining, mechanical fastening and fusion bonding. The primary means of joining of polymer composites is the mechanical fastening. In mechanical fastening, composites parts are joined by installing the bolt and nut in a hole. It is very important to design the bolted joint as this is the weakest region in the structure. Also, high stress localizes in and around the joint. The stress induced in a bolted joint is influenced by many factors such as the geometry of the joint, joint materials, bolt fastening torque etc. [3]. It has been revealed that the strength of a joint is affected by geometric variables such as width (w), edge distance (e), hole diameter (d) and laminate thickness (t) [4]. Several researchers have studied the failure behavior of bolted joints of synthetic fiber polymer composites both numerically and experimentally by analyzing the distribution of stress. The fastener pull-through failure in carbon/epoxy composites was investigated both experimentally and numerically [5-6]. Arnold *et al.* [7] optimized the different parameters of mechanically fastened composite joints. The joint strength was predicted by developing a new semi-empirical technique. Choi and Chun [2] predicted the load during failure of a mechanically fastened composite joint. The authors have developed a new method of prediction called 'failure area index'. The experimental values showed good agreement with the results obtained through numerical simulation. Chutima and Blackie [8] predicted the stress by developing 2-D finite element model in a joined specimen of carbon fiber reinforced composites. A 3-D finite element model was also developed to determine the strength and damage progression in joining of carbon fiber reinforced plastics [9-10] and graphite fiber reinforced plastics [11]. Experimental investigations on joining strength of carbon fiber reinforced plastics and glass fiber reinforced plastics were conducted for various laminate configurations, hole sizes, bolt tightening torque and geometric parameters (width, edge distance, hole diameter and laminate thickness) [12-13]. Senguttuvan and Lillymercy [14] experimentally determined the strength and mode of failure of riveted and bolted joint of glass fiber reinforced plastics by varying the number of rivet and bolt. However, a few works are available on joining behavior of green composites fastened by bolted joint. The green composites namely bamboo fiber reinforced polylactic acid (PLA) composites which is considered as an engineering material has been studied by many researchers by evaluating their mechanical properties [15-22], thermal properties [18,20], water resistance and interfacial adhesion characteristics [17] and crystallization behavior [22]. Fazita *et al.* [23] studied the recyclable and biodegradable behavior of bamboo/PLA composites. In the current study, the joining behavior of bamboo/PLA composites has been investigated by analyzing the compressive failure load. The bolt fastening torque and geometric parameters of bolted joint such as width to diameter ratio (*w/d*) and edge to diameter ratio (*e/d*) have been statistically analyzed and optimized to obtain the higher value of compressive failure load.

2. Materials and methods

2.1 Fabrication of composites

Bamboo fiber in woven form was purchased from Sri Lakshmi Group, Andhra Pradesh, India. PLA in pallets form was supplied by Natur Tec India Pvt. Ltd. The density of PLA is 1.24 g/cm^3. The melting temperature and glass transition temperature of PLA is 170°C and 58°C, respectively. The fabrication of the bamboo fiber reinforced PLA composites were carried out by film stacking method using hot compression. In order to remove the moisture, PLA pellets and the fibers were heated in an oven at 80°C for 4 hours. PLA films of 1.5 mm thickness were fabricated by hot compression of PLA pellets at 160°C. Initially, the pressure of 0.4 MPa was applied for 4 minutes and then increased to 3 MPa for another 2 minutes keeping the temperature constant. After allowed to cool the film for two hours under pressure, the film was removed from the mold. A total of four layers of fiber mats were stacked alternately between the PLA films. The stack of polymer films and fibers was compressed at a temperature of 180°C. Initially, 4 MPa pressure was applied for 8 minutes and then the pressure was increased to 6 MPa for 2 minutes. Composite was removed from the mold after 2 hours and allowed to cool. The thickness of the fabricated composite is 3 mm.

2.2 Experimentation

Table 1 shows the level of different parameters of bolted joint chosen for the purpose of investigation. The developed composites were cut as required size and hole was made with the help of vertical type milling machine under the dry condition, as shown in Figure 1(a). Drilling operations were carried out using a solid carbide drill bit (8-facet drill bit) of 8 mm in diameter (*d*) at 2000 RPM and 45 mm/min. The diameter of the bolt used for joining composite specimen was 8 mm with a pitch of 18 TPI. The bolt fastening torque was applied using a dedicated fixture mounted on a reaction type torque sensing device (Make: Syscon and Capacity: 100 kg-m). The torque sensing device with the developed fixture and the final bolted joined composite specimen is shown in Figure 1(b). The compressive failure load of the bolted joint is obtained by Universal Testing Machine at a cross-head speed of 2 mm/min. A fixture was designed and developed in order to hold and apply the compressive load on the bolted joined specimen, as shown in Figure 2.

Table 1. Levels of input process parameters

Factors	Unit	Levels		
		1	2	3
A: Width to diameter ratio (*w/d*)	-	2.5	3.25	4
B: Edge to diameter ratio (*e/d*)	-	1.2	1.8	2.5
C: Bolt fastening torque	N-m	10	15	20

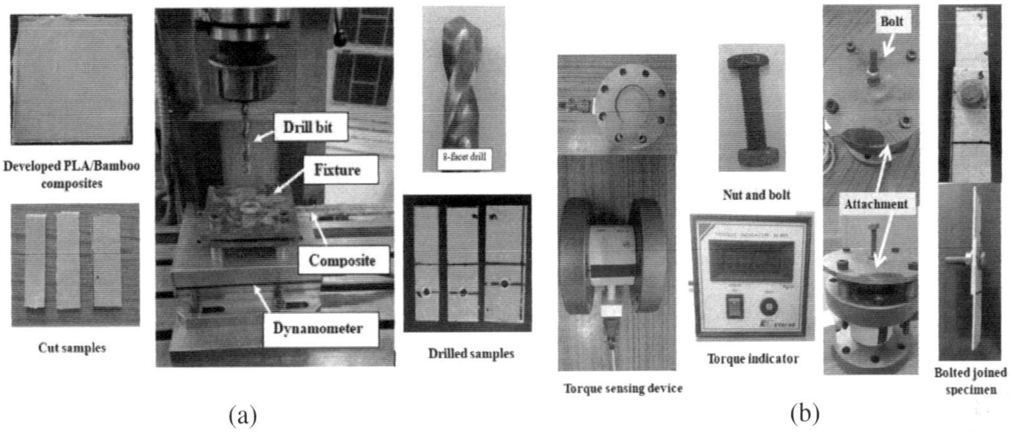

Figure 1. Experimental setup (a) drilling of composites and (b) applying bolt fastening torque

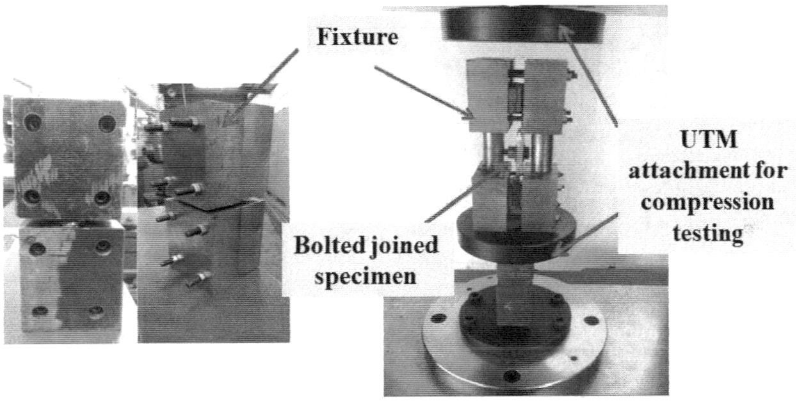

Figure 2. Measurement of compressive failure load

2.3 Design of experiments

It has been established that the bolt fastening torque and geometric variables of the bolted joint such as width to diameter ratio (*w/d*) and edge to diameter ratio (*e/d*) are important parameters that significantly affect the strength of bolted joint. In the current study, these three parameters at three different levels have been chosen to investigate the compressive failure load behavior of the composite joint. Orthogonal array L_9 was selected to carry out the experiments. The optimized parameters were found using Signal-to-Noise (*S/N*) ratio. The aim of the present study is to maximize the compressive failure load. Therefore, 'larger the better' quality characteristic was used for the compressive failure load. Analysis of variance (ANOVA) was also performed to obtain the relative significance of the process parameters affecting the compressive failure load. It also gives the percentage contribution of each process parameters on the compressive failure load.

IOP Conf. Series: Materials Science and Engineering **635** (2019) 012028 doi:10.1088/1757-899X/635/1/012028

3. Results and discussion

The experimental investigation was carried out to understand the effect of different geometric parameters of bolted joint on compressive failure load of the developed green composites. Table 2 presents the experimental layout and values of compressive failure load (both raw and calculated *S/N* ratios). The average values of raw data and *S/N* data are given in Table 3. The larger average values of raw data and *S/N* data at each level are shown in bold text. The main effect graphs of mean and *S/N* data are shown in Figure 3. Table 4 represents the optimal set of input parameters to obtain the maximum value of the compressive failure load. From the Taguchi's analysis, it was found that the higher level of *w/d* (4), medium level of *e/d* (1.8) and a higher level of bolt fastening torque (20 N-m) gives a higher value of compressive failure load. Confirmation experiments were performed at the optimum level of input parameters obtained through Taguchi's analysis. The compressive failure load was found within the confidence interval. The higher value of compressive failure load obtained at higher *w/d* and *e/d* is due to the increase in load carrying capacity of the joint with an increase in the width (*w*) and edge distance (*e*) keeping the diameter of the hole (*d*) constant [24]. The load is transferred in a bolted joint by two means: bearing mode and friction mode. In bearing mode, the load is transferred in the joined specimens by the bolt that bears against the hole. It gives the idea of forces acting on the hole and deformation as well. The bearing strength of the joined composite specimen is more when the values of *w/d* and *e/d* are more. The load that can bear by the overlapping area is high at higher *w* and smaller *d* as the cross-sectional area of the overlapping section is more. The highly stressed composite specimen can press the bolt and the mode of failure changes from tension to bearing [4]. The failure load tends to be zero when the ratio of *w/d* is unity [25]. In the bearing mode, one of the joined plates is subjected to tension during the transfer of load. This generated tension is neutralized by the bearing stress developed in the bolt and hole. Mobilization of bearing stress occurs due to the slipping of the plates relative to each other. Therefore, failure will occur at the edge during slipping of the plates if the edge distance (*e*) from the hole center is very small. Whereas the load is transferred in friction mode due to the induced tension between the plates as the force is applied during tightening of the bolt. It has been reported that the strength of the bolted joint is more when bolt fastening torque is high [26-27]. Higher fastening torque creates more friction between the bolt and specimen. This developed friction helps in releasing friction stress concentration in and around the bolted hole. Also, the washer that is used in bolted joint minimizes ply delamination. Hence, the specimen can carry the higher load when bolt fastening torque is high.

Table 2. The experimental layout of L_9 orthogonal array

Sl. No.	Factors			Compressive force (KN)	
	A	*B*	*C*	Raw data	*S/N* data
1	1	1	1	1.745	4.8359
2	1	2	2	2.575	8.2155
3	1	3	3	2.820	9.0050
4	2	1	2	3.288	10.3386
5	2	2	3	3.690	11.3405
6	2	3	1	2.863	9.1364
7	3	1	3	4.430	12.9281
8	3	2	1	3.980	11.9977
9	3	3	2	4.388	12.8453

A, B, C- Process Parameters
1, 2, 3- Levels of Process Parameters
S/N- Signal-to-Noise Ratio

Table 3. Response for mean data and S/N data

Data	Parameter designation	Average values		
		Level 1	Level 2	Level 3
Raw data	*A*	2.380	3.280	**4.266**
	B	3.154	3.415	**3.357**
	C	2.863	3.417	**3.647**
S/N data	*A*	7.352	10.272	**12.590**
	B	9.368	10.518	**10.329**
	C	8.657	10.467	**11.091**

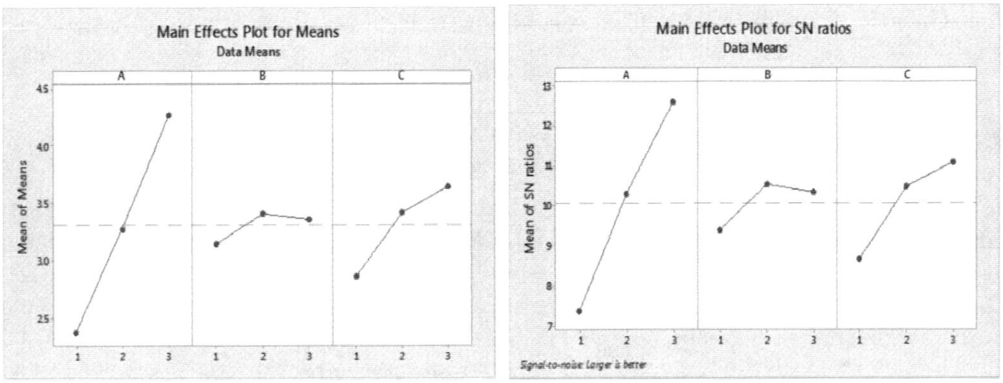

Figure 3. Main effect plot for means and *S/N* ratio

Table 4. Optimal setting of input parameters

Response	Factor *A*	Factor *B*	Factor *C*
Compressive failure load (kN)	Level 3 (*w/d* = 4)	Level 2 (*e/d* = 1.8)	Level 3 (20 N-m)

In Table 5, analysis of variance (ANOVA) of the experimental data for compressive failure load is presented. It is important to perform ANOVA to statistically analyze the influence of input parameters on the output response. From Table 5, it can be seen that the *F* value for *w/d* (factor *A*) is highest followed by bolt fastening torque (factor *C*) and *e/d* (factor *B*). This indicates that the *w/d* is the most significant model terms. It has the maximum influence on the compressive failure load. It was also observed in ANOVA that the *p*-value for *w/d* and the bolt fastening torque is lower than 0.05 at 95% confidence interval. This indicates that the *w/d* and bolt fastening torque are two significant parameters at a 95% confidence interval. R^2 value statically represents the closeness of the model to fit the data. The value of R^2 is 99.79%. This indicates that the fitness of the experimental data is quite satisfactory. The percentage contribution of *w/d* is more followed by the bolt fastening torque to obtain the higher compressive failure load as can be seen in Figure 4.

Table 5. ANOVA for compressive failure load

Source	DOF	Adj. SS	Adj. MS	*F*- Ratio	*P*-Value	% Contribution
A: *w/d*	2	5.33913	2.66957	398.28	**0.003***	82.910
B: *e/d*	2	0.11238	0.05619	8.38	0.107	1.745
C:Bolt fastening torque	2	0.97469	0.48734	72.71	**0.014***	15.135
Residual Error	2	0.01341	0.00670			0.208
Total	8	6.43961				100

DOF: Degree of Freedom
Adj. SS: Adjusted Sum of Square
Adj. MS: Adjusted Mean Sum of Square
*Significant at 95% Confidence Level (R^2 =99.79%).

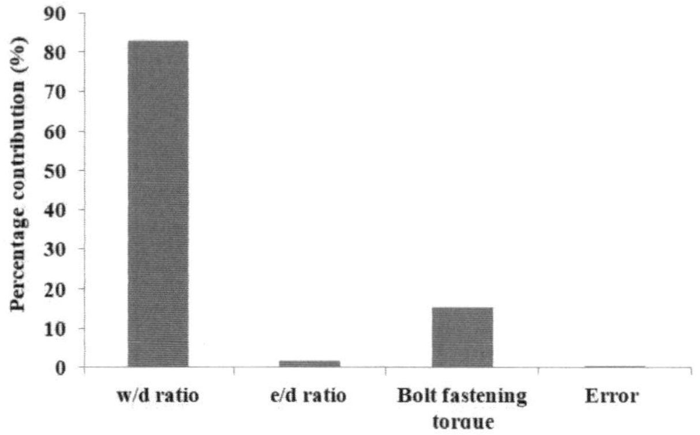

Figure 4. Percentage contribution of input parameters

4. Conclusions

The concluding remarks drawn from the present experimental endeavor are presented below:

- The optimum values of input parameters to obtain the higher value of compressive failure load are the higher level of *w/d* (4), medium level of *e/d* (1.8) and a higher level of bolt fastening torque (20 N-m).
- The compressive failure load increases with an increase in *w/d* and bolt fastening torque. At a high level of *w/d* and bolt fastening torque, the specimen can transfer more load by means of bearing and friction mode. The compressive failure load increases with increase in *e/d* up to a certain value and then it starts decreasing with further increase in *e/d*.
- The *w/d* and bolt fastening torque were found to be the significant parameters to obtain the higher value of compressive failure load.
- The percentage contribution of *w/d* is more followed by bolt fastening torque and *e/d* to obtain a higher value of compressive failure load.

References

[1] Ochi S 2012 Tensile properties of bamboo fiber reinforced biodegradable plastics *Int. J. Compos. Maters.* **2** 1.

[2] Choi J H and ChunY J 2003 Failure load prediction of mechanically fastened composite joints *J. Compos. Mater.* **37** 2163.

[3] Lie S T, Yu G and Zhao Z 2000 Analysis of mechanically fastened composite joints by boundary element methods *Composites Part B* **31**693.

[4] Camanho P P and Matthews F L 1997 Stress analysis and strength prediction of mechanically fastened joints in FRP: a review *Composites Part A* **28A** 529.

[5] Banbury A and Kelly D W 1999 A study of fastener pull-through failure of composite laminates. Part 1: Experimental *Compos. Struct.* **45** 241.

[6] Banbury A, Kelly D W and Jain L K 1999 A study of fastener pull-through failure of composite laminates. Part 2: Failure prediction *Compos. Struct.* **45** 255.

[7] Arnold W S, Marshall I H and Wood J 1990 Optimum design considerations for mechanically fastened composite joints *Compos. Struct.* **16** 85.

[8] Chutima S and Blackie A P 1996 Effect of pitch distance, row spacing, end distance and bolt diameter on multi-fastened composite joints *Composites Part A* **27** 105.

[9] Camanho P P and Matthews F L 1999 A progressive damage model for mechanically fastened joints in composite laminates *J. Compos. Mater.* **33** 2248.

[10] Kelly G and Hallström S 2004 Bearing strength of carbon fibre/epoxy laminates: effects of bolt-hole clearance *Composites Part B* **35** 331.

[11] Tserpes K I, Labeas G, Papanikos P and Kermanidis T 2002 Strength prediction of bolted joints in graphite/epoxy composite laminates *Composites Part B* **33** 521.

[12] Collings T A 1977 *The strength of bolted joints in multi-directional CFRP laminates Composites* **8** 43.

[13] Kretsis G and Matthews F L 1985 The strength of bolted joints in glass fibre/epoxy laminates *Composites* **16** 92.

[14] Senguttuvan N and Lillymercy J 2015 Joint strength analysis of single lap joint in glass fiber composite material *Int. J. Appl. Eng. Res.***10** 16535.

[15] Lee S H, Ohkita T and Kitagawa K 2004 Eco-composite from poly (lactic acid) and bamboo fiber *Holzforschung* **58** 529.

[16] Okubo K, Fujii T and Yamashita N 2005 Improvement of interfacial adhesion in bamboo polymer composite enhanced with micro-fibrillated cellulose *J. Soc. Mech. Eng. Int. J. Ser. A* **48** 199.

[17] Lee S H and Wang S 2006 Biodegradable polymers/bamboo fiber biocomposite with bio-based coupling agent *Composites Part A* **37** 80.

[18] Tokoro R, Vu D M, Okubo K, Tanaka T, Fujii T and Fujiura T 2008 How to improve mechanical properties of polylactic acid with bamboo fibers *J. Mater. Sci.* **43** 775.

[19] Okubo K, Fujii T and Thostenson E T 2009 Multi-scale hybrid biocomposite: processing and mechanical characterization of bamboo fiber reinforced PLA with microfibrillated cellulose *Composites Part A* **40** 469.

[20] Kang J T and Kim S H 2011 Improvement in the mechanical properties of polylactide and bamboo fiber biocomposites by fiber surface modification. *Macromol. Res.***19** 789.

[21] Zhang Q, Shi L, Nie J, Wang H and Yang D 2012 Study on poly (lactic acid)/natural fibers composites *J. Appl. Polym. Sci.* **125** E526.

[22] Li Y, Jiang L, Xiong C and Peng W 2015 Effect of different surface treatment for bamboo fiber on the crystallization behaviour and mechanical property of bamboo fiber/nanohydroxyapatite/poly (lactic-co-glycolic) composite *Ind. Eng. Chem. Res.* **54** 12017.

[23] Fazita M R, Jayaraman K, Bhattacharyya D, Hossain M S, Haafiz M K and HPS A K 2015 Disposal options of bamboo fabric-reinforced poly (lactic) acid composites for sustainable packaging: bodegradability and recyclability *Polymers* **7** 1476.

[24] Kelly G 2005 Load transfer in hybrid (bonded/bolted) composite single-lap joints *Compos. Struct.* **69** 35.

[25] Chang F K, Scott R A and Springer G S 1982 Strength of mechanically fastened composite joints. *J. Compos. Mater.* **16** 470.

[26] Karakuzu R, Gülem T and İçten B M 2006 Failure analysis of woven laminated glass–vinylester composites with pin-loaded hole *Compos. Struct.* **72** 27.

[27] Jadee K J and Othman A R 2011 Fiber reinforced composite structure with bolted joint–a review *Key Eng. Mater.* **471** 939.

IOP Conf. Series: Materials Science and Engineering **635** (2019) 012029 doi:10.1088/1757-899X/635/1/012029

A Study on Global and Cluster-wise Regression Model in the Automatic GMA welding

Tae-Jong Yun[1], Ji-Yeon Shim[2], Hong-Gun Kim[3], Zong-Liang Liang[1], Won-Bin Oh[1], Bo-Ram Lee[1], Min-Ho Park[4] and Ill-Soo Kim[1]

[1]Department of Mechanical Engineering, Mokpo National University, 1666 Youngsan-ro, Chungkye-myun, Muan-gun, Jeonnam, 534-729, South Korea
[2]Convergence Components & Agricultural Machinery Application Center, KITECH, Palbok-dong 2-ga, Deokjin-du, Jeonju-city, Jeollabuk-do, 561-202, South Korea
[3]Dept of Mechanical & Automotive Engineering, Jeonju University, 303 Choujiam-ro, Wangsan-gu, Jeonju, Jeonbuk-do, South Korea
[4]Research Institute of Medium & Small Shipbuilding, 55, Daebuljugeo 3-ro, Samho-eup, Yeongam-gun, Jeollanam-do, 58457, Korea

yzl143017@mokpo.ac.kr

Abstract. The automatic welding system is presently made use of high volume production industries even if the cost of the related equipment is justified by the large number of pieces to be made. Also, the detailed movement devices with the predetermined sequences of welding parameter and the use of timers to form the weld joints were required. A new mathematical model that predict the optimal welding parameters on a given bead geometry and accomplish the desired mechanical properties of the weldment to make the automatic GMA (Gas Metal Arc) welding process should be needed. The developed model should be employed a wide range of material thicknesses and be applicable for all welding positions as well. In addition, the algorithm must be available in the form of mathematical equations which can be programmed easily to the robot and give a high degree of confidence in predicting the bead dimensions. In this study, two regression models with global regression and cluster-wise regression are proposed to be applicable for prediction of optimal welding parameters on the bead reinforcement area. For development of the proposed regression models, an attempt has been done for applying to a several methods. A series experiments to research the effects of welding parameters on bead reinforcement area as a function of key output parameters for the lab-joint weld in the automatic GMA welding process was performed. Not only the fitting of these models was checked and compared by using a variance test (ANOVA), but also the prediction of bead reinforcement area using the developed regression models were carried out the basis of the additional experiments.

1. Introduction

The Metal Inert Gas (MIG) welding process, occasionally called GMA welding process, is a welding process that produces coalescence of metals by heating with a welding arc between a continuous filler metal electrode and the workpiece [1]. Even if the automatic welding system provides the same time saving and precision welding, it can only be applied for small-lot production and production of a single part [2]. A suitable process control algorithm that describe interaction of the welding parameters and their influence on optimal bead geometry in order to develop the automatic arc welding process, are required. Nevertheless, it is quite difficult task to apply them for various practical situations because relationship between the welding parameters and the bead dimensions is non-linear. Jeng et al. [3] applied for the laser butt welding process using a BP (Back Propagation) and LVQ (Learning Vector Quantization) neural networks, and proved that both neural networks were very useful in determining the suitable welding parameters and avoiding inappropriate welding design. Kim and Jun [4] have made use of a BP neural network for determining bead geometry in GMA welding process. The design parameters of the neural network model are chosen from the error analysis, and the proposed neural network model could be estimated bead geometry with reasonable accuracy. Li et al. [5] proposed a neural network model for on-line prediction of welding quality in GMA welding process.

Chandel [6] first applied regression analysis technique in GMA welding process and indicated the developed regression models derived from experimental results could be utilized for predicting the bead geometry fairly accurately. Yang, et al. [7] extended the related algorithm for prediction of weld deposit area and presented the effects of welding parameters on the weld deposit area. Juang and Tarng [8] demonstrated the selection of welding parameters by using a Taguchi method for obtaining the optimal bead geometry in the Tungsten Inert Gas (TIG) welding process of stainless.

To develop the automatic GMA welding process, the mathematical models should be required to select bead geometry as welding quality. No matter how, only a very few studies of prediction of welding parameters on the optimal bead reinforcement area for lab-joint weld in the automatic GMA welding process have been done.

This paper presents an intelligent model for the lab-joint weld in the automatic GMA welding process by regression algorithms for estimating the optimal bead reinforcement area and studying the effects of various welding parameters. Based on the experimental results, the two regression models with the global and cluster-wise regression analysis algorithms have been developed to predict the suitable bead reinforcement area. These two kinds of developed models are verified by data obtained from additional experiments with lab-joint welds and compared. Finally, predictive behaviors and advantages of each model are discussed.

2. Experimental Works

Experiments were designed for developing the two regression models which based on global and cluster-wise regression analysis algorithm for correlating independently controllable welding parameters. The experimental design provides the smallest number of treatment combinations with which the main effect of a factor and the interaction between the factors can be defined. Since the automatic GMA welding process is considered as a multi-parameter process, it's hard to find optimal welding parameters with lab-joint weld for good welding. According to previous studies [9], five welding parameters included welding voltage, arc current, welding speed, CTWD (Contact Tube Weld Distance) and welding angle were selected as the input parameters and the response was bead reinforcement area to control welding quality in this research. Figure 1 shows a schematic diagram for relationship between input and output parameters with lab-joint weld in the automatic GMA welding process.

Figure 1. A schematic diagram for relationship between input and output parameters

The bead reinforcement area is made use of studying the welding quality. A schematic diagram for measurement of bead reinforcement area with a lap-joint weld in the automatic GMA welding

IOP Conf. Series: Materials Science and Engineering **635** (2019) 012029 doi:10.1088/1757-899X/635/1/012029

process was presented in figure 2. In this study, the bead reinforcement area as welding quality was mainly considered.

Statistically designed experiments that are based upon full factorial techniques, reduce costs and supply the required information about the main and interaction effects on the response factors [10]. The design matrix that has 72 experimental welding runs was used where each row corresponds to one experimental run with two replications. The experimental results that included five welding parameters on bead reinforcement area were obtained by using a welding robot. In this study, the mean of these replications was considered output parameters to utilize the development of two regression models. The results of the experiment were employed on the basis of development of two regression models which based on global and cluster-wise regression analysis algorithms in the automatic GMA welding process.

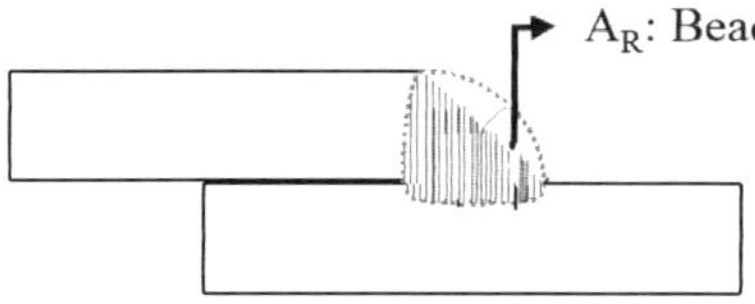

Figure 2. A schematic diagram for measurement of bead reinforcement area

3. Results and Discussion

3.1 Development of global regression model

In general, regression analysis is widely applied for understanding among which the independent variables are related to the dependent variable, and exploring the forms of these relationships. Many techniques to do regression analysis have been developed. Familiar methods such as linear regression and ordinary least squares regression are parametric, in that the regression function is defined in terms of a finite number of unknown parameters that are estimated from the data. The performance of regression analysis methods in practice depends on the form of the data generating process, and how it relates to the regression approach being used. Furthermore, the development of formalized approach for procedure optimization should be included to establish combination of welding parameters which would produce a good weld quality. Global regression analysis has been developed the basis of the experimental results. The following global regression model for bead reinforcement area was developed and presented as following:

$$\begin{aligned} A_R = {} & 6615 - 639.3V - 47.29I - 140.5S - 447.0C - 110.7A + 2.654VI + 7.885VS + 24.78VC + 6.167VA \\ & + 1.016IS + 3.362IC + 0.7873IA + 9.435SC + 2.360SA + 7.188CA - 0.05713VIS - 0.1860VIC - 0.04385VIA \\ & - 0.5257VSC - 0.1321VSA - 0.3978VCA - 0.0751521ISC - 0.01697ISA - 0.053201ICA - 0.1526SCA + \\ & 0.003966VISC + 0.000948VISA + 0.008489VSCA + 0.002933VICA + 0.001140ISCA - 0.000063VISCA \end{aligned} \quad (1)$$

The performance of the developed global regression model for predicting bead reinforcement area were showed figure 3. As indicated in figure 3, not only the straight line put on the measured bead reinforcement areas, but also the dash line represented the predicted results by the developed global regression model. It can be found that differences between the measured and predicted bead reinforcement areas were very small in cases of the whole trail numbers even if the maximum error was limited within 0.4mm. Figure 4 showed the error of the predicted bead reinforcement area using the developed global regression model. As shown in figure 4, it can also be indicated that distributions of the predicted bead reinforcement area were quite close to the best fit line so that the estimated results were reasonable reliable.

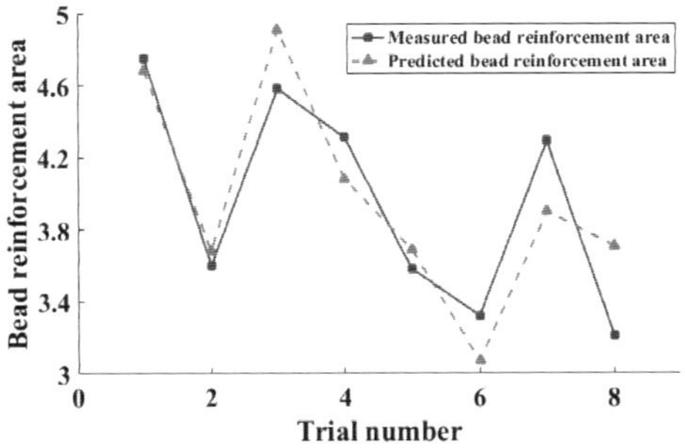

Figure 3. Performance of the developed global regression model for predicting bead reinforcement area

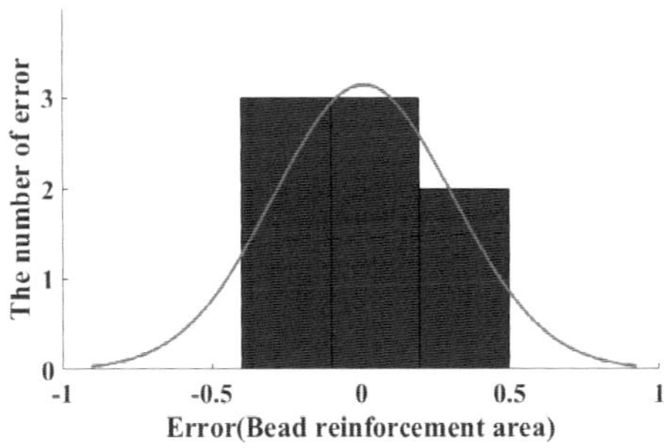

Figure 4. The error of the predicted bead reinforcement area using the developed global regression model

3.2 Development of cluster-wise regression model

Usually, accuracy of prediction of the output obtained by regression equations is depended on being higher or lower at the anchor points which used in the analysis. In this work, each welding parameter has been considered at its two levels for the global regression analysis. Hence, the error in prediction might be more or less deviate from their respectively expected values. In order to solve this problem, a cluster-wise regression analysis model was developed for this work. Cluster-wise regression analysis can be applied for dividing the input–output space into a number of clusters based on similarity. Once the clusters are obtained, the regression analysis can be carried out to determine the input–output relationships. Linear regression analysis involving main factors only can be adopted for this purpose. Initially, 72 sets of input–output data are randomly generated by selecting the input parameters lying within their respective ranges and using the output equations obtained above to determine the outputs. The technique is employed for identifying the cluster centres and grouping them into the clusters on the basis of the similarity measure. The minimum number of data sets in each cluster has been kept fixed to 8.3% (6 data sets out of 72) and it has been decided by the basis of the minimum number of data points required to carry out the linear regression analysis.

Moreover, the minimum number of data required for the regression analysis should be at least one greater than the number of input parameters. The parametric study is performed identifying the optimal number of cluster centres, data points that belong to each cluster centre and number of outliers by varying the T (threshold for similarity) from 0 to 1. When the clusters are identified by the basis of this cluster-wise input–output relationships, statistical regression analysis is performed. As the input–output relationships are established, the test scenarios are passed through the model. Furthermore, the minimum distance from the data point to an appropriate cluster centre in order to identify which cluster each of the test data points belongs is calculated. Once the cluster is identified for each test case, its outputs are predicted by using the linear regression equations corresponding to that particular cluster. Cluster-wise regression analysis for bead reinforcement area with lab-joint welds in the automatic GMA welding process carried out and the following equations are obtained for the different clusters as shown below:

- Cluster 1:

$$A_R = 25.70 - 0.3640V + 0.02700I - 0.3540S - 0.1240C + 0.007000A \quad (2)$$

- Cluster 2:

$$A_R = 25.93 - 0.3434V + 0.02648I - 0.3697S - 0.1141C + 0.006870A \quad (3)$$

- Cluster 3:

$$A_R = -3.3 + 0.581V + 0.0150I - 0.149S + 0.0206C + 0.0220A \quad (4)$$

- Cluster 4:

$$A_R = 20.23 - 0.0805V + 0.02884I - 0.3615S - 0.1150C + 0.01735A \quad (5)$$

- Cluster 5:

$$A_R = 25.452 - 0.355590V + 0.027222I - 0.35476S - 0.12352C + 0.00816A \quad (6)$$

The 8 test cases to check the performance of the developed cluster-wise regression model are passed through the regression equations. A particular test case is the first checked for its belongingness to a particular cluster by considering its Euclidean distance from the cluster centre. Table 1 indicates the belongingness of the test cases to different clusters.

Table 1. Belongings of test cases to different clusters in the automatic GMA welding

Cluster center	C1	C2	C3	C4	C5
Data set No.	5	6, 8	1, 3, 4	7	2

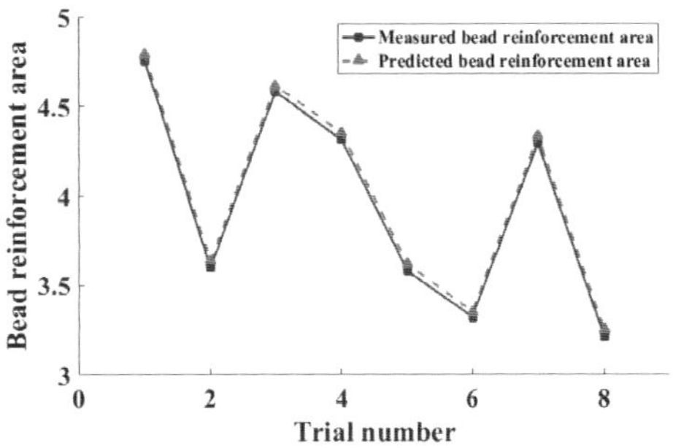

Figure 5. Performance of the developed cluster-wise regression model for predicting bead reinforcement area

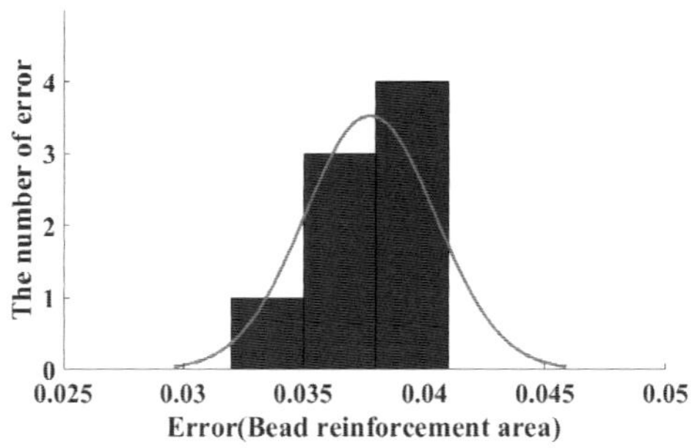

Figure 6. The error of the predicted bead reinforcement area with the developed cluster-wise regression model

Figure 5 represented performance of the developed cluster-wise regression model for predicting bead reinforcement area. As plotted in figure 5, performance of the developed regression model in order to estimate bead reinforcement area is excellent. It was noticed that differences between the measured and predicted bead reinforcement area was almost same in cases of the whole trail numbers even if the maximum error was limited within 0.04mm. Figure 6 demonstrated the error of the estimated bead reinforcement area with the developed cluster-wise regression model. Therefore, it can be concluded that the use of the developed cluster-wise regression model was able to predict bead reinforcement area with lab-joint weld for a given welding conditions.

3.3 Selection of the best regression model

To select the most accurate regression model for prediction of bead reinforcement area in the automatic GMA welding process, the 8 additional experiments were carried out. The convergence criterion for the developed regression models was determined by the average RMS error between the desired output value y_i and predicted output value y_i' for the prediction as following;

$$E_{RMS} = \sqrt{\frac{1}{n}\sum_{i=1}^{n} (y_i - y_i')^2} \quad (7)$$

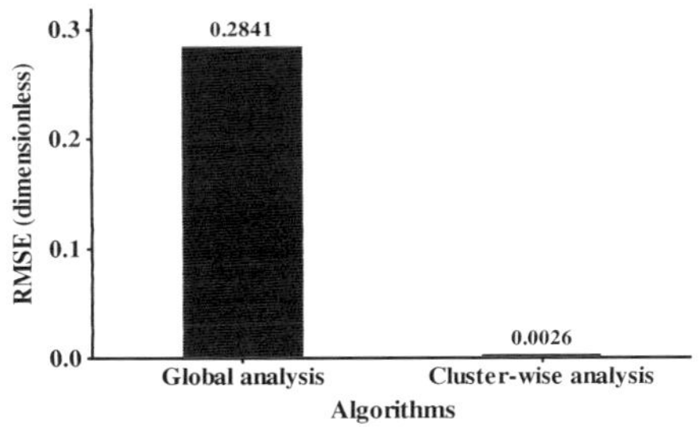

Figure 7. Comparison between the developed global and the developed cluster-wise regression models

Figure 7 plotted the comparison between the developed global and cluster-wise regression models in order to determine bead reinforcement area. The calculated values obtained using the developed cluster-wise regression model was universally lower than those by the developed global regression model as explained in figure 7. However, it is readily clear that the fitting on the experimental data of the cluster-wise regression model with the RMS value of 0.0021 is better than the developed global regression model to determine bead reinforcement area. No matter how, it was shown that the RMS values generated from the developed global regression model was still reasonably small to be accepted in most cases of practical applications.

4. Conclusions

The two regression models to determine the optimal welding parameters on the required bead reinforcement area and investigate the effects of welding parameters on the bead reinforcement area in lab-joint weld in the automatic GMA welding process has been proposed. Experimental results have been applied for developing the optimal algorithm to predict the optimal bead reinforcement area by a global and cluster-wise regression analyses in lab-joint weld in the automatic GMA welding process. The developed global and cluster-wise regression models were made comparing to the target value generated from additional experiment. Both of them were proved to be capable of predicting bead reinforcement area within an acceptable range of error. However, the developed cluster-wise regression model has yielded the slightly better predictions compared to the developed global regression model. A comprehensive analysis was further made for finding the optimal algorithm for prediction of bead reinforcement area. It can be concluded that not only the developed cluster-wise regression model has the least RMS in the aspects of bead reinforcement area, but also the selection of the most accurate regression model for prediction of bead reinforcement area was the cluster-wise regression model.

5. Acknowledgments

This work was supported by the Project of Technology Development for Industry Core (No.20002772, Smart Logistics & Production Automation System for Pipe Spool, Steel Out- Fitting and Block Logistics in Shipbuilding and Offshore Platform) funded By the Ministry of Trade, Industry & Energy (MOTIE, Korea)

6. References

[1] Marburger S J 1990 Welding automation and computer control *Welding: Theory and Practices, Elsevier Science Publishers* pp 209–233

[2] Heipe C R and Roper J R 1990 The geometry of gas tungsten arc gas metal arc and submerged arc weld beads *Welding: Theory and Practice, Elsevier Science Publishers* pp 1–33

[3] Jeng J Y, Mau T F and Leu S M 2000 Prediction of laser butt joint welding parameters using back-propagation and learning vector quantization networks *Journal of Materials Process Technology* 99 pp 207–218

[4] Kim I S and Jun K S 1999 Study on prediction of optimized penetration using the neural network and empirical models *Journal of Korean Society of Machine Tool Engineers* 81(5) pp 70–75
[5] Li X, Simpson S W and Rados M 2000 Neural networks for online prediction of quality in gas metal arc welding *Science and Technology of Welding and Joining* 5 pp 71–79
[6] Chandel R S 1988 Mathematical modeling of gas metal arc weld features *Modeling and Control of Casting and Welding Processes IV: Proceedings of the Fourth International Conference on Modeling of Casting and Welding Processes* pp 109–120
[7] Yang L J, Chandel R S and Bibby M J 1992 The effects of process variables on the bead height of submerged-arc weld deposits *Canadian Metallurgical Quarterly* 31(4) pp 289–297
[8] Juang S C and Tarng Y S 2002 Process parameter selection for optimizing the weld pool geometry in the Tungsten inert gas welding of stainless steel *Journal of Material Process Technology* 122 pp 33–37
[9] Liao T W, Li D M and Li Y M 1999 Detection of welding flaws from radiographic images with fuzzy clustering methods *Fuzzy Sets and Systems* 108 pp 145–158
[10] Vitek J M, David S A, Richey M W, Biffin J, Blundell N and Page C J 2001 Weld pool shape prediction in plasma augmented laser welded steel *Science and Technology of Welding and Joining* 6 pp 305–314

IOP Conf. Series: Materials Science and Engineering **635** (2019) 012030 doi:10.1088/1757-899X/635/1/012030

Optimization of dressing parameters of grinding wheel for 9CrSi tool steel using the taguchi method with grey relational analysis

Luu Anh Tung[1], Vu Ngoc Pi[1*], Vu Thi Lien[1], Tran Thi Hong[2], Le Xuan Hung[1], Banh Tien Long[3]

[1]Thai Nguyen University of Technology, Thai Nguyen city, Vietnam
[2]Nguyen Tat Thanh University, Ho Chi Minh city, Vietnam
[3]Ha Noi University of Science and Technology, Ha Noi city, Vietnam

*vungocpi@tnut.edu.vn

Abstract. This study aims to optimize the dressing parameters of grinding wheel for 9CrSi tool steel to minimize roughness average and flatness tolerance using Taguchi method and Grey Relational Analysis (GRA). These objectives are minimized by optimizing four four-level and two two-level dressing parameters in sixteen experiments based on an orthogonal array $L_{16}(4^4 \times 2^2)$. Taguchi method and GRA are combined to solve the multi-objective optimization problem and determine optimal dressing parameter level combination. Results show that, the optimal dressing parameters to obtain the minimal roughness average and flatness tolerance are coarse dressing depth of 0.025 mm, coarse dressing times of 3 times, fine dressing depth of 0.005 mm, fine dressing times of 2 times, non-feeding dressing of 3 times and dressing feed rate of 1.6 m/min, respectively. Experiments with the optimized dressing parameters have been done to verify the predict model. Results of the roughness average and flatness tolerance from the experiments match with those values of the models and satisfy practical requirements.

1. Introduction

Grinding process plays an important role in the quality of a surface finish [1-3]. This complex process depends on many factors such as grinding wheels, grinding parameters, dressing conditions. During grinding process, the surface of a grinding wheel is continuously worn and that reduces the accuracy of surface finish. To obtain a better grinding performance, the profiles and sharp levels of grinding wheel need to be stably maintained during grinding process. It is necessary to dress the grinding wheel to achieve the original requirements of the workpiece by using dressing tools such as a diamond. In the dressing process, dull grains are removed or severed to re-sharp and renew the cutting face of the grinding wheel.

Many studies have been done to understand the influence of dressing conditions on grind wheels and surface finish [4-16]. Normally, the objective of grinding process is to get a better surface finish (roughness, flatness tolerance, etc.) or high material removal rate of the workpiece. To obtain those objectives, many researchers have tried to optimize dressing parameters [17-25]. For example, the dressing speed ratio, the radial in-feed of the diamond roller dresser and the dress-out time are optimized to determine the grinding wheel lifetime, the cutting ability of the abrasive wheels, the cutting forces, and the roughness of the machined surfaces [23]. In the same study [25], the dressing depth of cut, the dressing cross feed rate, the drag angle of dresser, and the number of passes are investigated to minimize surface roughness. To optimize dressing parameters, many optimization methods are used such as simulated annealing method, neutral network method, Taguchi method and Grey Relational Analysis (GRA) method [17-27].

Although many published researches are available to optimize dressing parameters, there are still many parameters that have not been studied yet such as the effect of three dressing states i.e. the coarse dressing, the fine dressing and non-feeding dressing on the surface finish. In this study, six dressing parameters, including dressing feed rate, coarse dressing depth, coarse dressing times, fine dressing depth, fine dressing times, and non-feeding dressing are optimized to grind 9CrSi alloy steel workpiece on a surface grinding machine to minimize surface roughness and flatness tolerance. The coarse dressing depth, the fine dressing depth, the fine dressing times, and non-feeding dressing are set up in four four-level while the others two-level dressing parameters. Sixteen experiments are arranged based on an orthogonal array $L_{16}(4^4 \times 2^2)$. Two objectives i.e. the roughness average and the flatness tolerance are minimized by combination of Taguchi and GRA methods [28-32].

2. Methodology

2.1 Experimental machine and equipment

In this study, the grinding and dressing processes are implemented in a surface grinding machine as shown in Figure 1. The workpieces are made by 9CrSi tool steel (C: 0.85-0,95%, Si: 1.2-1.6%, Mn: 0.3-0.6%, Cr: ≤ 0.95-1.25%, P, S: ≤ 0.03%, Mo, W: ≤ 0.2%, Ni: ≤ 0.35%, V: ≤ 0.15%) with dimensions of 100×60×20 mm and hardness of 58-62 HRC. Lubricant is used in grinding process is Cantext Aquatex 3810 with flow of 10 l/min. All experiments are done on the devices that are listed in Table 1.

Figure 1. Surface grinding process.

Table 1. Experimental equipment.

Machine and Equipment	Specification
Surface grinding machine	MOTO – YOKOHAMA
Grinding wheel	Cn46TB2GV1.300.32.127.30 m/s
Dressing tool	3908-0088C type 2
Roughness measurement device	Mitutoyo 178-923-2A, SJ-201
Flatness tolerance measurement device	Mitutoyo CMM544

2.2. Design experiment

Grinding process parameters at each experiment are a cutting velocity of 26.7 m/sec, a depth of cut of 0.01 mm, a workpiece velocity of 10 m/min and an axial feed velocity of 8 mm/pass. Based on previous studies [17-27], six dressing parameters, including the dressing feed rate, the coarse dressing depth, the coarse dressing times, the fine dressing depth, the fine dressing times, and the non-feeding dressing are selected to investigate the roughness and flatness tolerance of the machined surfaces in this research. Each of these parameters is set up as shown in Table 2 in which there are four four-level factors and two two-level factors. The design of experiment (DOE) method has been used for the experiments consist of identical L16 orthogonal array fractional parametrical designs in seven first columns in Table 3.

Table 2. Dressing parameters and levels

Parameters	Unit	Levels			
		1	2	3	4
Dressing feed rate S	m/min	1.6	1.8	-	-
Coarse dressing depth t_{coarse}	mm	0.015	0.02	0.025	0.03
Coarse dressing times n_{coarse}	Times	0	1	2	3
Fine dressing depth t_{fine}	mm	0.005	0.01	-	-
Fine dressing times n_{fine}	Times	0	1	2	3
Non-feeding dressing $n_{non\text{-}feed}$	Times	0	1	2	3

Table 3. L_{16} Orthogonal array with factors and responses

No.	t_{coarse}	n_{coarse}	$n_{non\text{-}feed}$	n_{fine}	t_{fine}	S	Surface roughness average Ra (μm)			Flatness tolerance FT (μm)		
							Trial 1	Trial 2	Trial 3	Trial 1	Trial 2	Trial 3
1	0.015	0	0	0	0.005	1.6	-	-	-	-	-	-
2	0.015	1	1	1	0.005	1.8	1.154	1.117	1.229	27.4	27.6	26.35
3	0.015	2	2	2	0.01	1.6	0.745	0.825	0.842	24.65	24.4	25.4
4	0.015	3	3	3	0.01	1.8	0.652	0.517	0.596	17.15	19.4	19.6
5	0.02	0	1	2	0.01	1.8	1.225	1.042	1.252	16.15	16.7	17.4
6	0.02	1	0	3	0.01	1.6	1.267	1.322	1.35	16.35	16.6	17.55
	…											
15	0.03	2	1	3	0.005	1.6	0.635	0.769	0.677	18.3	20.35	19.65
16	0.03	3	0	2	0.005	1.8	0.771	0.874	0.892	16.3	18.65	17.1

2.3 Measurement

After each experiment, roughness is measured at three locations and these three roughness values are averaged. Flatness tolerance is determined by the difference of the maximum height and minimum height in 27 measured points on the workpiece. The roughness and flatness are measured three times at each experiment and their values are recorded in six last columns in Table 3.

3. Results and discussion

After all experiments are done as in Table 3, GRA [28-32] is used to solve multi-objective optimization problem in which the roughness average and flatness tolerance of the workpiece are minimized simultaneously. Then, Taguchi method is used to analysis and determine the optimal dressing parameters.

In the first step to optimize dressing parameters, based on the measured results, the grey relational grades ($\bar{\gamma}$) are generated as shown in Table 4.

Table 4. Grey relational co-efficient and grey grade values

No.	Grey relational co-efficient γ_i		$\bar{\gamma}$
	Ra	FT	
1	-	-	-
2	0.361	0.333	0.347
3	0.488	0.351	0.419
4	0.690	0.420	0.555
5	0.359	0.457	0.408
6	0.333	0.455	0.394
	...		
15	0.566	0.410	0.488
16	0.465	0.444	0.455

From Table 4, the higher grey relation grade implies the better product quality. Therefore, the factor effects can be estimated and the optimal level for each controllable factor determined. By observation of the grey relational grade corresponding to each experiment, the 10th experiment ($t_{coarse3}$, $n_{coarse2}$, $n_{non\text{-}feed4}$, n_{fine3}, t_{fine1}, S_1) has the maximum grey relational grade. That means that in this experiment, the S/N ratio is close to the normalized S/N ratio and it have better properties than other experiments. However, this result does not show optimal dressing parameters.

After that, in order to find these optimal factors, Taguchi method is used to determine the relation between the dressing parameters in different levels and grey relation grades as shown in Table 5. In addition, the effects of dressing parameters on the grey relation grades are plotted in Figure 2.

Table 5. Main effects on grey relation grades

Level	Dressing parameters					
	t_{coarse}	n_{coarse}	$n_{non\text{-}feed}$	n_{fine}	t_{fine}	S
1	0.4405	0.4652	0.4657	0.4694	**0.5216**	**0.6009**
2	0.4532	0.5317	0.4706	0.5401	0.4984	0.429
3	**0.6345**	0.4596	0.4457	**0.5705**		
4	0.4914	**0.5694**	**0.644**	0.4469		
Delta	0.194	0.1098	0.1982	0.1236	0.0231	0.172
Order of effect	2	5	1	4	6	3
Average of grey relation grades (T) is **0.509217**						

As observing in Table 5 and Figure 2, the optimal dressing parameters for surface grinding process to maximize the grey relation grade are $t_{coarse3}/n_{coarse4}/n_{non\text{-}feed4}/n_{fine3}/t_{fine1}/S_1$ or t_{coarse}=0.025mm, n_{coarse} = 3 times, $n_{non\text{-}feed}$ = 3 times, n_{fine} = 2 times, t_{fine} = 0.005 mm, S = 1.6 m/min.

Then, ANOVA table is established as presented in Table 6 to determine how much contribution of each dressing parameter on the grey relation grade. Based on the value of contribution in the last column in Table 6, the non-feeding dressing is the most significant parameter with a contribution of 26.5%, followed by the fine dressing times of 25%, and the coarse dressing depth of 18.7%. The fine dressing depth and coarse dressing times with contribution of 7.4% and 6.3% are the less significant. Since the contribution of the feed rate is very small, the influence on the surface finish of this parameter can be negligible.

IOP Conf. Series: Materials Science and Engineering **635** (2019) 012030 doi:10.1088/1757-899X/635/1/012030

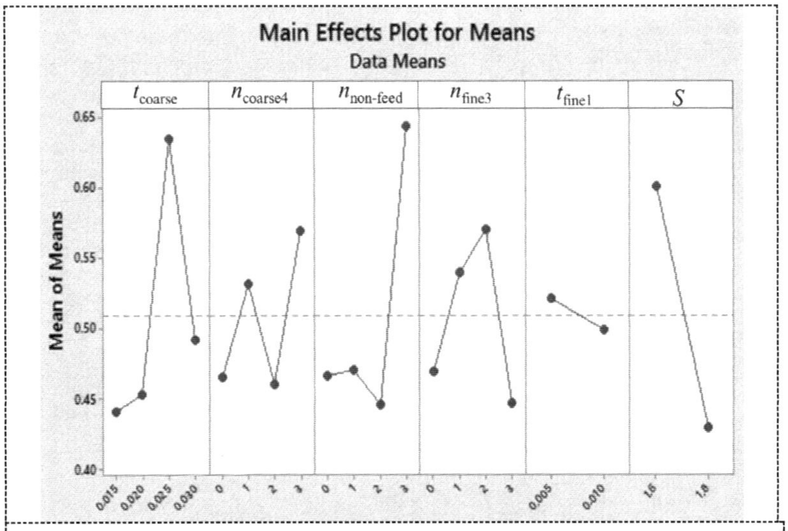

Figure 2. Main effects plot for means.

Table 6. Main effects on grey relation grades

Source	DF	SS	Adj SS	MS	SS'	F	C %
t_{coarse}	3	0.090798	0.137284	0.045761	0.076638	15.02	18.7
n_{coarse}	3	0.039872	0.07871	0.026237	0.025712	10.24	6.3
$n_{non\text{-}feed}$	3	0.122885	0.146906	0.048969	0.108725	15.05	26.5
n_{fine}	3	0.116591	0.085316	0.028439	0.102431	10.6	25.0
t_{fine}	1	0.035034	0.016319	0.016319	0.030314	7.42	7.4
S	(1)	(0.00472)	-	- pooled-	-	-	-
Error	1	0.00472	0.00472	0.00472			16.1
Total	14	0.409901					100.0

From above results, the predict modal of the optimal grey relation analysis are determined by the following equation:

$$\gamma_{op} = t_{coarse3} + n_{coarse4} + n_{non-feed4} + n_{fine3} + t_{fine1} - 4T \quad (1)$$

where T is the average of grey relation grades (T = 0.509217 as in Table 5), with the values of $t_{coarse3}$, $n_{coarse4}$, $n_{non\text{-}feed4}$, n_{fine3}, t_{fine1}, S_1 taken from Table 5. Using these optimal dressing parameters, the minimal values of roughness average and flatness tolerance are computed as:

$$(Ra, FT)_{\min} = \overline{t_{coarse3}} + \overline{n_{coarse4}} + \overline{n_{non-feed4}} + \overline{n_{fine3}} + \overline{t_{fine1}} - 4\overline{T} \quad (2)$$

where $\overline{t_{coarse3}}$, $\overline{n_{coarse4}}$, $\overline{n_{non-feed4}}$, $\overline{n_{fine3}}$, and $\overline{t_{fine1}}$ are mean values of the roughness average or the flatness tolerance corresponding with the coarse dressing depth at the level 3, the coarse dressing times at the level 4, the non-feed dressing at the level 4, the fine dressing depth at the level 3, and the fine dressing times at the level 1, respectively. Therefore, the predict values of the roughness average and the flatness tolerance are calculated as $(\mathrm{Ra})_{\min} = 0.3399\mu\mathrm{m}$ and $(\mathrm{FT})_{\min} = 10.3\mu\mathrm{m}$.

Finally, to evaluate the predict model, experiments have been done with the optimal dressing parameters in two times and their results are shown in Table 7. In this table, the differences of the surface finish properties between the predict model and experiment are small. Thus, the proposed method has been proven and it can be used to accurately predict both the roughness average and flatness tolerance simultaneously.

Table 7. The results from model and experiment

Properties of surface finish	Optimal parameters		
	Predict model	Experiment	Error (%)
	$t_{coarse3}$, $n_{coarse4}$, $n_{non\text{-}feed4}$, n_{fine3}, t_{fine1}, S_1	$t_{coarse3}$, $n_{coarse4}$, $n_{non\text{-}feed4}$, n_{fine3}, t_{fine1}, S_1	
Roughness average: Ra (μm)	0.3399	0.358	5.3
Flatness tolerance: FT(μm)	10.3	10.98	6.6
GRA value	0.9031	0.836	

4. Conclusion

This study has been done to optimize the dressing parameters of grinding wheel for 90CrSi tool steel to minimize the roughness average and the flatness tolerance using Taguchi method and Grey Relational Analysis. Sixteen surface grinding experiments are set up based on an orthogonal array $L_{16}(4^4 \times 2^2)$ for four four-level and two two-level parameters, including the dressing coarse dressing depth, the fine dressing depth, the fine dressing times, the non-feeding dressing, the dressing feed rate and, the coarse dressing times, respectively. Based on the results of GRA and Taguchi method, the optimal dressing parameters are determined to minimize the roughness average and flatness tolerance. Results show that the coarse dressing depth, the coarse dressing times, the fine dressing depth, the fine dressing times and the non-feed dressing are significant factors while the feed rate *S* is not important factor. The optimal dressing parameters that should be used for surface gringding process of 90CrSi alloy steel to obtain the minimal the roughness average and the flatness tolerance are t_{coarse}=0.025 mm, n_{coarse}= 3 times, $n_{non\text{-}feed}$= 3 times, n_{fine} = 2 times, t_{fine} = 0.005 mm, S = 1.6 m/min. In addition, the predict results agree with the experiment results, thus the proposed method in this study has been verified. This method can be further applied to optimize other mechanical manufacturing processes.

Acknowledgments

The work described in this paper was supported by Thai Nguyen University of Technology for the project N_0 T2016-46.

References

[1]. F Klocke 2009 *Manufacturing processes 2 – Grinding, honing, lapping* (Springer)
[2]. M C Shaw 1996 *Principles of Abrasive Processing* (Oxford University Press)
[3]. S Malkin 1989 Grinding technology: Theory and Applications of Machining with abrasives (England: Ellis Horwood Limited)
[4]. K Wegener, H W Hoffmeister, B Karpuschewski, F Kuster, W C Hahmann and M Rabiey 2011) *Conditioning and monitoring of grinding wheels* (CIRP Annals - Manufacturing Technology. 60) pp 757-777
[5]. T C Buttery, A Statham, J B Percival and M S Hamed 1979 Some effects of dressing on grinding performance (Wear. 55(2)) pp 195-219
[6]. X Chen 1995 Strategy for the selection of grinding wheel dressing conditions (Doctoral thesis, Liverpool John Moores University)
[7]. X Chen and W B Rowe 1996 Analysis and simulation of the grinding process Part I: Generation of the grinding wheel surface (International Journal of Machine Tools and Manufacture. 36(8)) pp 871-882
[8]. X Chen and W B Rowe 1996 Analysis and simulation of the grinding process. Part II: Mechanics of grinding (International Journal of Machine Tools and Manufacture. 36(8)) pp 883-896

[9]. D Chen, R Allanson, W B Rowe 1998 Life cycle model of the grinding process (Computers in Industry. 36(1-2)) pp 5-11

[10]. Prusak Z, J A Webster, and I D Marinescu 1997 Influence of dressing parameters on grinding performance of CBN/ Seeded Gel hybrid wheels in cylindrical grinding (International Journal of Production Research. 35(10)) pp 2899-2916

[11]. H Baseri, S M Rezaei, A Rahimi, and M Saadat 2008 Analysis of the disc dressing effects on grinding performance—part 1: simulation of the disc dressed wheel surface (Machining Science and Technology. 12(2)) pp 183-196

[12]. H Baseri, S M Rezaei, A Rahimi, and M Saadat 2008 Analysis of the disc dressing effects on grinding performance—part 2: effects of the wheel topographical parameters on the specific energy and workpiece surface roughness (Machining Science and Technology. 12(2)) pp 197-213

[13]. A Azizi, S M Rezaei and A Rahimi 2010 Study on the rotary cup dressing of CBN grinding wheel and the grinding performanc (The International Journal of Advanced Manufacturing Technology 47(9)) pp 1053-1063

[14]. M Novák, N Naprstkova, and H Kasuga 2016 Influence of Grinding Wheel Dressing on the Roughness of Final Surface and Cutting Force during GGG60 Grinding (Key Engineering Materials. 686) pp 218-223

[15]. P Walke, and N K Kamble 2017 A review on dressing and grinding operation (International Research Journal of Engineering and Technology 04(07)) pp2573-2575

[16]. J Palmer, H Ghadbeigi, D Novovic, and D Curtis 2018 An experimental study of the effects of dressing parameters on the topography of grinding wheels during roller dressing (Journal of Manufacturing Processes. 31) pp 348-35

[17]. S H Kim, and J H Ahn 1999 Decision of dressing interval and depth by the direct measurement of the grinding wheel surface (Journal of Materials Processing Technology. 88) pp 190 – 194

[18]. H Baseri 2012) Simulated annealing based optimization of dressing conditions for increasing the grinding performance (Int J Adv Manuf Technol. 59) pp 531–538

[19]. P Puerto, R Fernández, J Madariaga, J Arana and I Gallego 2013 Evolution of surface roughness in grinding and its relationship with the dressing parameters and the radial wear (Procedia Engineering. 63) pp 174-182

[20]. A Lefebvre, O Sinot, and A Torrance 2013 Optimization of dressing conditions for a resin-bonded diamond wheel by topography analysis (Machining Science and Technology. 17(2)) pp 312-324

[21]. H Deng, G Y Chen, C Zhou, S C Li, and M. J Zhang 2014 Processing parameter optimization for the laser dressing of bronze-bonded diamond wheels (Applied Surface Science. 290) pp 475-481

[22]. M K Sinha, D Setti, S Ghosh, and P V Rao 2014 An investigation into selection of optimum dressing parameters based on grinding wheel grit size (5th International & 26th All India Manufacturing Technology, Design and Research Conference),(India: IIT Guwahati, Assam)

[23]. I Aleksandrova 2016 Optimization of the dressing parameters in cylindrical grinding based on a generalized utility function (Chinese Journal of Mechanical Engineering. 29(1)) pp 63-73

[24]. S P Sanjay, and J B Yogesh 2017 Selection of levels of dressing process parameters by using TOPSIS technique for surface roughness of En-31 workpiece in CNC cylindrical grinding machine (IOP Conference Series: Materials Science and Engineering. 178(1)) pp 012-033

[25]. D Dadaso, Mohite, N Tiwari, and S Sontakke, Udayshankar Mishra 2017 Modelling and Optimization of Dressing Parameters of CNC Cylindrical Grinding Wheel for Minimum Surface Roughness (International Journal of Engineering Research and General Science. 5(4))

[26]. R Jeyapaul, P Shahabudeen, and K Krishnaiah 2005 Quality management research by considering multi-response problems in the Taguchi method - a review (Int J Adv Manuf Technol. 26) pp 1331–1337

[27]. W K Chan and T K L Tong 2007 Multi-criteria material selections and end-of-life product strategy: Grey relational analysis approach (Materials & Design. 28(5)) pp 1539-154

[28]. Liu, Sifeng; Yang, Yingjie, Forrest, Jeffrey 2017 Grey Data Analysis. Methods, Models and Applications (Singapore: Springer)

[29]. C L Lin 2004 Use of the Taguchi Method and Grey Relational Analysis to Optimize Turning Operations with Multiple Performance Characteristics (Materials and Manufacturing Processes. 19(2)) pp 209-220

[30]. P J Patil, and C R Patil 2016 Analysis of process parameters in surface grinding using single objective Taguchi and multi-objective grey relational grade (Perspectives in Science. 8) pp 367-369

[31]. R S Babu, V S Daniel, G S Subramanian, R S Narayanan, and S Shreerajan 2017 Grey Relational Analysis for multi objective Optimization of Cylindrical grinding process (SSRG International Journal of Mechanical Engineering – (ICCREST'17) - Special Issue)

[32]. A N Siddiquee, Z A Khan, and Z Mallick 2010 Grey relational analysis coupled with principal component analysis for optimisation design of the process parameters in in-feed centreless cylindrical grinding (The International Journal of Advanced Manufacturing Technology. 46(9)) pp 983-992

IOP Conf. Series: Materials Science and Engineering **635** (2019) 012031 doi:10.1088/1757-899X/635/1/012031

Reference position estimation for prosthetic elbow and wrist using EMG signals

Neelum Y S[1*], Z Kausar[1] and S A Usama[1]

[1]Department of Mechatronics Engineering, Air University, Islamabad, Pakistan

[1*]neelumyousafsattar@gmail.com

Abstract: Progress in Bio robotics is opening several doors to mimic the system and purpose of human limbs, thus allowing better prostheses for amputees. This paper presents EMG based Human Machine Interface framework for estimate of reference positions of prosthetic arm for patients suffering from above elbow amputation. In the present study, a set of information about the four basic motions regarding elbow flexion/extension along with the twist of radius and ulna referred as pronation/supination has been acquired from bicep of human arm. Upon muscle activation, the numerical data set points representing the electromyographic intentions were recorded using the Thalamic Labs product i.e. Myo Armband. To analyze these acquired signals, feature selection followed with feature extraction was done in order to classify the input data extracted from human muscle. Afterwards, the performance of Artificial Neural Network (ANN) for motion classification was evaluated. Utilizing the Computer Aided design skills in SolidWorks®, a 2-DOF prosthetic arm model was developed. A mathematical model of its kinematics based on Denavit-Hardenberg convention was presented in Matlab®. Forward kinematics model were analytically verified using Matlab® with that of PeterCorke® Robotic toolbox. In order to check the physical significance of proposed work, the information of forward kinematics from Matlab® linked with that of Solid works and was simulated with reference to the EMG intentions that were captured and classified to generate the reference positions of above mentioned four motions. The classification accuracy obtained from ANN i.e. 91.9% is found significant ($p<0.01$) for a group of ten healthy subjects.

1 Introduction

Evolution in the field of engineering has provided numerous solutions to complex problems. Advancements in the technology has revolutionized the concept of prosthesis[1]. Wooden prosthetic devices used for the lost functions of the amputated limbs are now being replaced by more versatile technological devices. These are not only light weighted but are easier to handle as compared to old ones. The new problem that has emerged is to control such devices according to the dire need of the user i.e. moving the prosthetic arm towards certain position with some specific speed and time duration[2, 3].

As the arm muscles contract, electrical signals are produced which are referred as electromyographic signals. In order to capture such information, electrodes are engaged on the surface of the contracting muscles. The signals generated are proportional to the amount of activity detected at the flexed muscle. In case of trans-humeral amputee, the only functional muscles for signal acquisition in arm are biceps, which make it suitable for the sensor to be set. As these fluctuations are different for each person at different position and time intervals, this becomes a disadvantage to work on control command generation, as the reference is non-linear and variable. For such a scenario, the implementation of machine learning techniques helps in determining the motion for the device. Most

IOP Conf. Series: Materials Science and Engineering **635** (2019) 012031 doi:10.1088/1757-899X/635/1/012031

of the work done for myoelectric prosthetic arm generates a command for go or not go for fixed type of motions [4]. This does not give accurate position of a prosthetic arm.

This paper presents a simple approach to estimate a reference position using EMG signals where prosthetic arm must reach. A prosthetic arm with 2 degree of freedom is used to verify the proposed method. A 2 degree of freedom arm is as simple as a 2-R configuration manipulator. It is proposed to first simulate the system and analyze how it behaves in real world rather than developing a prototype when one can virtually verify the work based on mathematical models and calculations.

The machine learning algorithms is proposed to be used as the use of computational methods to "learn" information rather is beneficial than relying on a mathematical model or equations which governs or represent any real-life problem. The algorithms improve their performance as the sample size increases [5]. By providing a better insight about the natural patterns, these algorithms help to make better predictions as well as decisions.

Furthermore, the method compromises of supervised & unsupervised learning. For a known set of data, a model is trained such that it makes predictions for the new or incoming data based on the evidence of uncertainties present in the trained set of input data and the known responses. For acquiring EMG data from bicep muscles, Myo arm is used [6] as it is a economical user-grade device that integrates an ARM Cortex-M4 unit with microcontroller, eight dry EMG electrodes, a micro nine-axis inertial measurement unit (IMU) and a module named Bluetooth Low Energy (BLE) module. These EMG armbands have reshaped the possibilities for myoelectric control applications, which are not restricted to traditional prosthetics market. Applications have combined wireless and accessible game like training initials for myoelectric prostheses [6,7], sport training systems[8, 9] and numerous other schemes. Supervised learning uses classification technique that predicts discrete responses whereas regression technique predicts continuous responses to develop predictive models[10, 11]. Feature selection plays an important role in data classification as well as the paradigm followed for data extraction. As soon as the information is received at the controller end, the robotic replacement of human arm is in specific position, as the healthy arm would be. For this purpose, the SolidWorks® model along with PeterCorke Toolbox® is used to simulate the reference positions generated using machine learning algorithm.

This paper presents the methodology and their results in two steps: First, EMG data acquisition &control command generation through data processing as given in section II, second the solid and kinematics modeling to generate reference positions using control commands and simulating the motion to the reference positions as given in section III.

2 EMG Data Acquistion and Processing

MYO armband is an application-based wireless wearable device which has eight EMG sensors. It recognizes muscle movements. The extraction process was done through the multiple chambers, which are integrated in the MYO armband. At first, the armband must be synchronized with computer. After that, different chambers showed peaks that it has collected at different positions respectively and are shown in Figure 2. To check whether the pods are activated or not, MYO diagnostics showed spikes on the pods when the sensors were triggered. As this research is specifically for people suffering from transhumeral amputation so MYO armband was placed on the biceps muscle to capture EMG signals as illustrated in Figure 1. No change in electrode position was followed. Changing the electrode positions cause disturbance as different muscles are activated which further gives information to predict the human arm motion. The change in electrode position also cause activation of wrong electrode while the arm is in motion. EMG signals were captured using the MYO Armband in a defined range of elbow and wrist motion. The signals were then acquired using MYO Data Capture. The experimental paradigm can be practiced as in table 1.

Table 1: Experimental Paradigm

No. of Subject(s)	10
No. of trials	4/person
Resting Time (For the Sensor)	5 min.
Time for signal extraction	20 sec (max)

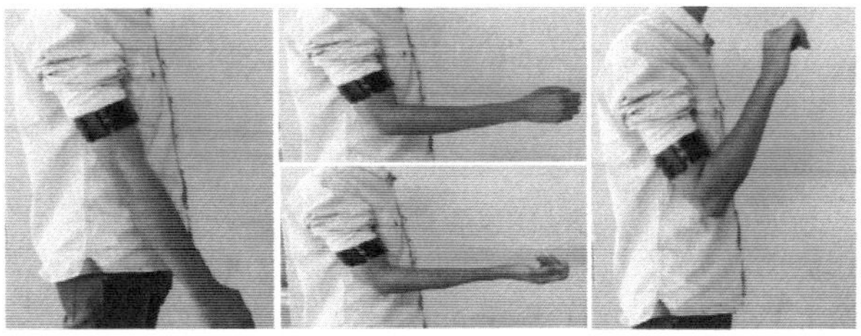

Figure 1. Sensor Placement at Biceps to Acquire Signals and Calibration of four different motions including Elbow Extension, Elbow Flexion, Wrist Pronation and Supination

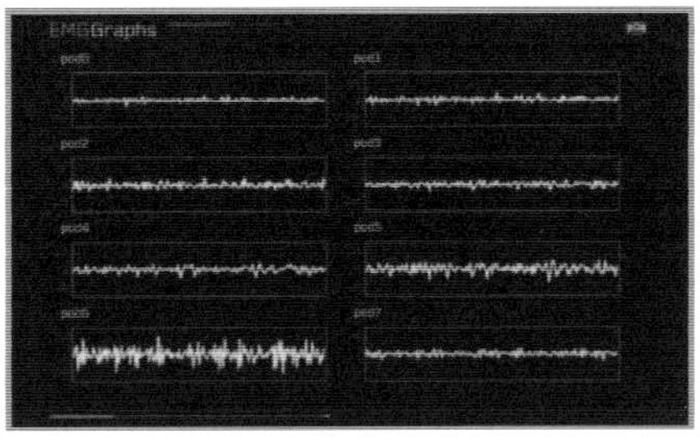

Figure 2. Active Signals at Myo Diagnostics

2.1 *Signal Analysis*

A Microsoft Excel® sheet was automatically generated as the sensor was synchronized followed by signal capturing application of armband as soon as it was executed. That numeric data was then analyzed in MATLAB® for further processing. Figure 3 represents the raw EMG data obtained from healthy subjects.

At every time instant, MYO armband delivers the electrical activity of muscle by modelling it in numeric value. Afterwards it is plotted together hence seen in a graphical form. Each impulse represents the different activity captured at that very instance on a certain electrode. About 8,000 set points were collected, plotted and analyzed for each of the defined class. These steps were frequently repeated on every patient. Several results were collected, and it can be said that the larger the data set, the higher the accuracies. The impulses are then marked at their highest point or peak in order to check if all the motions performed during signal acquisition were traceable or not. This can be established by simply counting the observable peaks.

2.2 *Feature Selection and Classification*

EMG signals extracted from muscles need advanced methods for detection, decomposition, processing, and classification. Figure 4 represents the entire algorithm that was responsible for the signal acquisition until control command generation. These acquired signals were raw data. They contained noise and a lot of disturbance. This was not used directly to get accurate results. Instead, after the detection of peaks, the features were extracted from the acquired signals. This featured data contributes to classify the signals. The feature set comprises of Mean Absolute Value, Variance, Waveform Length, Kurtosis and Peak among which only peaks & wave length showed different behavior. These two features were selected for classification. Different classifying techniques such as LDA (Linear Discriminant Analysis), SVM (Support Vector Machine), ANN (Artificial Neural Networking) are used in different applications to obtain good accuracies. Among such techniques, ANN is precisely experimented to obtain greater accuracy and to provide with more effective offline training and accurate classification.

IOP Conf. Series: Materials Science and Engineering **635** (2019) 012031 doi:10.1088/1757-899X/635/1/012031

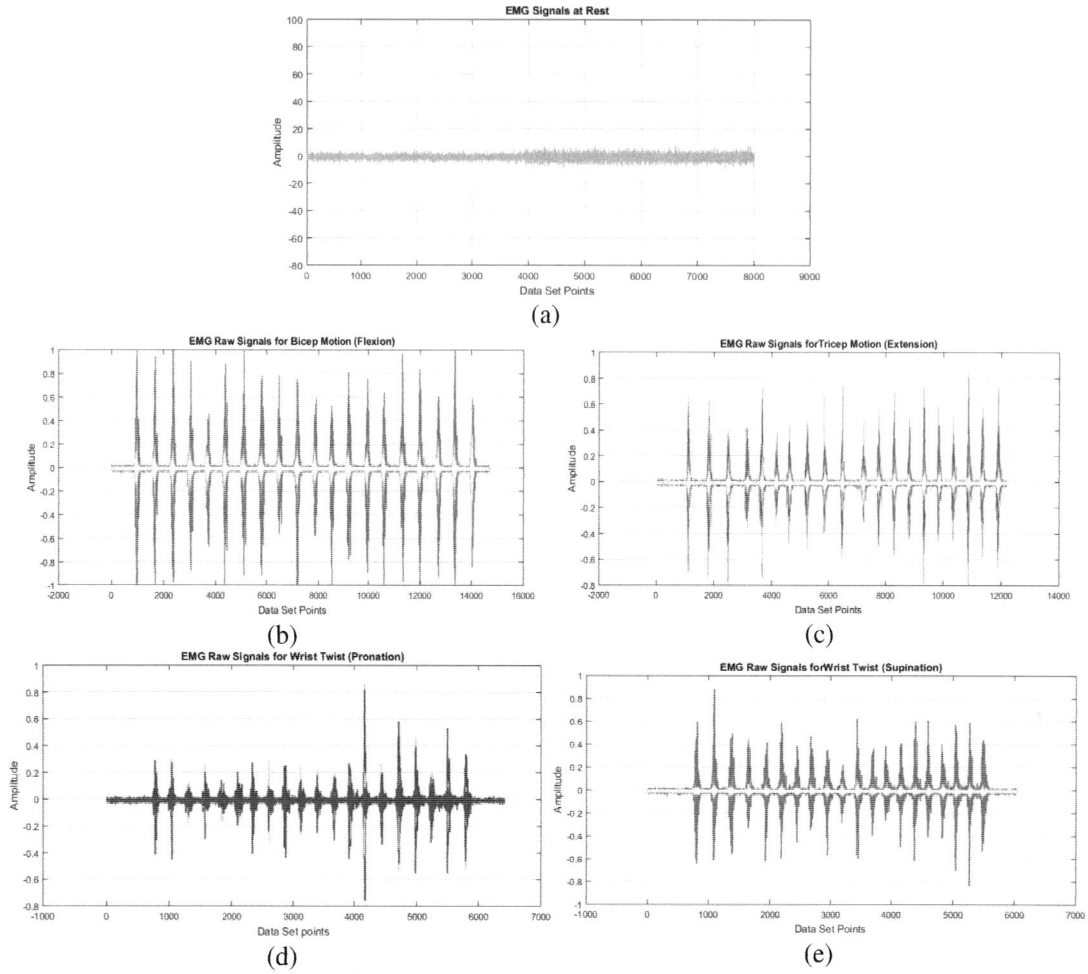

Figure 3. Raw electromyographic intentions of a healthy subject at a) Rest position b) Elbow Flexion c) Elbow extension d) Wrist Pronation e) Wrist Supination

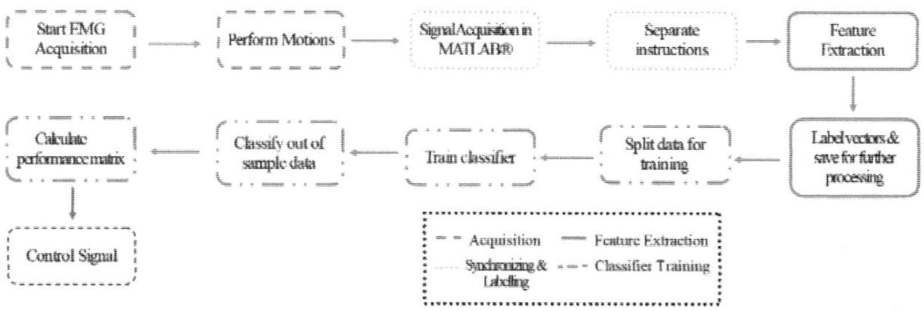

Figure 4. EMG Signal Acquisition and Processing algorithm

2.3 *Artifical Neural Network(ANN)*

ANN is a technique that utilizes multiple neuron layers to map data from one distribution to another to allow for better and optimized classification [12]. This technique is implemented to yield state of the art results while using several layers giving rise to a deep neural network in its place. Figure 5 is a step by step representation of working boundaries of Artificial Neural Network [3].

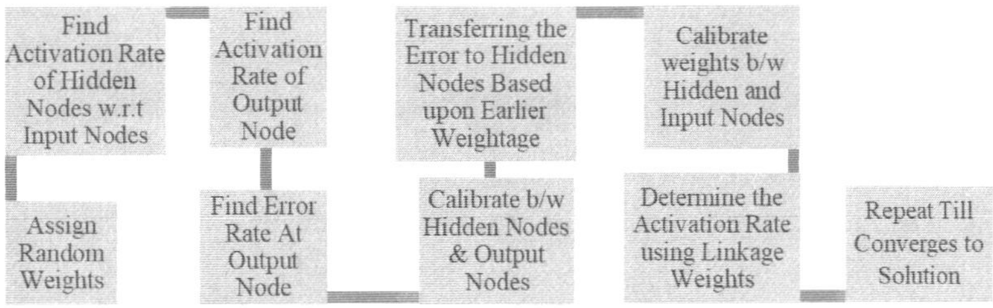

Figure 5. Algorithm for Artificial Neural Networking

A set of data was entered which passed through hidden nodes/layers before giving a certain output. At first, it passed through hidden nodes, which comprises of different activation functions. They decided based on certain input or some predicted value. Mostly, activation functions are non-linear. For this research, Relu was employed as an activation function for neurons, the weights were initialized using the Xavier distribution, the network utilized the Adam optimizer function for gradient descent and categorical cross entropy was used as a loss measure. For repeated validation, these activations were then directed back as input and cross check. This iterative process is referred as *Training*.

In the training section, the exact class for each record was known by the supervised training & the output nodes therefore gave correct value i.e. "1" for the node corresponding to that class, and "0" for the others. Error term was calculated from the correct output term. After labelling the data as inputs & output by one hot encoding method, furthermore, these labelled signals were cruised to the classification steps as discussed earlier. These were then fed to the classifiers. In order to actuate the prosthetic device, *control commands* were generated. These commands are founded on the classified signals. As soon as the motion is captured, the classifier comes into action categorizing the type of activity based on previous training, that needs to be completed and hence a control signal is generated by means of a controller.

The intentions are classified by muscle contractions. The controller classifies the signals by separating such intentions i.e. if the arm is moving upwards or downwards. It then commands the motors to move in clockwise direction or anti clockwise direction. The pronation & supination command generation was carried out in same way, but it was observed that the signals for this motion changed as the elbow joint was displaced more than 30 degrees. For such control commands to be generated, the offline training was done so that in real time the classifier will know what motion is to be performed.

3 Elbow and Wrist Modeling

In order to find a reference position and verification of the proposed methodology, mathematical modeling is required. For such a purpose, the SolidWorks model along with PeterCorke Toolbox® comes in action providing the reference position by using the information of forward kinematics. A SolidWorks model was developed in SolidWorks® for the system. Simulation of different motion types of the elbow and wrist are shown in Figure 6. This SolidWorks model window was used for the visual verification of the reference positions generated as given earlier in section II-D. The SolidWorks model receives the generated reference values from workspace that is imported from the kinematics model developed in Matlab, and both the virtual windows were linked together.

Forward kinematics is a method commonly used to help in computing the position of end effector [13]. The desired position would not be just suggested by electromyographic signals but also through kinematics equations. In this research, a two degree of freedom (DOF) system (for elbow and wrist joint one each) has been considered. The purpose of computing forward kinematics model is to determine the collective effect of the entire set of joint variables. A generally used convention for selecting frames of reference in robotic applications is the Denavit-Hartenberg, or DH convention [14]. The DH parameters for the proposed prosthetic arm design including the elbow and wrist are illustrated in Table 2. MATLAB® was used to verify five reference positions, given in table 3.

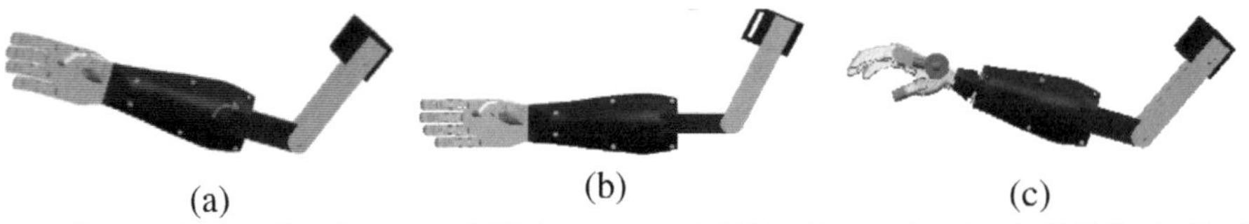

(a) (b) (c)

Figure 6. a) Elbow Flexion b) Elbow and Wrist at rest c) Wrist Pronation in SolidWorks®. A virtual representation of the device for how it is actuated physically. MATLAB® providing the coordinates at certain angle with reference to forward kinematics already linked with Solid works® motion to study the error.

Table 2: DH Parameters

Link Length a_i(mm)	**Link Twist α_i (deg)**	**Joint Distance d_i(mm)**	**Joint Angle θ_i(deg)**
0	90^0	d_1	θ_1
0	90^0	d_2	θ_2
0	-90^0	0	θ_3

Table 3: Reference positions of the prosthetic elbow and wrist for different motion types

Arm Part	**Motion Type**	**Angular Position**
Elbow	Extension	-80°
	Flexion	80°
	Rest	0°
Wrist	Pronation	-85°
	Supination	85°

These reference positions were generated for rest and four types of motions. The generated results in PeterCork are shown in Figure 7 for two selective positions.The Solid Works® design was imported in Matlab® for processing through SimMechanics® which was used to produce a mechanical model. It contains all the physical aspects such as weight, gravity etc. so that the real environment parameters are virtually tested/validated.

The transformation matrix represents the general mathematical form of special linkages within a robot [15]. The general transformation equation is written as (1).

$$
{}^{0}_{i}T = \begin{bmatrix} c\theta_i & -s\theta_i c\alpha_i & s\theta_i s\alpha_i & a_i c\theta_i \\ s\theta_i & c\theta_i c\alpha_i & -c\theta_i s\alpha_i & a_i s\theta_i \\ 0 & s\alpha_i & c\alpha_i & d_i \\ 0 & 0 & 0 & 1 \end{bmatrix} \quad (1)
$$

According to the DH-table the link transformation matrices of proposed two DOF prosthetic arm (elbow and wrist) is given in (2).

$$
{}^{0}_{2}T = \begin{bmatrix} c_2c_1 & s_1 & s_1c_1 & s_1d_2 \\ s_1c_2 & -c_1 & s_1s_1 & -c_1d_2 \\ s_2 & 0 & -c_1 & d_1 \\ 0 & 0 & 0 & 1 \end{bmatrix} \quad (2)
$$

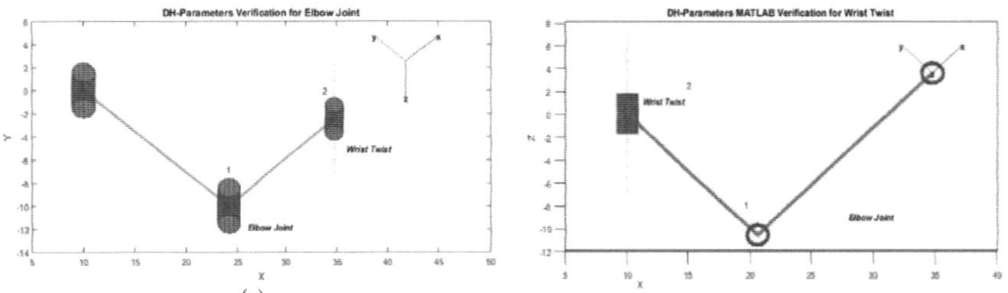

(a) (b)

IOP Conf. Series: Materials Science and Engineering **635** (2019) 012031 doi:10.1088/1757-899X/635/1/012031

Figure 7. a) Elbow Joint at 80o representing Flexion b) Wrist Joint at 85 o represnting Pronation in PeterCorke Robotics Toolbox Matlab®

4 Conclusions

The electromyographic signals were acquired (extracted & recorded) for elbow joint extension and flexion movement along with the wrist twist i.e. supination and pronation. After analyzing the obtained data, feature selection played an important role in data classification. Dissimilar features and their combinations facilitated to classify the acquired raw signals. Artificial neural network (ANN) was used for offline training. A CAD model of two DOF prosthetic arm was developed in Solid works® and its Kinematic Model was built in Matlab®. Afterwards both were linked to check the output of the system virtually rather than developing a real-life model. It verified that the arm joints move and reach desired position as soon as the control signals were attained. Unlike forearm, upper arm has minor muscle activity & motions for the elbow joint and forearm twist can't be predicted very easily. Therefore, signal processing is a big challenge for above elbow amputation, but more experimentation & better feature and classifier selection may help in resolving such problem.

References

[1] D. A. Bennett, J. E. Mitchell, D. Truex, and M. Goldfarb, "Design of a Myoelectric Transhumeral Prosthesis," IEEE/ASME Transactions on Mechatronics, vol. 21, pp. 1868-1879, 2016.

[2] C. A. Calderon, C. Ramirez, V. Barros, and G. Punin, "Design and Deployment of Grasp Control System applied to robotic hand prosthesis," IEEE Latin America Transactions, vol. 15, pp. 181-188, 2017.

[3] A. Phinyomark and E. Scheme, "A feature extraction issue for myoelectric control based on wearable EMG sensors," in Sensors Applications Symposium (SAS), 2018 IEEE, 2018, pp. 1-6.

[4] A. Stango, F. Negro, D. Farina, "Spatial correlation of high density EMG signals provides features robust to electrode number and shift in pattern recognition for myocontrol", IEEE Trans Neural Syst Rehabil Eng. vol. 23(2), pp.189–198, 2015.

[5] D. S. Childress, "Closed-loop control in prosthetic systems: historical perspective," Annals of biomedical engineering, vol. 8, pp. 293-303, 1980.

[6] S. S. Esfahlani, B. Muresan, A. Sanaei, and G. Wilson, "Validity of the Kinect and Myo armband in a serious game for assessing upper limb movement," Entertainment Computing, 2018.

[7] A. Tabor, S. Bateman, and E. Scheme, "Game-Based Myoelectric Training," in Proceedings of the 2016 Annual Symposium on Computer-Human Interaction in Play Companion Extended Abstracts, 2016, pp. 299-306.

[8] J. Ding, R.-Z. Lin, and Z.-Y. Lin, "Service robot system with integration of wearable Myo armband for specialized hand gesture human–computer interfaces for people with disabilities with mobility problems," Computers & Electrical Engineering, 2018.

[9] V. Becker, P. Oldrati, L. Barrios, and G. Sörös, "TouchSense: Classifying and Measuring the Force of Finger Touches with an Electromyography Armband," 2018.

[10] C. Loconsole, G. D. Cascarano, A. Brunetti, G. F. Trotta, G. Losavio, V. Bevilacqua, et al., "A model-free technique based on computer vision and sEMG for classification in Parkinson's disease by using computer-assisted handwriting analysis," Pattern Recognition Letters, 2018.

[11] X. Jiang, L.-K. Merhi, Z. G. Xiao, and C. Menon, "Exploration of Force Myography and surface Electromyography in hand gesture classification," Medical Engineering & Physics, vol. 41, pp. 63-73, 2017.

[12] M. A. Oskoei and H. Hu, "Myoelectric control systems—A survey," Biomedical Signal Processing and Control, vol. 2, pp. 275-294, 2007.
[13] N. Y. Sattar, U. Farooq, Z. Kausar, and L. A. Khan, "Inverse Kinematic Model and simulation of a Prosthetic Upper Limb," in Control System, Computing and Engineering (ICCSCE), 2015 IEEE International Conference on, 2015, pp. 526-531.
[14] J. J. Craig, Introduction to robotics: mechanics and control vol. 3: Pearson/Prentice Hall Upper Saddle River, NJ, USA:, 2005.
[15] S. Niku, Introduction to robotics: John Wiley & Sons, 2010.

IOP Conf. Series: Materials Science and Engineering **635** (2019) 012032 doi:10.1088/1757-899X/635/1/012032

A hybrid process planning for energy-efficient machining: Application of predictive analytics

S-J Shin[1]

[1] Division of Interdisciplinary Industrial Studies, Hanyang University, 222 Wanshimni-ro, Seongdong-gu, Seoul, 04763, Republic of Korea

sjshin@hanyang.ac.kr

Abstract. Computer-aided process planning for energy-efficient machining is essential as energy consumption becomes a major environmental metric in the metal cutting industry. This paper introduces a process planning approach that enables energy prediction in the process planning phase through incorporating Generative Process Planning (GPP) and Variant Process Planning (VPP), called Hybrid Process Planning. GPP is used to provide decision making algorithms in computers by generating energy prediction models specific for machining conditions. VPP is adopted to reuse existing process plans with inclusion of such prediction models so that process planners can anticipate the energy values to be consumed in machine tools. Particularly, the present approach builds upon predictive analytics to efficiently handle sensor-level data collected from real machining operations, and create energy prediction models by using a machine-learning technique.

1. Introduction

In metal cutting processes, process planning is mandatory to determine individual machining operations and strategies needed to fabricate a part [1]. Due to the multi-functionality of machine tools and the complexity of process planning, Computer-Aided Process Planning (CAPP), which uses computers for process planning, becomes essential to gain better speed, cost and time [1]. Recently, environmentally conscious (eco-) manufacturing receives much attention to enhance environmental performance [2] and energy consumption particularly becomes an important environmental metric [3]. Thus, eco-CAPP systems considering the energy aspect are vital for realizing energy-efficient machining.

However, few studies have been made for providing eco-CAPP systems. The previous studies including [3] [4] [5] and [6] have contributed to embody energy-efficient machining through providing logical procedures in the domain of CAPP. The methods introduced in those studies are mainly based on Generative Process Planning (GPP). As GPP uses the algorithms implemented in computers to create process plans automatically without human intervention [7], it requires algorithm coding to decide their process parameters using energy prediction models. However, GPP has not been much propagated to industries because it serves the algorithms restricted under limited experimental conditions [8]. Unless such algorithms dynamically calibrate their models into changes in machining conditions, they lack in providing predictive results accurately for a huge number of machining conditions.

Meanwhile, Variant Process Planning (VPP) is based on Group Technology (GT), which identifies part families and their standard plans with regard to their classification attributes [7]. If an inputted part can be classified with a code number and its relevant standard plan exists in a standard process plan database, this standard plan would be retrieved through similarity search. If a part doesn't have relevant standard plans, the most-similar plan would be retrieved. Due to this simplicity, VPP is more usable in industries; however, it also has a drawback [8]. VPP hardly provides predictive algorithms for improving performance because it only reuses existing standard plans.

This paper introduces a CAPP approach for energy-efficient machining based on Hybrid Process Planning (HPP), which integrates advantages of GPP and VPP. This paper also contains a case study to demonstrate the feasibility of the proposed approach. The proposed approach: 1) uses sensor-level machine-monitoring data collected from previous machining executions for creating energy models, 2) applies GPP to create machine-specific energy models that can predict energy up to the level of tool paths, and 3) advances traditional VPP to enable model-based process planning through providing standard process plans together with such energy models. Therefore, the proposed approach makes CAPP systems feasible so that process planners proactively forecast energy consumption during process planning. The present work is an attempt of predictive analytics in that it aims at implementing a data-driven prediction environment to learn the past knowledge from sensor-level data.

Section 2 introduces the concept of the proposed approach. Sections 3 and 4, respectively, explain the methods of energy predictive modelling and model-based process planning. Section 5 shows a case study, and Section 6 concludes the remark. Note that the concept of HPP has been introduced like [8]. It is also noted that STEP-compliant data interface for Numerical Controller (STEP-NC) [9] is chosen as the data interface for process planning and MTConnect [10] as the data interface for machine-monitoring.

2. Proposed approach

Figure 1 shows the concept of the proposed approach. It comprises the two stages: preparatory and operation (these are the generic stages in VPP). The preparatory stage pre-identifies standard process plans with generating energy models before their use in the operation stage. This stage adopts the concept of GPP. On the other hand, as the operation stage builds upon the traditional VPP, it retrieves and edits standard plans with the use of energy models during the process planning phase. Here, an energy model is defined as a numerical function that characterizes the relationship between x variables (process parameters: feedrate, spindle speed, cutting depth and cutting width) and y variable (energy) at a certain set of Machining Configuration (MC) parameters. An energy model provides the ability to predict energy at the level of tool paths. The MC parameters are the classifier for identifying a specific machining condition where an energy model can be applied. Note that the machining data consist of process plan, Numerical Control (NC) program and machine-monitoring data.

3. Methods for preparatory stage

Section 3 explains technical methods for individual functions in the preparatory stage, which consist of data elements identification, data synchronization, prediction modelling, and model accumulation.

3.1. Data elements identification

This function defines the data elements that should be extracted from machining data to acquire the data instances for energy modelling. Specifically, this function identifies the data elements regarding x variables, y variable and a set of MC parameters while the data elements can be designated among process plan, NC program and machine-monitoring data. We first define a list of MC parameters (M) as below. Here, *command* stands for an NC block code such as G00 or G01, and *trajectory* does the specific purpose of an NC block code.

$$M = \{ m_1, m_2, m_3, m_4, m_5, m_6, m_7 \mid machine, material, operation, feature, strategy, command, trajectory \}$$

Then, we define x variables as the four process parameters - feedrate (x_1), spindle speed (x_2), cutting depth (x_3) and cutting width (x_4) - and y variable as delta energy as below. Here, we use the concept of delta energy, which multiplies power with a sampling time interval of measuring power [11]. It comes from that a power meter typically measures not energy directly but power at each timestamp because energy-unit is a scalar quantity [12].

$$(X, y)_M = \{x_1, x_2, x_3, x_4, y \mid feedrate, spindle\ speed, cutting\ depth, cutting\ width, delta\ energy\}$$

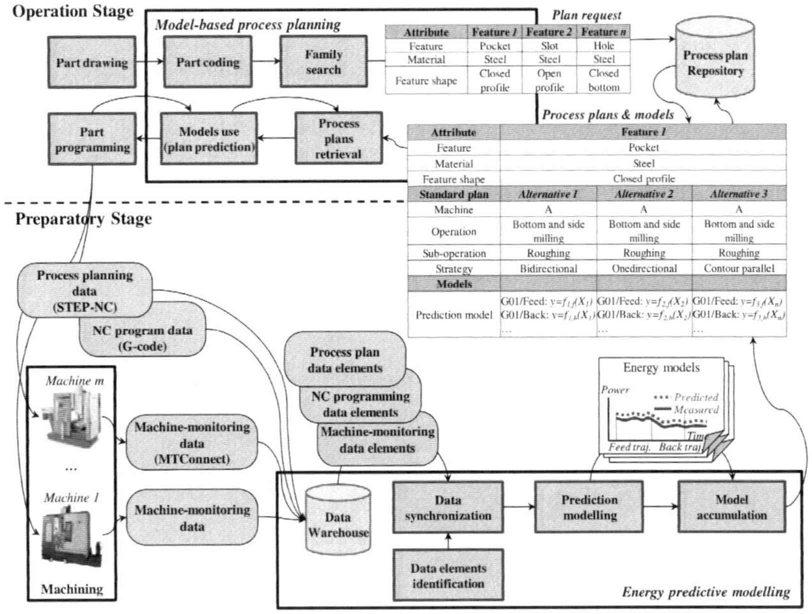

Figure 1. Concept of the proposed approach

3.2. Data synchronization

This function extracts and collects data element instances, and synchronizes machine-monitoring data with process plan and NC program data to generate training datasets for machine-learning analysis. These data instances are extracted from entities or attributes designated in machining data. We can extract *machine*, *material*, *operation*, *feature*, *strategy*, *feedrate*, *spindle speed*, *cutting depth* and *cutting width* from a STEP-NC part program; *command* and *trajectory* from an NC program; while, *delta energy* from an MTConnect document, which records sensor-level data about a machine's movements and actions [13]. As described in Section 3.1, the delta energy is indirectly acquired from *wattage* attribute in an MTConnect document.

Data synchronization comprises time and context synchronizations. Time synchronization aligns machining data instances on timestamps. Context synchronization configures input-and-output datasets aligning with sets of MC parameters. These synchronizations can be achieved by backward connection. The first connecting point is the *position* attribute in an MTConnect document. Each position instance can be matched with its corresponding NC block because this NC block obviously commands the tool movement associated with the position. Such NC block can be the next connecting point with a STEP-NC part program because NC blocks are generated based on the STEP-NC program where machining operations and tool path strategies are specified. In such way, training datasets can be prepared, formalized into *{M: X, y}*.

3.3. Energy modelling

This function generates energy models through machine-learning analysis that uses training datasets. Machine-learning analysis learns training datasets to acquire knowledge. This energy modelling relates to supervised learning because input and output variables are supervised by humans. This supervised learning derives a numerical function $y=f(X)+\varepsilon$ for an energy model through minimizing the error term (ε), which can be achieved by computing and adjusting appropriate coefficients repeatedly. Here, we use polynomial regression, which is commonly-used and applicable for supervised learning. Figure 2 describes the structure of an energy model, which consists of a set of MC parameters and its corresponding energy function derived by regression.

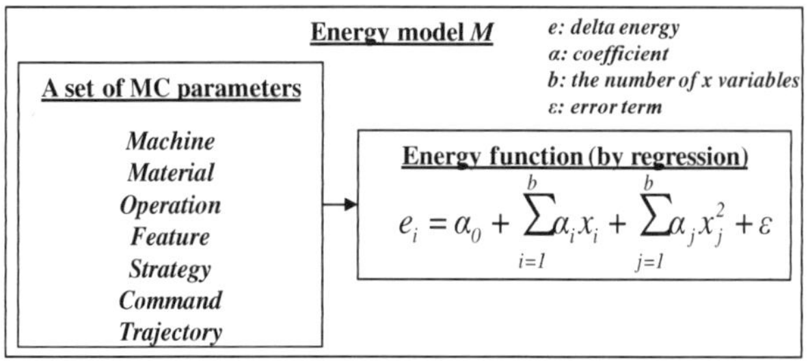

Figure 2. Structure of an energy model

3.4. Model accumulation
One problem of machine-learning analysis is that the models for the machining conditions that have not been run in machines cannot be created. It comes from the high dependency on training datasets, which do not exist in such machining conditions. For this case, this function creates alternative energy models based on existing energy models. Existing models can be reused to substitute for non-existent models through three kinds of methods: *cloning*, *blending* and *competition*.

Some machining strategies already own their energy models by the methods explained from Section 3.1 to Section 3.3, but other strategies may not own. Each strategy is composed of not only similar but also different tool paths with other strategies. For example, the feed path in bidirectional can be treated to be similar with that of unidirectional due to their similarity in cutting pattern and contact volume. This similarity allows to make correlations among strategies. Hence, we can prepare a correlation matrix, which records the mutual similarity in energy demands in terms of high, middle and low levels. These levels can be determined by preliminary energy analysis using machining data. High-level correlation has priority of creating alternative models than middle or low level correlation.

Figure 3 explains *cloning*, *blending* and *competition*. In Figure 3(a), *cloning* copies and pastes a high-level energy model. In Figure 3(b), *blending* chooses or synthesizes high-level models existent in multiple strategies. In Figure 3(c), *competition* selects the most-similar model when several high-level models exist in multiple strategies. This competition can be determined by likeliness or default setup. In such way, alternative energy models can be created for the machining conditions that do not own training datasets. Lastly, we can generate and accumulate machine-learned and alternative energy models in a knowledge base, *process plan repository*.

4. Methods for operation stage
Section 4 explains technical methods for individual functions in the operation stage, which consist of part coding, family search, process plans retrieval and models use. These methods build upon GT. Model-based process planning is possible through the integration of GT with the use of energy models.

4.1. Part coding
Once a part drawing is inputted, this function generates numerical codes, which identifies design attributes by numerical digits for efficient similarity search. A part can configure multiple sets of design attributes with respect to machining features. For problem simplification, we define the design attributes as below.

$$A = \{\, a_1, a_2,\ a_3 \mid feature, material,\ feautre\ shape \}$$

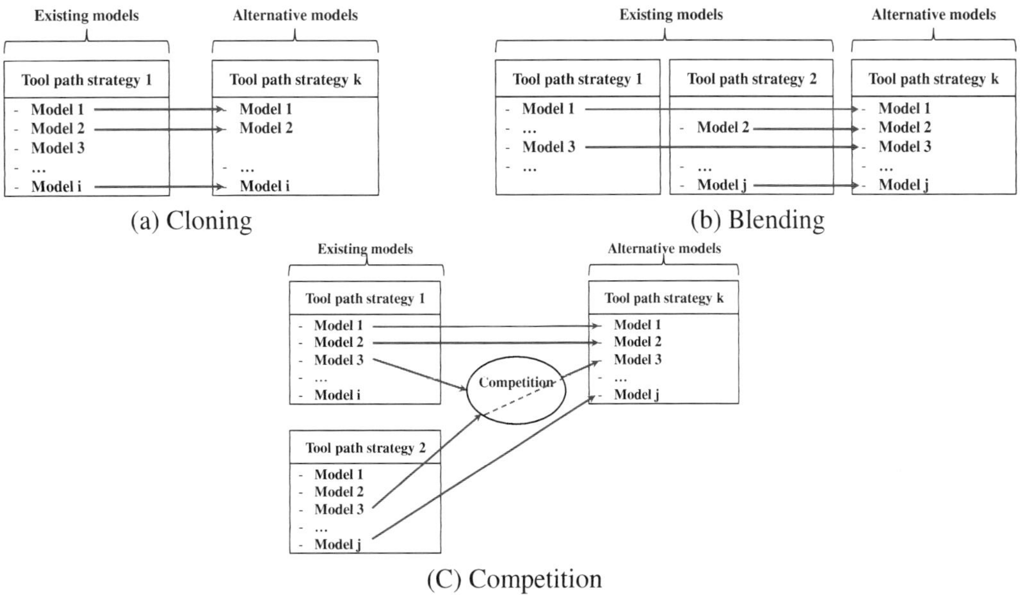

Figure 3. Methods of creating alternative models

4.2. Family search

This function searches the most-relevant process plan from the process plan repository by the input of the numerical digits. The similarity search can be applicable by the nearest neighbour search, which calculates Euclidean distance between an inputted case and existing cases [14], as expressed in Equation (1). We assume that all the weight factors are identical for problem simplification. Calculating similarity distances allows to find the most-similar process plan that scores the minimum distance in their similarities.

$$Similarity(a_i, a_e) = \sqrt{\sum_{i=1}^{m} w_i (a_i - a_e)^2} \qquad (1),$$

where, a_i: an inputted case, a_e: an existing case, m: the number of attributes, w: weight factor

4.3. Process plan retrieval

This function retrieves the most-similar process plan found from the process plan repository. The process plan information contain machine, (sub-) operation and strategy, as showed in Figure 4. Moreover, the retrieved process plan directly links with energy models for model-based process planning. In Figure 4, Alternatives 1 and 2, respectively, suggest two selective bidirectional and contour parallel strategies.

4.4. Models use

This function allows process planners to anticipate energy values during their process planning and part programming. Once process planners decide tool path movements and generate an NC program, sets of MC parameters are fully identifiable aligning with a sequence of tool paths. Thus, energy models can be combined with regard to sequential sets of MC parameters. Each energy model can calculate a predicted energy value whenever process parameters are inputted. In such way, the total energy consumed during the execution of an NC program can be calculated through summing each area bounded by a moving time and a predicted energy value. The moving time on a tool path can be obtained from the path length divided by cutting speed.

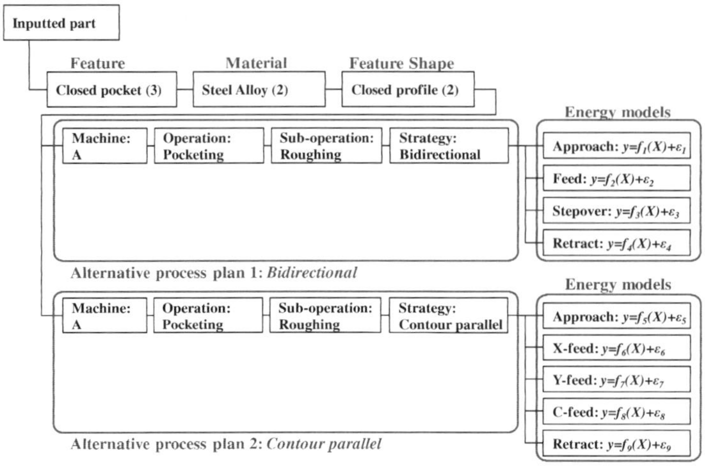

Figure 4. An example of standard process plan and energy model retrieval

5. Case study

5.1. Experiment

Table 1 lists twelve sets of process parameters for fabricating twelve parts that contain thirteen machining features. Feedrate (x_1), spindle speed (x_2) and cutting depth (x_3) are randomly determined within allowable ranges, and identically assigned on each trial. Cutting width (x_4) is fixed as the cutting tool diameter. The experimental setup is as follows: machine (Mori Seiki NVD 1500 DCG), Computerized Numerical Controller (CNC) (Fanuc 0i), workpiece (steel 1018, 10.16*10.16*1.27cm), cutting tool (flat end mill, 8mm diameter, 4 number of flutes), and power meter (system insights high speed power meter).

We perform actual machining based on the experimental setup. We collect twelve pairs of STEP-NC and NC programs and their associating MTConnect documents. Machine (m_1) and material (m_2) are fixed as described above while operation (m_3), feature (m_4), strategy (m_5), command (m_6) and trajectory (m_7) vary in terms of the entities and attributes coded in individual STEP-NC and NC programs.

Table 1. List of process parameters

Trial	Feedrate (x_1) (mm/tooth)	Spindle speed (x_2) (RPM)	Cutting depth (x_3) (mm)
1	0.0127	1500	1.5
2	0.0127	2000	1.5
3	0.0127	1750	1.0
4	0.0229	1750	1.0
5	0.0127	1750	2.0
6	0.0178	1500	1.0
7	0.0178	2000	1.0
8	0.0178	2000	2.0
9	0.0178	1750	1.5
10	0.0076	1750	1.5
11	0.0152	1750	1.5
12	0.0127	1750	1.5

5.2. Data collection and synchronization

Machine-monitoring data (MTConnect documents) comes from two sources. The CNC generates data instances including position and NC block, depending on task processing or events. The power meter measures power at 100 Hz frequency, independently with the CNC. Time synchronization combines these data instances from the two sources on the same timestamps, specially merged at average 0.365 second intervals. Then, STEP-NC and NC data instances are synchronized with MTConnect data instances on the same time intervals. The context synchronization generates individual datasets with regard to sets of MC parameters, formalized into *{M: X, y}*.

In such way, we collect fifty-one datasets, which correspond to the number of total sets of MC parameters on twelve trials. For better performance in energy models, we carry out data cleaning. We exclude the erroneous power data instances null or less than 1500 watt. It comes from that the machine spends minimum 1500 watt during turned-on. We exclude power data instances lowermost 0.5% or uppermost 0.5% among the entire dataset. It causes from our outlier decision.

5.3. Energy prediction modelling

We generate fifty-one polynomial regression functions for energy models using KNIME, a data mining tool [15]. These models contain different coefficients in terms of m_3, m_4, m_5, m_6 and m_7. Table 2 presents the comparison result between the measured and predicted energy values on the twelve trials. Here, Root Mean Square Error (RMSE) means the average difference between individual measured and predicted energy values; meanwhile, Relative Total Error (RTE) measures percentages of the total measured and predicted energy difference. It is observable that energy models make a good performance to anticipate energy values within absolute 1.08% RTE. It conjectures from that energy models are accurately learned to ensure the rigidity of prediction through decomposing energy mechanism granularly up to the level of tool paths.

Table 2. Comparison between measured and predicted energy

Trial	Measured energy (kJ)	Predicted energy (kJ)	RMSE (J)	RTE (%)
1	13980.5	13929.1	28.67	-0.37
2	11410.0	11442.1	29.72	0.28
3	19571.3	19494.0	23.62	-0.40
4	9859.9	9853.1	28.39	-0.07
5	9973.1	10037.9	34.50	0.65
6	13392.9	13444.0	25.30	0.38
7	11079.3	11111.9	26.26	0.29
8	6038.8	5973.8	41.10	-1.08
9	9764.7	9734.8	32.58	-0.31
10	19312.8	19329.6	21.82	0.09
11	10821.5	10903.4	30.12	0.76
12	12607.7	12557.4	29.98	-0.40

5.4. Model accumulation

We produced three closed pocket features by bidirectional and contour parallel strategies but not unidirectional nor contour bidirectional. Nevertheless, the energy models for the two latter strategies can be alternatively created using the method explained in Section 3.4. In Figure 5(a), the approach and retract tool paths on unidirectional can clone those from the common strategy. The linear feed path on unidirectional can chose that of bidirectional through competition (specifically, likeliness) because they have high-level correlation and make the closest tool path patterns. In Figure 5(b), the energy models for contour bidirectional can be created from those of bidirectional or contour parallel through blending.

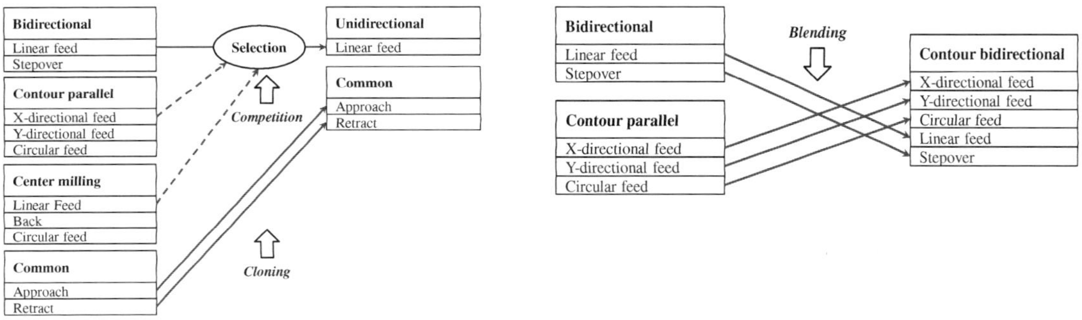

(a) Unidirectional (b) Contour bidirectional

Figure 5. Cases of creating alternative models

6. Conclusion

This paper proposed a HPP-based process planning approach to enable energy prediction in the process planning phase through incorporating GPP and VPP. The proposed approach could create machine-specific models that can predict energy up to the level of tool paths, and use such energy models to implement model-based eco-process planning.

However, the present work has limitations. It only provides energy prediction models, but does not cover optimal process parameters for minimizing energy automatically. It excludes productivity performance including machining time and surface roughness, which are still significant in eco-process planning. We plan to advance the proposed approach toward multi-objective optimization through accommodating energy minimization and productivity performance.

7. Acknowledgments

This work was supported by Basic Research Program in Science and Engineering through the Ministry of Education of the Republic of Korea and National Research Foundation of Korea (No. NRF-2018R1D1A1B07047100).

8. References

[1] Groover MP, Zimmers Jr EW 1984 *CAD/CAM: Computer-aided design and manufacturing* (Englewood: Prentice-Hall)

[2] Kara S and Li W 2011 Unit process energy consumption models for material removal processes *CIRP Annals – Manufacturing Technology* **60** 37–40

[3] Newman ST, Nassehi A, Imani-Asrai R and Dhokia V 2012 Energy efficient process planning for CNC machining *CIRP Journal of Manufacturing Science and Technology* **5** 127-136

[4] Zhao F, Murray VR, Ramani K and Sutherland JW 2012 Toward the development of process plans with reduced environmental impacts *Frontiers of Mechanical Engineering* **7(3)** 231-246.

[5] Peng T and Xu X 2014 A holistic approach to achieving energy efficiency for interoperable machining systems *International Journal of Sustainable Engineering* **7(2)** 111-129.

[6] Wang S, Lu X, Li XX and Li WD 2015 A systematic approach of process planning and scheduling optimization for sustainable machining *Journal of Cleaner Production* **87** 914-929.

[7] Amstead BH, Ostwald PF and Begeman ML 1987 *Manufacturing process* (New York: John Wiley & Sons)

[8] Elinson A, Nau DS, Herrmann JW, Minis IE and Singh G 1997 Toward hybrid variant/generative process planning *ASME Design Engineering Technical Conference* (Sacramento: U.S.A.)

[9] Xu XW, Wang H, Mao J, Newman ST, Kramer TR, Proctor FM and Michaloski JL 2005 STEP-NC compliant NC research: the search for intelligent CAD/CAPP/CAM/CNC integration *International Journal of Production Research* **43(17)** 3703-3743

[10] Shin SJ, Woo JY, Kim DB, Kumaraguru S and Rachuri S (2015) Developing a virtual machining model to generate MTConnect machine-monitoring data from STEP-NC *International Journal of Production Research* **54(15)** 4487-4505

[11] Bhinge R, Park J, Law KH, Dornfeld DA, Helu M and Rachuri S 2017 Toward a generalized energy prediction model for machine tools *Journal of Manufacturing Science and Engineering* **139(4)** 041013.

[12] Campatelli G, Lorenzini L and Scippa A 2014 Optimization of process parameters using a response surface method for minimizing power consumption in the milling of carbon steel *Journal of Cleaner Production* **66** 309-316

[13] Vijayaraghavan A and Dornfeld AD 2010 Automated energy monitoring of machine tools *CIRP Annals - Manufacturing Technology* **59** 21–24

[14] Zezula P, Amato G, Dohnal V and Batko M 2006 *Similarity Search - The metric space approach* (New York: Springer)

[15] Berthold MR, Cebron N, Dill F, Gabriel TR, Kotter T, Meinl T, Ohl P, Sieb C, Thiel K and Wiswedel B 2008 KNIME: The Kostanz information miner *Data Analysis, Machine Learning and Applications* (Berlin: Springer-Verlag)

IOP Conf. Series: Materials Science and Engineering **635** (2019) 012033 doi:10.1088/1757-899X/635/1/012033

An Integrated Management System of Man-machine Communication for Tailings Pond Based on RSCIC

Guangming Yu[1,2,*], Yingnian Yu[3], Junwei Liu[1,2], Ju Qiu[1,2], Yanhua Zuo[1,2], Xiankun Zeng[1,2], Xin Dong[4], Weiquan Wang[5]

1. School of Civil Engineering, Qingdao University of Technology, Qingdao 266033, China
2. Cooperative Innovation Center of Engineering Construction and Safety in Shandong Blue Economic Zone, Qingdao University of Technology, Qingdao 266033, China
3 Qingdao Company, Zhongqi Jiaojian Group Co., Ltd., Qingdao 266300, China
4 Shandong Zhaojin Company, Zhaoyuan 265400, China
5. Fuyunmengku Iron Ore Co. Ltd, Urumqi 830022, China

[*]Corresponding author: yu-guangming@263.net

Abstract. Tailings pond plays a very important role in the construction of mine engineering. At the same time, tailings pond also produces huge pollution to the environment. If dam break occurs, it will cause loss of people and property. In this paper, a tailings pond management system based on RSCIC is studied, the monitoring content of the system is determined, the more innovative system and hardware are emphasized, the program flow chart of digital processing is put forward. Adopting scientific and effective means to manage the tailing dam automatically can not only strengthen the safety protection of the tailing dam, but also provide a certain theoretical basis for the construction of other digital tailings dam projects.

1. Introduction

Tailings pond is a place for storing mine residue or industrial waste residue, which is built by intercepting the terrain with poor terrain [1]. According to incomplete statistics, there are about 8,000 tailings dams of different sizes in China. However, it is inevitable that there will be over-exploitation and even non-compliance with the requirements of mining, leading to the tailings dam failure straight up. In recent years, dozens of large tailings dam accidents have occurred in our country, which bring great damage to the safety of people's lives and property as well as the living environment. For example. In 2006, 17 people were dead and 5 injured in the dam break accident of a tailing pond in Shaanxi Province. In 2008, 128 people were dead and 35 injured in the dam break accident of a tailing pond of a mining company in Shanxi Province [2].

With the occurrence of several large tailings dam accidents, the government has increased its investment in the safety of tailings dam. With the deepening of the research theory and practical application of tailing dam safety in various scientific research institutions, people know more and more about tailings dam. It has been able to realize the real-time on-line monitoring of the tailings pond, and preliminary realization of the prototype of the digital tailings pond.

At present, our country has tailing ponds to realize the dam body stability and the safe production monitoring, although has separated from the artificial survey method, but will be affected by many factors the disturbance and the restriction. For example, the workload of measurement is large, the accuracy of measurement is not accurate, the function is single, and so on. In this situation, it is necessary to explore a new tailings pond management system. Therefore, on the basis of the embryonic form of the digital tailings pond, this paper introduces the RSCIC system, through the characteristics of information, intelligence, real-time and network of the RSCIC system, carries on the comprehensive and effective control to the tailings pond. In the five letters of RSCIC, *R* stands for reconnaissance, *S* for surveillance, *C* stands for communication, *I* stands for intelligence, *C* stands for command.

2. Implementation and innovation of RSCIC system

At present, the monitoring system of tailings pond has been used in many practical projects. For example, the use of high-density resistivity method to monitor the infiltration line, the shortcomings of this monitoring mechanism is mostly manual monitoring, and a long period of work; using GNSS, GPS technology to monitor the deformation of the tailings dam. But for other parameters of the monitoring effect is not significant; the use of remote data transmission network based on RS- 485 to monitor the seepage of the tailings pond, the function is relatively simple [3-5]. Therefore, the development of an automatic management system with higher efficiency and accuracy is a vital part of doing a good job in disaster monitoring of tailings ponds.

The purpose of RSCIC system is to reveal the particularity of tailings dam structure, the complexity of deformation and the danger of dam break, to study the instability mechanism of tailings dam, and to study and establish the theory and system of monitoring and early warning of tailings dam based on RSCIC. Submitting the risk early warning method of tailings dam based on RSCIC and compiling software provide some theoretical support for the safe operation of tailings dam, thus can promote its development through more scientific and innovative technical means.

The RSCIC system (Fig. 1) consists of five modules that are both continuous and related to each other. For example, from the design, construction, completion and production of tailing dam, it is necessary to investigate the geographical and natural position, topographic and geological conditions, the shape of dam body, the arrangement of monitoring points on the dam body and the communication with the control and command center. All the data into the database, the need for intelligent analysis of data and among the rules of change and then put forward its evolution process early warning or production recommendations; all of the above modules are under the command of the command module.

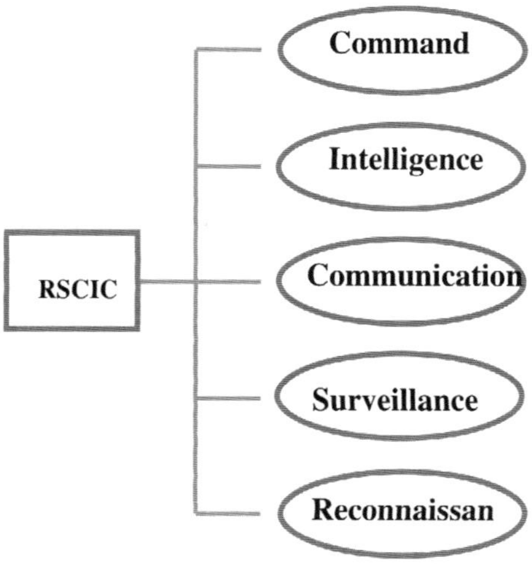

Fig. 1. Composition of RSCIC management system

The innovative and progressive nature of the RSCIC system is as follows.

(1) Real-time monitoring: the real-time monitoring system of tailings pond can monitor the technical indexes such as rainfall, soakage line, water level, dry beach length, dam displacement, pore water pressure and so on.

(2) Video monitoring: scanning and recording the key parts of the dam body through the electronic monitoring system, forming a grid three-dimensional monitoring system at different angles and different spatial positions, so as to ensure the safe operation of the tailings dam.

Under the condition that the tailings dam has certain flood regulating capacity, the system can measure the indexes such as water level, rainfall, dam deformation, soakage line and so on. And these data are collected through the real-time online monitoring system [6]. In order to keep the tailing dam under a higher safety standard, the established system should cooperate with each other, realize the on-line monitoring and real-time analysis of the tailing dam, and reduce the hazard of the tailing dam to the minimum.

3. Working principle of RSCIC system

System mode of operation: Reconnaissance system should plan the layout of the monitoring system before collecting basic data. Information obtained by surveillance systems (such as tailings dam monitoring instrument, the water level warning, wireless transmission network, etc.). This information is transmitted to the main control platform through the communication system. The intelligent control platform provides the location of the tailings dam alarm to the manager through data analysis, the scene of mine tailing movement, velocity and seepage field in dam body. The manager further monitors the site through the control system and implements flood control measures according to the scene and tailings dam safety specifications. The specific working principle process is shown in Fig. 2.

Fig. 2. Working principle flowchart of Rscic system

3.1. Reconnaissance system

Investigation is the process from site selection, design to completion to on-line monitoring of tailings pond. The investigation of tailing pond is of great significance to the stability and safety of tailing pond. Whether the selected location of the tailings pond conforms to the safety regulations, the geological structure and lithologic characteristics of the selected location should be drilled and sampled. At the same time, the flood discharge capacity, water storage capacity and various factors which may cause danger to the downstream residents should be taken into account. In addition to the above aspects, we can also use professional equipment (Fig. 3) to survey and analyze the displacement, rainfall and flood level indicators, and collect data in real time. The real time analysis of the initial dam and the accumulative dam in the formation process of the tailings pond, etc [7].

The design and construction standards for tailings ponds should specify the safety parameters needed for operation, such as the total storage capacity, the final dam height and the final stacking height, etc.

IOP Conf. Series: Materials Science and Engineering **635** (2019) 012033 doi:10.1088/1757-899X/635/1/012033

(a) Static level

(b) Inclinometer

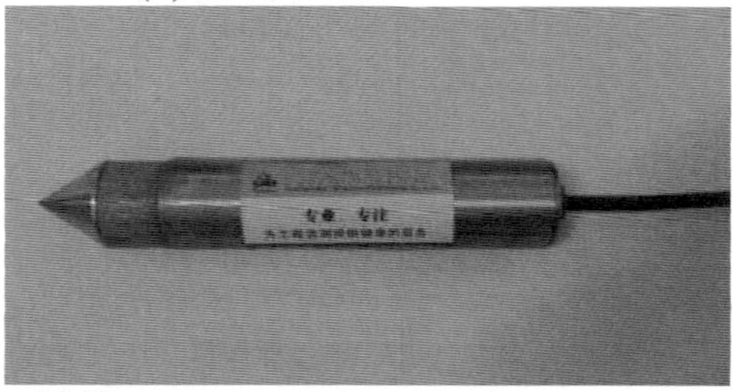

(c) Pore water pressure gauge

Fig. 3. Dynamic monitoring equipment

3.2. Surveillance system

As shown in Fig. 4, the main observation and monitoring projects have saturation line position, dam body displacement, dry beach length, pond water level and rainfall, etc. After the collected data is transported to the central Management Center, the plan and profile of the monitoring point can be observed in the main control platform of the central Management Center, which can be alerted in real time when the data is over-timed.

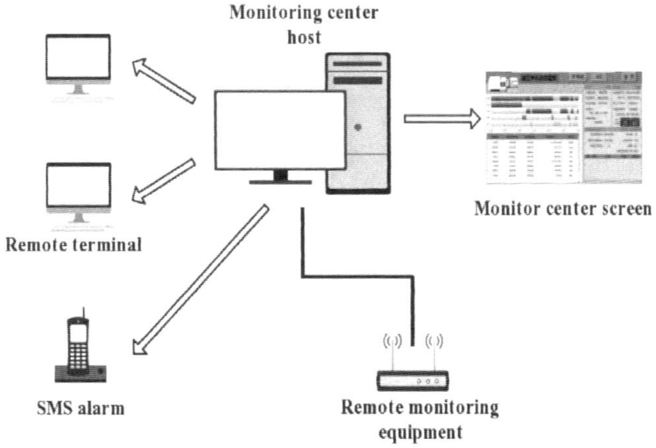

Fig. 4. Monitoring system structure diagram

3.3. Communication system

The purpose of communication is to collect and transmit the data observed by tailings pond to the management department. The communication system structure is shown in Fig. 5, which mainly includes 1 main line cable, several branch cables, data acquisition box and so on. The monitoring device transmits the received data to the automatic acquisition box through the sub-line, and then the GPRS transmission module and the INTERNET network are transmitted to the management center through the main line.

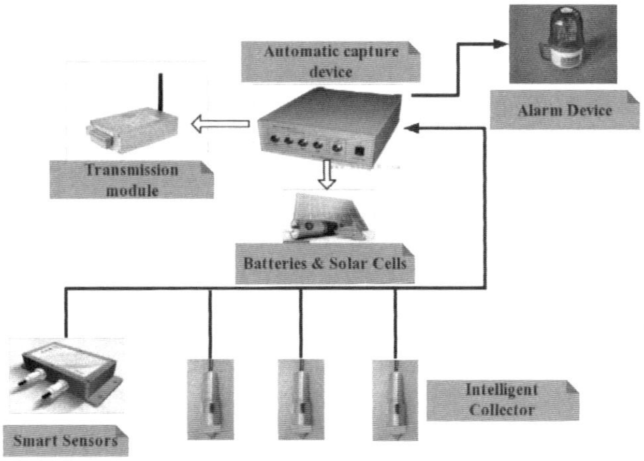

Fig. 5. Communication system structure diagram

3.4. Intelligent system

The working process of the intelligent system is shown in Fig. 6. When all kinds of data collected are transported to the management center, the data report and calculation results obtained after analysis are classified and stored in the server. The controller gets the security information and technical data from the filtered data according to his own needs.

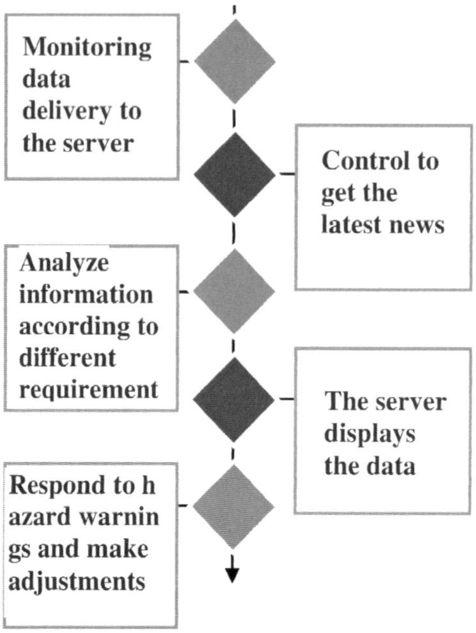

Fig. 6. Intelligent system work flow chart

As the safety monitoring system of tailings pond is still unable to realize fully intelligent monitoring and inspection, it requires us to combine the manual inspection and compare and analyze the data obtained from the two to obtain more specific and scientific safety results [8].

3.5. Command system

The dynamic data collected by the instrument can be alerted to the danger after computer processing, which can be controlled from the source. If the infiltration line is too high, the mining can be suspended. If the length of dry beach is insufficient, it can change the position of the drainage pipeline or build a dam. The workflow of the command system is shown in Fig. 7 to ensure that the received data is real-time dynamic data, and automated security analysis is carried out to ensure the smooth operation of the warning system.

According to the analysis and judgment data, the displacement of the tailings dam is judged qualitatively, so as to be able to grasp the force of the tailings dam, evacuate the dangerous people on the basis of the early warning information, and take effective reinforcement measures to prevent the occurrence of greater damage.

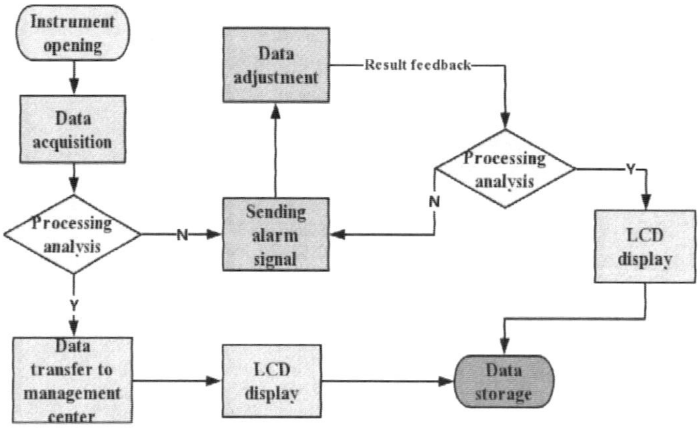

Fig. 7. Flowchart of the command system

IOP Conf. Series: Materials Science and Engineering **635** (2019) 012033 doi:10.1088/1757-899X/635/1/012033

3.6. Overall optimization of RSCIC system

The monitoring system of tailings pond based on RSCIC integrates multiple work on the basis of communication technology. The dynamic monitoring system of RSCIC automation can not only improve the working ability of managers, but also show the advantages of the age of big data in the rapid automation.

4. Implementation steps of RSCIC system

4.1 In-site investigation stage

The work flow chart of RSCIC in the exploration phase as shown in Fig. 8, must first master the tailings dam site, from the natural conditions, geographical environment, historical extreme weather conditions investigation as follows.

(1) Firstly, we should understand the engineering geological conditions of the tailing pond, such as rock hardness and mechanical properties, seismic grade and so on, and make qualitative and quantitative evaluation of some necessary investigation factors, so as to make preparations in advance.

(2) To investigate the hydrogeological conditions of the site, such as the depth of groundwater and the possible seepage flow, so as to make a comprehensive evaluation of the dam body.

(3) After ascertaining the possible adverse effects of bad geological conditions, we should take different protective measures according to different phenomena, determine the required indexes and data, and correctly classify the grade of rock and soil.

(4) According to the characteristics of different soils, stability analysis is carried out according to different standards.

(5) To provide effective and scientific safety measures. In the later design, the information of the location of the monitoring instruments should be taken into account in the design of the tailings dam, especially the monitoring pipeline channels should be left when the overflow wells and overflow culverts are poured.

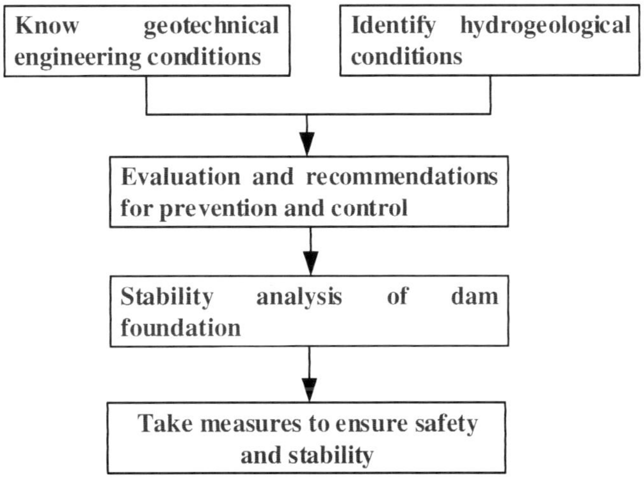

Fig. 8. Flowchart of the exploration stage

4.2 Selection of monitoring indexes for surveillance system

(1) The water level of the tailings dam is relatively stable because of the waste residue accumulated in the tailings dam for a long time. The waste residue removed from the tailings dam will inevitably percolate in the course of flowing. At the same time, it is also possible that the pond water level will rise rapidly due to heavy rainfall during the flood season. In this life cycle, the tailings performance may be changed due to some reasons other than human activities. All of these are the reasons that lead to the abnormal seepage of tailings dam. Therefore, the reasonable control of the location of the soakage line is the most important task in the monitoring task.

(2) The monitoring of pond water level is used to check whether the flood control capacity of tailings pond can meet the safety requirements. Specifically, a complete design document will give the

specified design flood level, the highest flood level and the regulation of flood depth, these indicators must meet the design requirements. Therefore, accurate control of the height of the pond water level can effectively avoid the occurrence of accidents, but also conducive to the safety inspection of the relevant departments.

(3) Once the tailing dam breaks, it will be disastrous and irreparable. Therefore, it is necessary to establish a disaster evolution process, which can intuitively respond to indicators, can take appropriate measures in time to prevent dam deformation and other disasters, so as to avoid or reduce the occurrence of disasters.

(4) In the quantitative evaluation of the flood control capacity of the tailings pond, it is necessary to determine the beach top elevation and the allowable dry beach elevation under the designed maximum flood level. Because of the limitation of technical means, the calculated results of total dry beach length and flood regulation dry beach length are not convincing enough. Therefore, in the automatic monitoring process, it is necessary to be equipped with an accurate and rapid determination method.

4.3 Security management of communication system

The communication system of RSCIC consists of several branch cables that collect information through data sensors and then connect the cables to the automatic collection box. For the sake of safety, all the cables are covered with PPR pipes and buried in the ground to ensure the safety in use. In addition from the point of view of economy and stability the data collected from the different profiles of the field are processed and sent to the computer room of the monitoring center for automatic preservation. All authorized users can monitor or analyze the data through the Internet or remote login surveillance system.

4.4 Early warning control of intelligent system

The intelligent system of RSCIC automatically analyzes the information it collects and alerts it when it exceeds the preset value. Therefore, the establishment of a sound early warning model and the preparation of scientific prediction software, can better predict the stability of the tailings dam, the development of safe and reliable control measures [9]. The main functions include real-time data display and query, remote data access, automatic flood routing, different levels of automatic early warning, video recording and real-time monitoring, as well as the daily safety monitoring of tailings pond informatization.

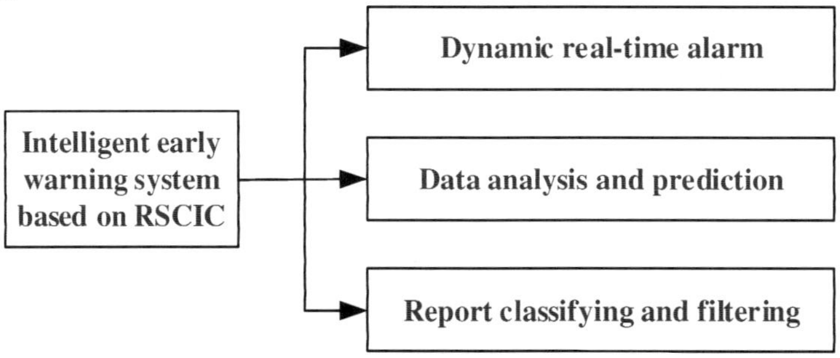

Fig. 9. Organization chart of intelligent system

The organization chart of the intelligent system is shown in Fig. 9, and the main components are as follows.

(1) Timely alarm: The system will automatically identify the hazard information and make the corresponding judgment. The information can be sent to the controller at the first time through GPRS, WIFI, SMS and many other ways.

(2) Data analysis and prediction: Through comprehensive analysis and comparison of the relevant data, such as water level of dam pond, soakage line, large displacement of dam body, rainfall and seepage flow of dam body, the correlation among various data is calculated, and the health condition of dam body is comprehensively judged.

(3) Statistical reports: Classification and presentation of various data to facilitate extraction of useful data. Support the export and print.

4.5 Integration of the command system

The command system of RSCIC integrates the above four stages, this system can consult the data of water level, rainfall, depth of infiltration line, deformation of dam and so on needed by tailings management in real time. On the basis of theoretical analysis and data comparison, work measures are optimized.

5. Conclusions

This paper mainly studies the application of RSCIC system in the digital management of tailings ponds, giving a brief account of the present situation of the research on the monitoring technology of tailings in our country. Secondly, the equipment and working principle of RSCIC system are introduced, and the running system and working flow of RSCIC system are described in detail. The automatic safety monitoring of tailings pond is a part of digital mine construction, it is also the need of modern management of enterprises.

The establishment of RSCIC system can not only protect the lives and property of the downstream people, but also reduce the environmental pollution caused by dam break. Therefore, it not only promotes the long-term stability of society, but also accumulates valuable experience for the research and development of automatic safety monitoring and early warning system of tailings pond.

Acknowledgments

The reaearch was jointly supported by the National Natural Science Foundation of China (51674150) and the Construction Science and Technology Projects of Urumqi (2016002).

References

[1]. Zhang Z.Q., Hu J., 2010. Automatic monitoring system for safety of tailings dam. *Mining Engineering*, vol. 8, no. 02, pp. 53-55.

[2]. Wang C., Wang Y.H., Li Q.M., Zhou Z., Shi Z.D., Meng X.S., 2008. Research on an online mine tailings dam optical fiber surveillance technology. *Shandong Science*, vol. 21, no. 06, pp. 4-8.

[3]. Zhang L.， Li Y.Q.， Sun H., 2011. Applications of online deformation survey monitoring system in tailing pond of metal mine. *Computer Engineering & Software*, vol. 32, no. 01, pp. 61-63.

[4]. Zhan J.J.， Xiao T.Y., 2013. Research and application of monitoring system for tailings dam based on GPS data. *Journal of Safety Science and Technology*, vol. 9, no. 04, pp. 5-10.

[5]. Li Z.K., Zhang Y.P., 2008. Research and design of tailing dam seepage monitoring system based on RS-485. *Mining and Metallurgy*, vol. 36, no. 03, pp. 77-80.

[6]. Zeng Q.W., Xie D.R., Su J.D., Yuan M., 2010. Risk analysis of dam failing of the tailings reservoir. *Industrial Safety and Environmental Protection*, vol. 36, no. 01, pp. 44-46.

[7]. Li Q.M., Liao G.L., Wang Y.H., Wang Q., 2007. Study on assessment methods for safety in controlling flood and possibility of cracks of tailing dams. *Journal of Safety Science and Technology*, no. 05, pp. 24-29.

[8]. Wu Q., Chen Q., 2008. An analysis of environmental effects induced by environmental problems in mines. *Hydrogeology & Engineering Geology*, no. 05, pp. 81-85.

[9]. Yu G.M., Song C.W, Pan Y.Z., Li L., Li R., Lu S.B., 2014. Review of new progress in tailing dam safety in foreign research and current state with development trent in China. *Chinese Journal of Rock Mechanics and Engineering*, vol. 33, no. S1, pp. 3238-3248.

IOP Conf. Series: Materials Science and Engineering **635** (2019) 012034 doi:10.1088/1757-899X/635/1/012034

Effect of process parameters to flexural strength of 3D printed anatomical bone part

T Tajarernmuang[1,2,a], C Sawangrat[3,*,b], and S Jomjunyong[3,c]

[1]Biomedical Engineering Institute, Chiang Mai University, Chiang Mai, 50200, Thailand

[2]Graduate School, Chiang Mai University, Chiang Mai, 50200, Thailand

[3]Department of Industrial Engineering, Faculty of Engineering, Chiang Mai University, Chiang Mai, 50200, Thailand

*Corresponding author e-mail: [b]choncharoen@eng.cmu.ac.th, [a]tanapong.tajarernmuang@gmail.com, [c]sermkiatj@gmail.com

Abstract. Training and practice on synthetic anatomical bone before perform real medical operation is useful for medical student and recently become part of medical procedure. Since there are various factors to considered, the attempt to imitate anatomical bone from medical image data of patient, using technology that based on Fused Deposition Modelling (FDM) is alternatively developed. Since durability and qualities of 3D printed model depend on process parameters which referred to 3D printer setting, this study aims to investigate effect of essential 3D printer setting parameters involve with mechanical properties of 3D printed anatomical bone part, focused on infill density and layer height, and determine regression model for mechanical strength prediction from selected printing parameters by using experimental design and statistical analysis that based on Design of Experiments (DOE). 2-levelled full factorial experiment was designed, 3 factors including layer height, infill density, and print speed were taken into account. After that, selected 27 Mid-section of femur bone from 3D medical image data were fabricated from filament of Poly-lactic acid (PLA). The models were underwent a three-point bending testing to investigate mechanical properties and statistical analyse to investigate the relationship of parameters.

1. Introduction

For medical student, training and practice on synthetic anatomical bone before perform real medical operation is useful and recently become part of medical procedure. [1-3]. Thus, the attempt to fabricate solid 3D anatomical model from medical image data of patient using a technology based on Fused Deposition Modeling (FDM) is alternatively developed [4]. FDM is one of additive manufacturing trends that work based on a layer-by-layer fabrication basis [5-7]. According to working principle, printing material for FDM must be able to liquefy above certain temperature and solidify upon cooling, for example, metal and thermoplastic polymer [8, 9]. The fabrication process of 3D printed anatomical model typically consists of several sequences [10, 11]; however, the quality and durability of model is mostly depended on 3D printer setting parameter adjustment [12, 13]. When focused on basic 3D printer setting, Infill density is a setting that determines a percent of material filling inside, layer height determine thickness of outer layer, and print speed was a speed of moving nozzle head. All of them consequently influence strength of an entire 3D printing model [13, 14]. However, geometric detail on each bone is difference, resulting in varies mechanical properties of the 3D printed final product [12, 15]. Thus, the investigation of mechanical properties of bone-shape 3D printed model is required. Regarding the development of 3D printed anatomical bone, this study aims to investigate the effect and relationship between mechanical strength of anatomical model to withstand given load from transverse direction, referred to flexural strength, and process parameters, consisting of 3 parameters; layer height, infill density, and print speed, for additional information of 3D printed anatomical model and related study.

2. Materials and Methods

In this study, the experimental method was consisted of 3 part; test specimen preparation, design of experiment, and mechanical testing

2.1. Test specimen preparation

Medical imaging data of mid-shaft of human left femur (size 33x52x150 mm) from Computer Tomography (CT) scan was selected from open-source digital library. After image segmenting, test specimens related to number of observation were fabricated in vertical direction. In addition, summary of 3D printer setting were represented in Table 1. The 3D printer was based on FDM and printing material in this study was focused on filament of PLA.

Table 1 Summary of 3D printer setting adjustment

Process parameter	Value
Nozzle temperature (°C)	210
Bed temperature (°C)	60
Printing orientation	Triangle
Infill density (%)	40, 60, 80
Layer height (mm)	0.2, 0.15, 0.1
Print speed (mm/s)	50, 75, 100
Wall thickness (mm)	1.2

2.2. Design of experiment

2-levelled full factorial experimental design, considering 3 factors; infill density, layer height, and print speed was selected. Each experimental runs was repeated 3 times.

2.3. Mechanical testing

3-point bending test was conducted by Instron5566 (0.2 mm/s head-cross speed, 10kN load). This study considered that shape of 3D printed model was identical to bone [16]. Thus, the equations based on beam theory that follow related study [17] consisted of bending moment, flexural stress, flexural strain, and flexural modulus were applied and calculated accordingly to equation 1, 2, 3, and 4

$$M = \frac{F \cdot L}{4} \quad (1)$$

$$\sigma = \frac{Mc}{I} \quad (2)$$

$$\epsilon = \frac{12 \cdot c}{L^2} \cdot \Delta \quad (3)$$

$$E = \frac{\sigma}{\epsilon} \quad (4)$$

Where F is applied load, L is the distance between two bottom support (100 mm), c is a distance of the chosen point from neutral axis, I is the cross-sectional moment of inertia based on the geometric entity (c and I are 52.92 mm. and 175385.1 g/mm^2, respectively), Δ is the vertical displacement at the point of load application.

2.4. *Statistical analysis*

After mechanical properties were calculated, Analysis of Variance (ANOVA) from software Minitab16 was used to analyze interaction of the process parameters.

3. Results and Discussion

Test specimens of 3D printed anatomical bone were fabricated as represented in Figure 1 relation to the full factorial experiment designed from Minitab2.

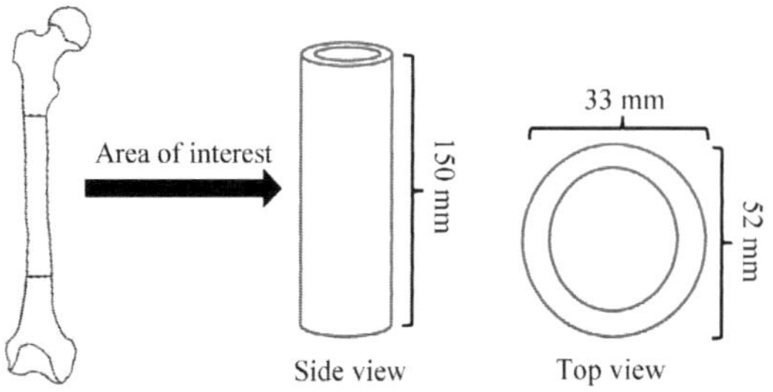

Figure 1. Test specimens of 3D printed anatomical bone

After testing, mechanical properties including flexural strength and Flexural modulus were represented in Table 2 and the statistical data was represented in Table 3.

Table 2 Flexural strength and Flexural modulus of anatomical test specimens

Group			Flexural strength		Flexural modulus	
Infill density (%)	Layer height (mm)	Speed (mm/s)	Mean (MPa)	STD	Mean (MPa)	STD
40	0.10	50	34.67	2.13	136.87	13.04
40	0.10	100	13.85	0.34	90.99	7.95
40	0.20	50	14.53	1.25	38.74	1.52
40	0.20	100	14.95	5.13	56.37	39.22
60	0.15	75	30.78	0.72	107.28	1.06
80	0.10	50	36.88	3.26	125.94	6.60
80	0.10	100	29.50	0.07	181.87	16.79
80	0.20	50	26.60	0.73	105.82	3.76
80	0.20	100	21.63	8.94	165.10	13.68

Table 3 Analysis of Variance (ANOVA) for Flexural strength

Source	DF	Seq. SS	Adj. SS	Adj. MS	F-value	P-value
Main effect	3	1478.73	1478.73	492.909	33.27	0.000
Infill density	1	490.36	490.36	490.364	33.09	0.000
Layer height	1	506.89	506.89	506.886	34.21	0.000
Print speed	1	481.48	481.48	481.476	32.49	0.000
2-ways interactions	3	278.00	278.00	92.665	6.25	0.004
Infill density*Layer height	1	0.67	0.67	0.665	0.04	0.835

Infill density*Print speed	1	9.22	9.22	9.221	0.62	0.440
Layer height*Print speed	1	268.11	268.11	268.110	18.09	0.000
3-ways interactions	1	92.81	92.81	92.812	6.26	0.022
Infill density*Layer height*Print speed	1	92.81	92.81	92.812	6.26	0.022
Curvature	1	121.78	121.78	121.776	8.22	0.010
Residual error	18	266.71	266.71	14.817		
Pure error	18	266.71	266.71	14.817		
Total	26	2238.02				

R-Square = 88.08% R-Square (pred.) = 73.19% R-Square (adj.) = 82.79%

The result of ANOVA showed that infill density, layer height, and print speed significantly influence flexural strength of the 3D printed model (p-value = 0.000), the interaction between them was also statistically significant (p-value = 0.022). Standardized effect plot of interaction between the process parameters were represented in Figure 2 and summary residual plots in Figure 3.

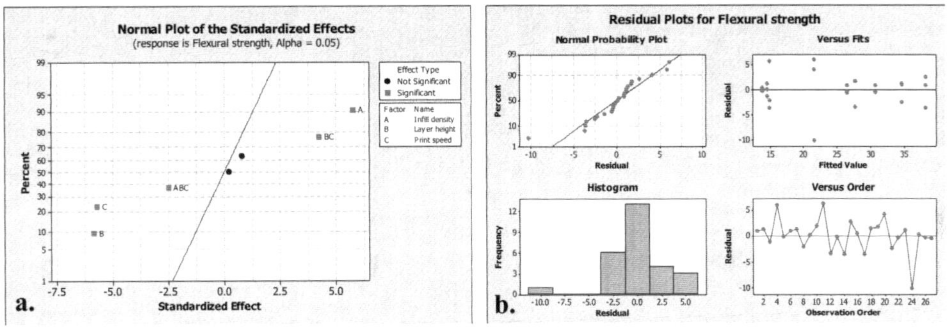

Figure 2. Standardize effects plot of process parameters to flexural strength of the model (a), 4-in-1 Residual plots of process parameters (b)

In Figure 2(a), infill density (factor A) possess positive standardized effect meanwhile layer height (factor B) and print speed (factor C) possess negative standardized effect, in which clarified that when increase in infill density, decrease in layer height and print speed, maximum flexural strength was higher. The result was also in agreement with related study [14]. In Figure 2(b), the points of observation data were dispersed linearly in normal probability plot and were dispersed randomly in residual versus fits and residual versus order plots, these patterns showed that the parameters were normal and independent distribution. Accordingly, the regression model of 3 process parameters including infill density, layer height, and print speed could be described as

$$Y = 24.022+4.520X_1-4.596X_2-4.479X_3+0.167X_1X_2+0.620X_1X_3+3.342X_2X_3-1.967X_1X_2X_3$$

When X_1 is infill density, X_2 is layer height, and X_3 is print speed.

After 3-point bending test, the sample specimens were dissected and scanning electrons microscope (SEMs) was used to observe the difference of fracture between each group. In Figure 3, traces of transgranular fracture were seen in all groups of specimen. Rough fracture scars were seen in specimens with 40% infill density, 0.2 mm layer height and 100 mm/s print speed. The scars were slightly small when observe in high infill density, higher layer height, and low print speed. According to size of crack, outer layer or first layer of wall thickness of all specimens was affected from external force the most, and then the pattern of fracture changed in next layers. These behaviors corresponded to characteristic of thermoplastic material [18] that the model become more brittle when increase infill density, increase print speed or decrease layer height. Thus, when brittle model receive external force from transverses side, a little rough scar was occurred on the fractured surface. More brittle specimens also related to higher flexural modulus from the result of 3-point bending test as well. In addition, inner layer of specimen contacted with many support structures that referred to triangle infill pattern in this study. When a number of support structures inside were increased due to high infill density setting, the received force could be good transferred through the contact area, resulting in several small cracks occurred through the layer.

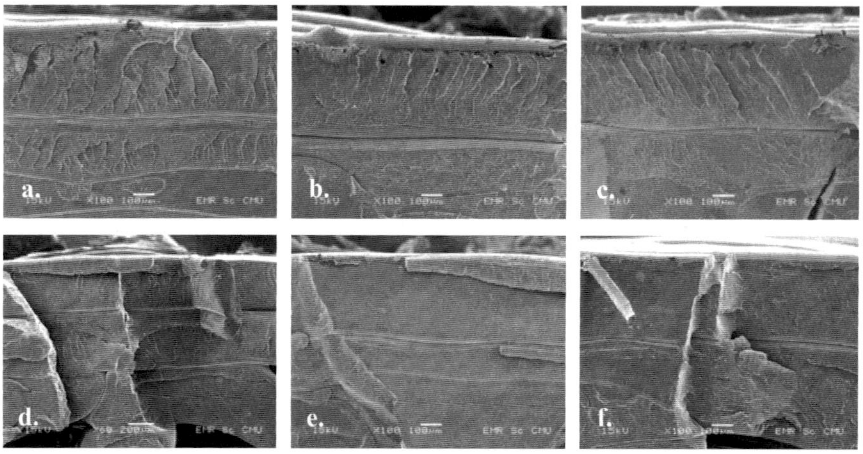

Figure 3. Fracture surfaces of 3D printed specimens with 40% infill density, 0.1 mm layer height and 50 mm/s print speed (a), 40% infill density, 0.2 mm layer height and 50 mm/s print speed (b), 40% infill density, 0.2mm layer height and 100 mm/s print speed (c), 60% infill density, 0.15 mm layer height and 75 mm/s print speed (d), 80% infill density, 0.2 mm layer height and 50 mm/s print speed (e), and 80% infill density, 0.2 mm layer height and 100 mm/s print speed (f)

Since 3D printed anatomical bone model typically use with bone saw and bone drill to perform medical training such as testing bone fixator [19, 20], when use a high infill density, the model will become too stiff that the machine is not able to cut or break. However, when use a low infill density, large cracks would be occurred and the fixator would be failed to attach the 3D printed model. Thus, using moderate infill setting and focus on the other setting should prevent the rough large fracture that lead to a low tolerance 3D printed model.

4. Conclusions

This study presented the investigation of effect and relationship of process parameters including layer height, infill density, and print speed to flexural strength of 3D printed anatomical models. The result revealed that when layer height and print speed setting were decreased and infill density setting was increased, average maximum flexural strength and flexural modulus were higher relatively and the model also represented more brittleness characteristic. The author suggested that further progressive studies such as reinforced printing material development or analytical mathematic model should be considered.

Acknowledgment

The author would like to express gratitude to Asst. Prof. Dr. Dumnoensun Pruksakorn from Department of Orthopedic, Faculty of Medicine, Chiang Mai University, for his advice and guidance, to Graduate School, Chiang Mai University and Biomedical Engineering Institute, Chiang Mai University, for facility support.

References

[1] Berman, B. (2012). 3-D printing: The new industrial revolution. *Business Horizons, 55*(2), 155-162.

IOP Conf. Series: Materials Science and Engineering **635** (2019) 012034 doi:10.1088/1757-899X/635/1/012034

[2] Azer, S. A., and Azer, S. (2016). 3D Anatomy Models and Impact on Learning: A Review of the Quality of the Literature. *Health Professions Education, 2*(2), 80-98..

[3] AbouHashem, Y., Dayal, M., Savanah, S., and Štrkalj, G. (2015). The application of 3D printing in anatomy education. *Medical Education Online, 20*, 10.3402/meo.v3420.29847.

[4] Smith, M. L., and Jones, J. F. X. (2017). Dual-Extrusion 3D Printing of Anatomical Models for Education. Anat Sci Educ.

[5] Cui, X., Boland, T., D'Lima, D. D., and Lotz, M. K. (2012). Thermal inkjet printing in tissue engineering and regenerative medicine. *Recent Pat Drug Deliv Formul, 6*(2), 149-155.

[6] Hoy, M. B. (2013). 3D printing: making things at the library. *Med Ref Serv Q, 32*(1), 94-99.

[7] Zhai, Y., Lados, D., and Lagoy, J. (2014). Additive Manufacturing: Making Imagination the Major Limitation (Vol. 66).

[8] Lim, L. T., Auras, R., and Rubino, M. (2008). Processing technologies for poly(lactic acid). *Progress in Polymer Science, 33*(8), 820-852.

[9] Hwang, S., Reyes, E., Moon, K.-s., Rumpf, R., and Kim, N. (2015). Thermo-mechanical Characterization of Metal/Polymer Composite Filaments and Printing Parameter Study for Fused Deposition Modeling in the 3D Printing Process. *Journal of Electronic Materials, 44*(3), 771-777.

[10] Mitsouras, D., Liacouras, P., Imanzadeh, A., Giannopoulos, A. A., Cai, T., Kumamaru, K. K., and Rybicki, F. J. (2015). Medical 3D Printing for the Radiologist. *Radiographics, 35*(7), 1965-1988.

[11] Wang, C.-S., Wang, W.-H. A., and Lin, M.-C. (2010). STL rapid prototyping bio-CAD model for CT medical image segmentation. *Computers in Industry, 61*(3), 187-197.

[12] Sood, A. K., Ohdar, R. K., and Mahapatra, S. S. (2010). Parametric appraisal of mechanical property of fused deposition modelling processed parts. *Materials & Design, 31*(1), 287-295.

[13] Luzanin, O., Movrin, D., and Plancak, M. (2014). *Effect of layer thickness, deposition angle, and infill on maximum flexural force in FDM-built specimens* (Vol. 39).

[14] Fernandez-vicente, M., Calle, W., Ferrándiz, S., and Conejero, A. (2016). *Effect of Infill Parameters on Tensile Mechanical Behavior in Desktop 3D Printing* (Vol. 3).

[15] Lee, C. S., Kim, S. G., Kim, H. J., and Ahn, S. H. (2007). Measurement of anisotropic compressive strength of rapid prototyping parts. *Journal of Materials Processing Technology, 187-188*, 627-630.

[16] Oksztulska-Kolanek, E., Znorko, B., Michalowska, M., and Pawlak, K. (2016). The Biomechanical Testing for the Assessment of Bone Quality in an Experimental Model of Chronic Kidney Disease. *Nephron, 132*(1), 51-58.

[17] Sharir, A., Barak, M. M., and Shahar, R. (2008). Whole bone mechanics and mechanical testing. *The Veterinary Journal, 177*(1), 8-17.

[18] Ashby, M. F. (2011). Chapter 4 - Material Property Charts *Materials Selection in Mechanical Design (Fourth Edition)* (pp. 57-96). Oxford: Butterworth-Heinemann.

[19] Elfar, J., Menorca, R. M., Reed, J. D., and Stanbury, S. (2014). Composite bone models in orthopaedic surgery research and education. *J Am Acad Orthop Surg, 22*(2), 111-120.

[20] Lin, H.-H., Lonic, D., and Lo, L.-J. (2018). 3D printing in orthognathic surgery – A literature review. *Journal of the Formosan Medical Association, 117*(7), 547-558.

Digital Fabrication of a 3D Object using Fixtureless Assembly

Ratchatin Chancharoen and Chaiwuth Sithiwichankit

Department of Mechanical Engineering, Chulalongkorn University, Thailand

E-mail: Ratchatin.C@chula.ac.th

Abstract. Digital fabrication is the kind of the manufacturing process expected to be the next industrial revolution due to its capability converting a digital model into a physical object with fully-automated operation of manufacturing. Although the assembly operation of mass-produced components gives relatively higher cost effective, it is still unprepared for the digital fabrication in practise, unlike the subtractive or additive operation. This research proposes the minimal-required robotic system and process enabling the digital fabrication using the assembly operation done at fixtureless location. The robotic system comprises a SCARA robot, a vacuum gripper, a ram extruder, a USB camera, and MATLAB® on a personal computer. The calibrations giving the required kinematic correlations, and the pose estimation capacitating the fixtureless location operation, are also described in this paper. The necessary process are product designing, product cell-decomposition, and product assembly. In result, the fully-automated assembly of a 3D object is successfully done at fixtureless location in about three minutes and haft which is short compared with using the subtractive or additive operation. This research gives promise of applying the digital fabrication using the assembly operation at fixtureless location for practical applications, such as building and repairing of furniture, constructive structure, and etc.

1. Introduction

Digital fabrication is a term describing the manufacturing process building a physical 3D object from the corresponding 3D digital model using various kinds of fully-automated manufacturing operation [1]. According to various aspects of its benefit, e.g., customization, automation, transportation, storage, and etc., the digital fabrication is expected to be the next industrial revolution [1,2]. Among many kinds of the fully-automated manufacturing operation, the assembly operation gives relatively high cost-effective, if the assembling components are massively produced. With low time consumption, the assembly operation also provides high productivity which is a great advantage, especially for the fabrication of a large product [3,4]. However, the assembly operation is still unprepared in practise for the digital fabrication [5,6], while the digital fabrication using the subtractive or additive operation are widely used [7]. Consequently, the digital fabrication of large product which is suitable for the assembly operation due to long-time consumption, is done using the operation of the subtractive and additive in some researches [8].

This research aims to propose the minimal-required robotic system and process for the digital fabrication using the assembly operation. Since the existing assembly system involves the guidance system [9], the machine vision enabling the assembly operation done at fixtureless location is also included in this research. In this paper, the proposed robotic system is described in Section 2, the necessary calibration of the kinematic correlations is explained in Section 3, the routine of the pose estimation done by the machine vision system is presented in Section 4, the proposed process of

IOP Conf. Series: Materials Science and Engineering **635** (2019) 012035 doi:10.1088/1757-899X/635/1/012035

digital fabrication is described in Section 5, the assembly operation done by the proposed is demonstrated in Section 6, and the conclusion is in Section 7.

2. Robotic system

The robotic system consists of a manipulator, a gripper, a system of joining component, a machine vision system, and a control system of assembly operation. First, the manipulator is a SCARA robot, Yamaha YK350XC connecting to its controller, Yamaha RCX142. Next, the gripper is a single cup vacuum gripper [10], while the pressure at the cup is controlled by two solenoid-valves whose the logic is switched by the digital signal from the robot controller. In this case, the type of joining is adhesive bounding [11]. The system of joining component consequently comprises a ram extruder [12], and its control unit. including an Arduino MEGA board, a Ramp 1.4 3D printing shield, and a Polulu A4988 driver. The extruder is actuated by a NEMA 11 stepper motor through a lightweight ball-screw drive. For the machine vision system used to estimate the pose of assembling base and components, the sensing device is a USB camera [13], and the sensed result is inspected by image processing and analysis programs on MATLAB® running on a personal computer. To control the assembly operation, an operation-control program developed and run on MATLAB® is used. Thus, MATLAB® transfers the operation information with the robot controller, the extruder controller, and the USB camera via serial communication [14]. The layout of the entire robotic system is demonstrated in Figure 1, while the gripper, the extruder, and the USB camera are all equipped to the robot tip as the assembling head shown in Figure 2.

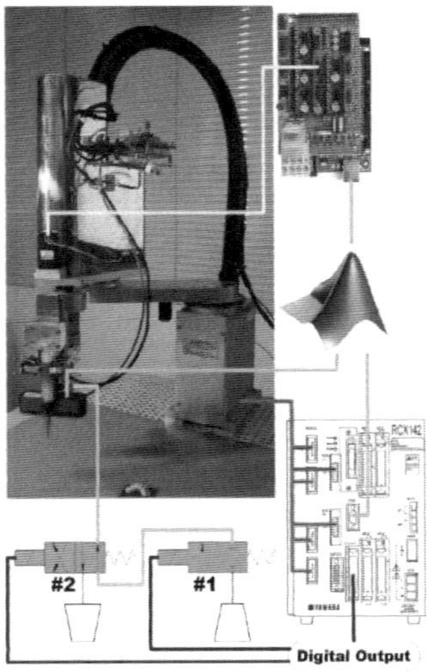

Figure 1. Robotic system

Figure 2. Assembling head

3. Kinematic calibration

The kinematic correlations between the robot tip and all the devices of the assembling head, are constant and necessary, but unknown. The frames for kinematic analysis are attached of the robot base ({B}), the robot tip ({T}), the gripper ({G}), the extruder ({E}), the camera ({C}), the pose-estimating objects ({O}) which are the assembling base and components. The frames of the robot and the assembling head are shown in Figure 3(a) and Figure 3(b) respectively. First, the camera calibration is done to determine the lens-distortion coefficients and the intrinsic parameters of the camera. The lens-distortion calibration of Heikkila and Silven [15], and the intrinsic parameters calibration of Zhang Z. [16] are used in order. Next, the correlation between the robot tip frame ({T}) and the camera frame ({C}) is calibrated following the approach of Zhang H. [17]. Then, the pure-translation correlation of the robot tip frame ({T}) and the gripper frame ({G}) locating at the center of gripping, is calibrated by gripping and releasing a sample object of same shape and size with the gripper, in the robot workspace, then estimating the object pose with respect to the robot tip frame ({T}) as the representative. Lastly, the pure-translation correlation of the robot tip frame ({T}) and the extruder frame ({E}) locating at the tip of the extruder, is determined by extruding the adhesive in the robot workspace, then estimating the pose of the extruded adhesive with respect to the robot tip frame ({T}) to represent the extruder pose.

(a) (b)

Figure 3. Coordinate frames

4. Pose estimation

The machine vision system provides the pose estimation of the assembling base and components, so the assembly operation can be done at a fixtureless location. There are five steps done for this purpose. The input and result of the pose estimation are shown in Figure 4. First, image acquisition captures an image of the workspace by the USB camera, and acquired the image into MATLAB®. Next, image undistortion correct the imperfection caused by the camera lens-distortion of the image using the coefficients obtained from the camera calibration. Then, image segmentation distinguishes the assembling base and components from the background in the undistorted image using the color thresholding [18]. So, the binary image representing the assembling base and components with white regions, and the background with the black region, is given. After that, features extraction is done on the binary image to obtain the position with respect to the image coordinate, of the concerned features. In this case, blob analysis [19] and the corners detection approach of Shi and Tomasi [20] are used, so the area, the centroid position, and the corners positions, of each white region are determined. Finally, the pose estimation project the extracted positions into the real world using the inverse calibration projection [21]. With known extrinsic parameters of the projection frame, these positions can be mapped into the robot tip frame ({T}) and the robot base frame ({B}) respectively. Consequently, the position and orientation of the assembling base and components are obtained

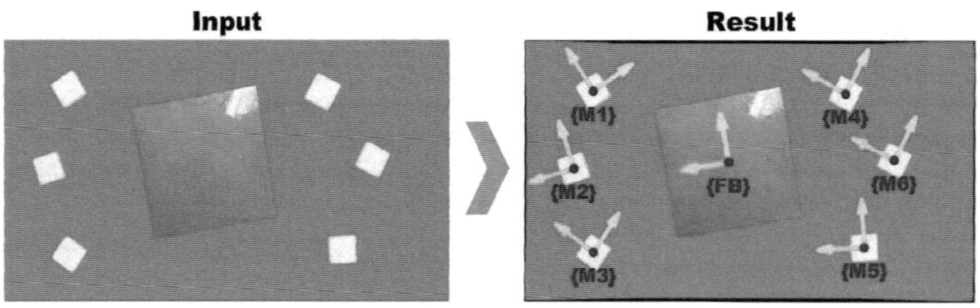

Figure 4 Pose estimation

5. Process

The process of the digital fabrication starts with product designing in which a 3D digital model of a product is built using CAD. Unlike the subtractive and additive which are the processing operation [11], the assembly operation joins discrete items together as the assembling components. Therefore, the cell decomposition [22] have to be applied to the designed model before the operation-control program is generated. In the decomposed model, the pose of each cell represents the pose of the corresponding assembling component. These poses are used to inform the poses of the assembling components relative to the assembling base in the operation-control program, while the pose of the assembling base and components are specified as the variables obtained by the pose estimation during the assembly operation. After the operation-control program is generated, the program is run on MATLAB® until the program end, meanwhile the assembly operation is fully-automated performed.

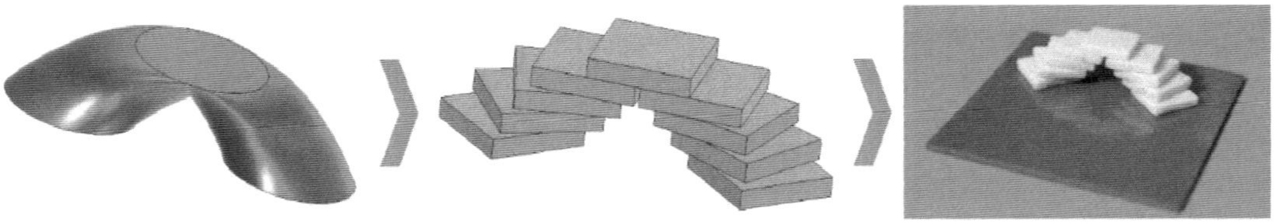

Figure 5. Digital fabrication process

6. Assembly demonstration

The curved-bridge model demonstrated in Figure 5, is fabricated using the proposed system and process. So, the performance of the proposed is observed. The physical model is successfully fabricated, while the fully-automated assembly operation is smoothly completed in about three minutes and half. Since the assembling base and components are placed at unknown location in the workspace, the assembly operation is exactly done at the fixtureless location. Furthermore, all four degrees of freedom of the robot are used to manipulate the assembling components. Together with the machine vision system, the robot can move the gripper and the extruder to their operating locations correctly. Lastly, the gripper grips and releases the assembling components firmly, while the extruder deposit the adhesive with the amount slightly deviated from the desirable.

7. Conclusion

The digital fabrication using the assembly operation at fixtureless location, can be practically done with the proposed robotic system and process. Lacking any part of the system or process, cause the failure. In the aspect of the product quality, the critical issues are the quality of the assembled joints, and the pose estimation noised by the surrounding light. Compared to the digital fabrication using the operation of the subtractive or additive, the presented digital fabrication gives lower time consumption. For the calibration approaches, all of them provides the accurate correlations of satisfying. Finally, this research gives the promise of applying the digital fabrication using the assembly operation at fixtureless location for the practical applications of building or repairing large product which cannot be moved easily, such as furniture, constructive structure, and etc.

8. References

[1] Lau M, Mitani J, Igarashi T. Digital fabrication. Computer. 2012 Dec 1;45(12):76-9.

[2] Smith RC, Iversen OS, Hjorth M. Design thinking for digital fabrication in education. International Journal of Child-Computer Interaction. 2015 Sep 1;5:20-8.

[3] Kalpakjian S, Schmid S. Manufacturing, Engineering and Technology SI 6th Edition-Serope Kalpakjian and Stephen Schmid: Manufacturing, Engineering and Technology. Digital Designs; 2006.

[4] Lim S, Buswell RA, Le TT, Austin SA, Gibb AG, Thorpe T. Developments in construction-scale additive manufacturing processes. Automation in construction. 2012 Jan 1;21:262-8.

[5] Keating S, Oxman N. Compound fabrication: A multi-functional robotic platform for digital design and fabrication. Robotics and Computer-Integrated Manufacturing. 2013 Dec 1;29(6):439-48.

[6] Willmann J, Knauss M, Bonwetsch T, Apolinarska AA, Gramazio F, Kohler M. Robotic timber construction—Expanding additive fabrication to new dimensions. Automation in construction. 2016 Jan 1;61:16-23.

[7] O'Neill B. Three-Dimensional Digital Fabrication. Inkjet Technology for Digital Fabrication. 2012 Nov 26:325-42.

[8] Yoshida H, Igarashi T, Obuchi Y, Takami Y, Sato J, Araki M, Miki M, Nagata K, Sakai K, Igarashi S. Architecture-scale human-assisted additive manufacturing. ACM Transactions on Graphics (TOG). 2015 Jul 27;34(4):88.

[9] Radi M, Reinhart G. Industrial haptic robot guidance system for assembly processes. In 2009 IEEE International Workshop on Haptic Audio visual Environments and Games 2009 Nov 7 (pp. 69-74). IEEE.

[10] Lundström G. Industrial robot grippers. Industrial Robot: An International Journal. 1974 Feb 1;1(2):72-82.

[11] Groover MP. Fundamentals of modern manufacturing: materials processes, and systems. John Wiley & Sons; 2007 Jun 14.

[12] Perrot A, Mélinge Y, Rangeard D, Micaelli F, Estellé P, Lanos C. Use of ram extruder as a combined rheo-tribometer to study the behaviour of high yield stress fluids at low strain rate. Rheologica acta. 2012 Aug 1;51(8):743-54.

[13] Kadkhodayan A, Jiang X, Menon C. Continuous prediction of finger movements using force myography. Journal of Medical and Biological Engineering. 2016 Aug 1;36(4):594-604.

[14] Cai Y, Warwick TP, Rane SG, Masserrat E. Digital serial communication device testing and its implications on automatic test equipment architecture. In null 2000 Oct 3 (p. 600). IEEE.

[15] Heikkila J, Silven O. A four-step camera calibration procedure with implicit image correction. In Computer Vision and Pattern Recognition, 1997. Proceedings., 1997 IEEE Computer Society Conference on 1997 Jun 17 (pp. 1106-1112). IEEE.

[16] Zhang Z. A flexible new technique for camera calibration. IEEE Transactions on pattern analysis and machine intelligence. 2000 Dec 1;22.

[17] Zhang H. Hand/eye calibration for electronic assembly robots. IEEE Transactions on Robotics and Automation. 1998 Aug;14(4):612-6.

[18] Chaabane SB, Sayadi M, Fnaiech F, Brassart E. Color image segmentation using automatic thresholding and the fuzzy C-means techniques. In Electrotechnical Conference, 2008. MELECON 2008. The 14th IEEE Mediterranean 2008 May 5 (pp. 857-861). IEEE.

[19] Mustafa NB, Fuad NA, Ahmed SK, Abidin AA, Ali Z, Yit WB, Sharrif ZA. Image processing of an agriculture produce: Determination of size and ripeness of a banana. In2008 International Symposium on Information Technology 2008 Aug 26 (Vol. 1, pp. 1-7). IEEE.

[20] Shi J, Tomasi C. Good Features to Track, 1994. The proceedings of the IEEE Conference on Computer Vision and Pattern Recognition on 1994 (pp. 593-600). IEEE.

[21] Martynov I, Kamarainen JK, Lensu L. Projector calibration by “inverse camera calibration”. In Scandinavian Conference on Image Analysis 2011 May 23 (pp. 536-544). Springer, Berlin, Heidelberg.

[22] Kada M. Scale-dependent simplification of 3D building models based on cell decomposition and primitive instancing. In International Conference on Spatial Information Theory 2007 Sep 19 (pp. 222-237). Springer, Berlin, Heidelberg.

Acknowledgments

This work was a part of the intelligent 3D printing for development project, which was supported by the Development into the 2nd Century Fund, Chulalongkorn University.